Principles and Technical Aspects of PCR Amplification

Elizabeth van Pelt-Verkuil
Alex van Belkum
John P. Hays

Principles and Technical Aspects of PCR Amplification

Springer

Elizabeth van Pelt-Verkuil
Hogeschool Leiden, Leiden
The Netherlands

Alex van Belkum
Erasmus MC, Rotterdam
The Netherlands

John P. Hays
Erasmus MC, Rotterdam
The Netherlands

ISBN 978-90-481-7579-6 e-ISBN 978-1-4020-6241-4
DOI 10.1007/978-1-4020-6241-4

Softcover reprint of the hardcover 1st edition 2008

Printed on acid-free paper

9 8 7 6 5 4 3 2 1

springer.com

Foreword

This book aims to provide an introduction to: (i) the concept of PCR, (ii) PCR technologies, (iii) PCR applications and (iv) PCR quality, with particular emphasis being placed upon PCR applications and techniques relevant to the clinical laboratory. This book will without doubt be useful as a reference work pertaining to technical aspects of PCR for all bio-medical students, research technicians, medics and scientists interested in the PCR technique and it's applications *per se*.

Wherever possible, the authors have tried to provide figures, scientific publications, or references to commercially available products, in order to illustrate any particularly important concepts or comments. Indeed, all commercial PCR biotechnology companies offer information about their products on internet sites and in online technical manuals. These online resources will be invaluable for any readers requiring more detailed PCR protocols.

The authors have provided references for many PCR concepts and applications that are directly useful to the clinical laboratory audience. These references provide a starting point for a more detailed investigation into the PCR techniques and concepts mentioned. In particular, further detailed information may be acquired by (i) referring to the reference section of cited publications, and (ii) by referring to other (more recent) publications published by the cited author(s).

Great efforts have been made to include descriptions of the vast majority of PCR applications and technologies that currently exist, though the dynamic nature of the PCR field means that no book can ever be regarded as totally inclusive of all the PCR refinements that have been (and are currently being) developed.

The writing of this book has been facilitated by the Hogeschool Leiden (The Netherlands), the Department of Medical Microbiology and Infectious Diseases at Erasmus MC (Rotterdam, The Netherlands) and an unrestricted financial grant donated by Roche Molecular Diagnostics (Almere, The Netherlands).

Finally, the authors would like to wish everyone success in their PCR-related studies!

April 2007

Elizabeth van Pelt-Verkuil
Alex van Belkum
John P. Hays

Contents

Chapter 1

The Polymerase Chain Reaction

1.1 An Overview of the PCR Process

The polymerase chain reaction (PCR) allows the specific and exponential synthesis of a predetermined DNA region via the use of two small, specifically designed fragments of DNA (primers or oligonucleotides), which form the two termini of the nucleic acid molecule to be amplified. PCR amplification reactions in general are highly specific, specificity being determined by the correct hybridisation of primer specific sequences to complementary sequences present on the target DNA molecule to be amplified. PCR primers comprise specific nucleotide sequences which are designed to hybridise to either the parallel or anti-parallel strand of the target DNA molecule, and as such since primers need to be precisely complementary to their target sequences, some sequence data from the terminal ends of the DNA is required for primer design (Fig. 1.1). Once hybridised to the target DNA, the primers provide the double stranded 3′-hydroxyl terminus required by thermostable DNA dependent DNA polymerases to begin the synthesis of a new DNA strand (complementary to the strand to which the primer has hybridised). Moreover, because PCR uses two primers (one designed for each strand of the DNA molecule to be amplified), repeated cycles of primer hybridisation (annealing) and disassociation allows DNA amplification in the 5′ to 3′ direction on both strands to occur, with the primers effectively acting as Okazaki fragments [Marinus, 1976].

PCR amplification is in fact a cyclical process where the sample DNA is initially denatured in order to unwind and separate the DNA double helix into single strands. This is usually achieved by heating the DNA sample in an aqueous environment, usually at a temperature of 94°C for 30 seconds to 5 minutes. Hybridisation of the specific oligonucleotide primers to each strand is then achieved by lowering the temperature of the reaction mix to the annealing temperature (T_m) which is usually set between 40°C and 65°C (dependent on the design of the oligonucleotide sequences used as primers). After the primer hybridisation step, the temperature is raised to approximately 72°C, (an optimal temperature for thermostable DNA polymerase mediated DNA strand replication), and the whole cycle is then repeated a pre-determined number of times. After each cycle of replication, each newly synthesised double stranded DNA molecule (known as an amplimer or amplicon)

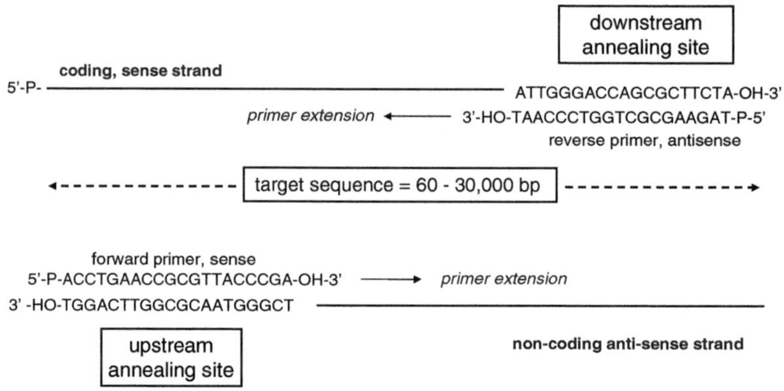

Fig. 1.1 The key components of PCR are the primers and the thermostable DNA polymerase. The forward and reverse primers determine the PCR specificity and the length of the amplification products. Primers need to be precisely complementary to their respective target sequences. Both the sense and anti-sense strands of the DNA double helix are copied and therefore amplified during PCR thermocycling. After target DNA disassociation, the primers are allowed to anneal to the target DNA so that strand amplification via a thermostable DNA polymerase can occur. The primer that hybridises nearest the "ATG" start codon of the gene (fragment) to be amplified is called the upstream or forward primer, whilst the primer which hybridises closest to the "stop" codon is referred to as the downstream or reverse primer

contains terminal sequences, which are complementary to the primer sequences used (Fig. 1.2). This process allows each amplimer to serve as a template for replication in subsequent rounds of PCR cycling, resulting in a theoretical doubling (exponential amplification) of the number of target molecules during each cycle. Some of the fundamental principles introduced above are detailed in a large body of international scientific literature (e.g. [Jain, 2002; Lubeck and Hoorfar, 2003; Klein, 2002; Wolk et al., 2001; Foy and Parkes, 2001; Erlich, 1999; Kiechle, 1999; Lisby, 1999]).

1.2 Before PCR and Beyond

Before PCR, molecular biologists utilised nucleic acid sequence data (sequence motifs) to design "hybridisation" probes for use in assays for the detection and identification of specific RNA and DNA fragments (moieties). These "hybridisation" assays were used to determine the presence/absence of specific RNA or DNA sequences within complex mixtures of nucleic acids, via the use of specifically designed (semi-synthetic) complementary DNA or RNA molecules (nucleic acid probes) which had been equipped with radioactive labels for detection purposes. The target nucleic acid population was initially attached to a solid carrier phase and

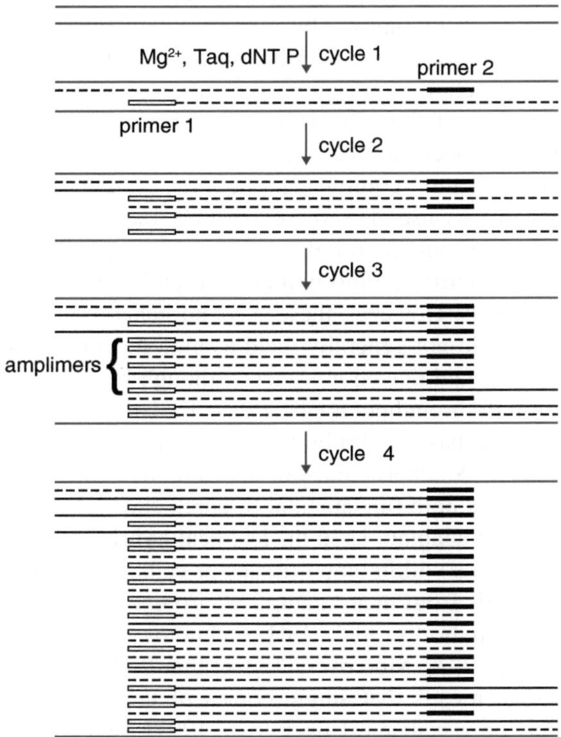

Fig. 1.2 Schematic representation indicating the principle of PCR and its key components. The duplication of a region within a DNA target molecule is facilitated by the specific hybridisation of two different oligonucleotide primers (primer 1 and primer 2). A thermostable DNA-dependent DNA polymerase recognises these primers and extends the DNA strand in the 5′ to 3′ direction while consuming dNTPs. Repeated cycles of strand separation by heat denaturation (melting), primer hybridisation (annealing) and new DNA strand synthesis (elongation) results in the exponential amplification of a specific DNA region, as defined by the user designed primers. In general, the minimum number of cycles of PCR performed is 20 cycles with 50 cycles usually being regarded as an upper limit

then hybridised with specific labelled probe. After stringent hybridisation and extensive washing procedures, the presence of the target DNA fragment could be determined by the presence/absence of the radioactively labelled probe on the solid carrier phase. Alternatively, direct visualisation of the probe and target molecule was achieved via electron microscopy, with hybridisation being quantified on the basis of the different widths of double stranded (hybridised) versus single stranded (non-hybridised) nucleic acids. The most convenient of these hybridisation test systems utilised filter hybridisation (where the target DNA extract was first immobilised on nitrocellulose or nylon filters), in combination with post-hybridisation autoradiography or scintillography. This format greatly increased test sensitivity and vastly improved the technical reliability and speed of the hybridisation procedure, allowing the detection of picogram quantities of target material. However, one major disadvantage of these hybridisation systems was the requirement for radioactively labelled probes, not least because working with radioactive materials is hazardous for your health, requires legal permits, correct disposal systems, and is relatively expensive to use. For these reasons, radioactive labels have been largely replaced by various (non-radioactive) chemical labels, facilitating the development of colorimetric, chemo-luminescent and chemo-fluorescent hybridisation detection methods. However, these "second generation" chemical-labelling and detection systems do not generally yield as high a degree of sensitivity as the original radio-labelling and detection systems, though both systems are amenable to automation and high throughput applications. To date, a variety of elegant techniques based upon the basic hybridisation principle have been developed (e.g. sandwich hybridisation, Southern- and Northern-blot hybridisations, etc.) and these are frequently applied in both fundamental research and clinical diagnostics.

Newer advances and adaptations in nucleic acid hybridisation have allowed the detection of nucleic acids within target tissues or cells *per se*, without the need to extract the nucleic acids from their "natural environment". Many elegant applications for this "*in situ* hybridisation" (ISH) technology have been developed and refined [Brigati et al., 1983]. Moreover, the fact that ISH assays can be coupled to microscopic analysis, means that single copies of target regions present within individual cells may be visualised [Unger et al., 1991]. The sensitivity of such ISH tests may be up to 1,000x greater than traditional filter hybridisation techniques. Currently, most ISH tests are coupled to amplification techniques which include a visualisation step involving the gradual accumulation of a reporter stain (such as biotin-tyramine) in the presence of probe hybridisation (Fig. 1.3). The introduction of a few additional steps to the detection protocol facilitates the use of (multiplex) ISH testing protocols using multiple probes and specimens. However, the disadvantage of ISH/stain-accumulation coupled protocols is a reduction in the ability to accurately localise the focal point of probe hybridisation within a cell due to diffusion of amplimers and stain(s) within the cellular matrix. Loss of target or target inaccessibility are well-known complications of ISH protocols and separate optimisation reactions are required for each cell type and ISH probe used in order to achieve the best results.

The need to detect very small numbers of clinically relevant molecules has significantly increased over the past few years. For example, the detection of low-titre viral infections, minimal residual disease in leukaemia patients, point mutations in genes or genetic aberrations in tumours etc., all require highly sensitive methodologies. This has led to the development of novel approaches specifically aimed at the amplification of target (gene) sequences prior to detection, such that sensitivity issues related to hybridisation/probe detection protocols are no longer the limiting step of DNA and RNA detection protocols.

The PCR technique provided the first practical solution to overcoming the limitations of sensitivity which were inherent in early nucleic acid hybridisation and detection

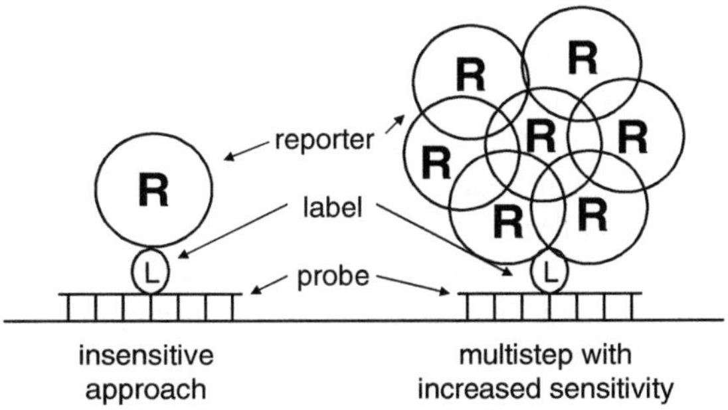

Fig. 1.3 The sensitivity of DNA hybridisation tests may be increased by the accumulation of a stained or fluorescent reporter molecule at a given location. This is facilitated by a reactive label that is coupled to the hybridisation probe. A gradual increase in stain accumulation over time increases ISH sensitivity by simply increasing the signal intensity. The specificity of the interaction between a (synthetic) DNA probe and its target is dependant upon complementary hybridisation between the DNA probe and its target sequence

protocols. The leading principle of PCR lies in the fact that extremely small amounts of target DNA can be specifically amplified to large amounts of synthetic DNA *in vitro*. Essentially, the continuous, exponential, semi-conservative replication of a well-defined DNA region present in the template DNA leads to the accumulation of large quantities of newly synthesised but specific target DNA product. The methodological principle behind the modern form of PCR amplification had actually been demonstrated before the development of PCR thermocycling *per se*, the process initially relying on a non-thermostable DNA dependent DNA polymerase isolated from *Escherichia coli* (the Klenow fragment), which (due to its non-thermostable nature) became heat inactivated after every PCR cycle. The introduction of thermostable DNA polymerases, (which were identified more than 10 years before the first PCR process was described [Chien et al., 1976]), meant that DNA dependent DNA polymerases were no longer inactivated after every cycle of PCR amplification, and hence the laborious task of adding fresh enzyme after every amplification cycle was no longer necessary, thereby reducing both the hands-on technical time required to perform an assay and the likelihood of (cross) contamination. The only extra requirement was a machine that could automatically heat and cool the samples (i.e. a PCR thermocycler), which eliminated the need for several separate water baths all set at different temperatures. In fact, this form of PCR-based amplification via automated thermocycling has now become almost universally popular for the detection of specific RNA and DNA molecules within clinical specimens, being extremely popular and invaluable in both the routine scientific laboratory and in the field of scientific research *per se*.

The development of a convenient PCR amplification methodology has itself provided an impetus for the development of techniques, which are based upon nucleic acid amplification and detection reactions. In general, the most common of these currently used in the scientific laboratory may be divided into three main categories:

1. *Target amplification*: The initial amount of target nucleic acid (or a specific region within this target nucleic acid) is exponentially amplified using an optimised reaction mix and a thermostable DNA dependent polymerase (e.g. PCR). Other variants (related to PCR but using different reaction mixes and enzymes) include – the ligase chain reaction (LCR) [Laffler et al., 1993], transcription amplification systems (TAS) [Kwoh et al., 1989] and the nucleic acid sequence based amplification (NASBA) technique [Malek et al., 1994; Cook, 2003]. The latter two techniques may be especially useful in amplifying RNA instead of DNA, and involve the generation of DNA/RNA intermediates.
2. *Probe amplification*: This technology depends on an initial hybridisation between a probe molecule and the target nucleic acid, where the probe contains additional sequence elements that enable detection of the probe via a specific interaction with these additional sequence elements, e.g. the Qβ-replicase system which exponentially amplifies RNA using the bacteriophage Qβ RNA dependent RNA polymerase. Other systems involving "replicable probes" and "branched DNA assays" (which enable the Christmas tree-like "networking" of multiple probes), have been produced and developed into commercially available diagnostic products. Though these are currently much less widely available or used than the "target amplification" systems mentioned above [Qian and Lloyd, 2003; Blok and Kramer, 1997].
3. *Signal amplification*: This is the common term for all of the techniques that amplify a signal previously generated by either the target amplification systems or probe amplification systems described above. Signal amplification systems frequently utilise chemical reactions that result in an increase in the accumulation of coloured, fluorescent or luminescent product (i.e. signal) after target amplification, thereby resulting in an increase in test sensitivity [e.g. Ness et al., 2003].

This book is specifically dedicated to the polymerase chain reaction (developed in 1985 by Saiki et al. [1985]), and to the large multitude of additional applications and formats which have been described for PCR over the past 2 decades. The goal of this book is to provide the reader with an overview of the technique, development and application of PCR technology today, rather than providing an all-encompassing and intimately detailed account of all aspects of PCR (as if this were possible!). To this extent, the PCR process is first compared to DNA replication, and different PCR test formats. Further, the identification of amplified PCR products is discussed in detail, and the design and execution of PCR tests (including quality control and trouble shooting) are also discussed.

References

Blok HJ, Kramer FR. 1997. Amplifiable hybridization probes containing a molecular switch. Mol Cell Probes 11:187–194.

Brigati DJ, Myerson D, Leary JJ, Spalholz B, Travis SZ, Fong CK, Hsiung GD, Ward DC. 1983. Detection of viral genomes in cultured cells and paraffin-embedded tissue sections using biotin-labeled hybridization probes. Virol 126:32–50.

Chien A, Edgar DB, Trela JM. 1976. Deoxyribonucleic acid polymerase from the extreme thermophile *Thermus aquaticus*. J Bacteriol 127:1550–1557.

Cook N. 2003. The use of NASBA for the detection of microbial pathogens in food and environmental samples. J Microbiol Methods 53:165–174.

Erlich HA. 1999. Principles and applications of the polymerase chain reaction. Rev Immunogenet 1:127–134.

Foy CA, Parkes HC. 2001. Emerging homogeneous DNA-based technologies in the clinical laboratory. Clin Chem 47:990–1000.

Jain KK. 2002. Current trends in molecular diagnostics. Med Device Technol 13:14–18.

Kiechle FL. 1999. DNA technology in the clinical laboratory. Arch Pathol Lab Med 123:1151–1153.

Klein D. 2002. Quantification using real-time PCR technology: applications and limitations. Trens Mol Med 8:257–260.

Kwoh DY, Davis GR, Whitfield KM, Chappelle HL, DiMichele LJ, Gingeras TR. 1989. Transcription based amplification system and detection of amplified human immunodeficiency virus type I with a bead-based sandwich hybridization format. Proc Natl Acad Sci USA 86:1173–1177.

Laffler TG, Carrino JJ, Marshall RL. 1993. The ligase chain reaction in DNA-based diagnosis. Ann Biol Clin 51:821–826.

Lisby G. 1999. Application of nucleic acid amplification in clinical microbiology. Mol Biotechnol 12:75–99.

Lubeck PS, Hoorfar J. 2003. PCR technology and applications to zoonotic food-borne bacterial pathogens. Methods Mol Biol 216:65–84.

Malek L, Sooknanan R, Compton J. 1994. Nucleic acid sequence based amplification (NASBA). Methods Mol Biol 28:253–260.

Marinus MG. 1976. Adenine methylation of Okazaki fragments in *Escherichia coli*. J Bacteriol 128:853–854.

Ness JM, Akhtar RS, Latham CB, Roth KA. 2003. Combined tyramide signal amplification and quantum dots for sensitive and photostable immunofluorescence detection. J Histochem Cytochem 51:981–987.

Qian X, Lloyd RV. 2003. Recent developments in signal amplification methods for in situ hybridisation. Diagn Mol Pathol 12:1–13.

Saiki RK, Scharf S, Faloona F, Mullis KB, Horn GT, Erlich HA, Arnheim N. 1985. Enzymatic amplification of beta-globin genomic sequences and restriction site analysis for diagnosis of sickle cell anemia. Science 230:1350–1354.

Unger ER, Hammer ML, Chenggis ML. 1991. Comparison of 35S and biotin as labels for in situ hybridization: usa of an HPV model system. J Histochem Cytochem 39:145–150.

Wolk D, Mitchel S, Patel R. 2001. Principles of molecular microbiology testing methods. Infect Dis Clin North Am 15:1157–1204.

Chapter 2

A Brief Comparison Between *In Vivo* DNA Replication and *In Vitro* PCR Amplification

In principle, PCR generates large quantities of DNA from a minute amount of nucleic acid starting material using a methodology similar to (but much simpler than) that seen in living cells. For living cells, *in vivo* DNA synthesis is dependent upon a well defined but complex set of enzymes and co-factors, which have evolved to act in a concerted fashion during the synthetic phase (S-phase) of the cell cycle. In comparison, PCR facilitates in vitro DNA synthesis in a much simpler fashion, making use of a smaller set of defined ingredients and reaction conditions involving relatively high temperatures. The range of factors contributing to successful PCR amplification is reviewed below.

2.1 Nucleic Acid Targets

2.1.1 DNA

In vivo DNA duplication, which is essentially a form of limited DNA amplification, is performed under the direction of a select and diverse set of structural proteins, enzymes and additional co-factors (a detailed description of which is beyond the scope of this book and interested readers are referred to the international literature for a more in-depth look into this particular topic, e.g. Shcherbakova et al., 2003; Goren and Cedar, 2003; Kelman, 2000; Nasheuer et al., 2002; Nasmyth, 2002). In eukaryotes, the DNA molecule is intimately associated with positively charged proteins, which are strongly electrostatically bound to the phosphate moieties on the DNA chain. These "histone" proteins associate into octameric complexes, bind to approximately 400 base pairs (bp) of genomic DNA, and constitute approximately half of the mass of the eukaryotic chromosome. These histones do not completely disassociate from the DNA during replication. An additional level of complexity also exists in eukaryotic DNA, in that further folding processes induce secondary (300 nm) and tertiary (700 nm) order folding on the template DNA, thereby generating tightly coiled DNA with a high molecular density [Stewart, 1997] (Fig. 2.1). For prokaryotes, the situation regarding the *in vivo* state of DNA

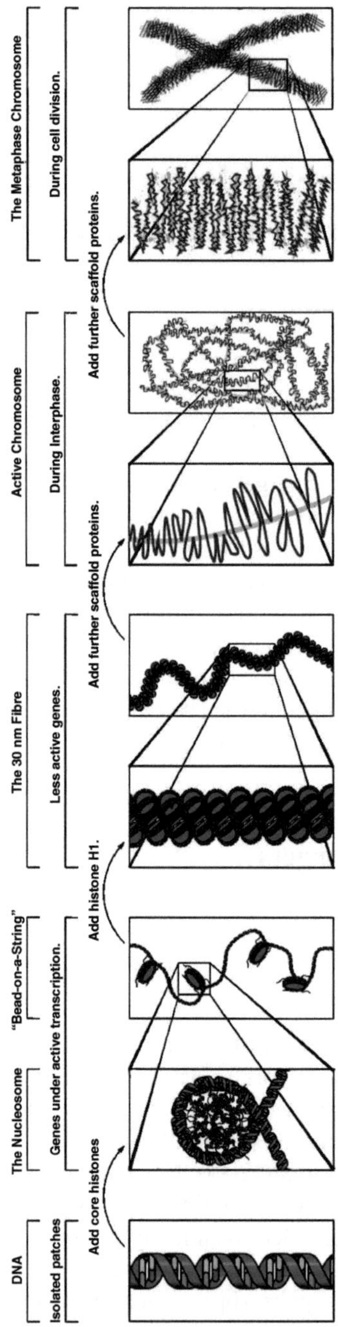

Fig. 2.1 *In vivo*, the double stranded DNA molecule is organized into a highly condensed structure with an impressive molecular density. The primary double helix structure is embedded into spherical nucleosomes (11 nm in diameter) via an interaction with chaperone or histone proteins. These nucleosomes are in turn ordered into 30 nm chromatin fibers which are looped into macrostructures called chromatin. This chromatin may be further condensed into heterochromatin, providing the framework that generates the final eukaryotic chromosomal structure. What is shown is a beads-on-a-string model of chromatin with a linker length of 20 nucleotide pairs in three dimensions (From http://en.wikipedia.org/wiki/Chromatin)

is generally considered to be somewhat simpler, with microorganisms lacking a visible nucleus and containing only one (circular) chromosomal copy per cell. However, it is known that prokaryotic DNA is also associated with a variety of histone-like proteins, which intimately interact with the cellular DNA, protecting it and conferring complex structure (though this structure is generally less ordered or regular than that associated with eukaryotic chromosomes). Therefore, one of the major requirements for DNA replication in *in vivo* systems is the systematic uncoiling of specific regions of genomic DNA in order to expose single strands of DNA ready for replication. In both eukaryotic and prokaryotic kingdoms this process is achieved via specific DNA topoisomerases, helicases and gyrases, which act together to help uncoil the double stranded DNA.

In contrast, the uncoiling of DNA during PCR amplification is not enzymatically controlled but is in fact achieved using a far less complicated procedure, i.e. the DNA to be amplified is usually chemically extracted from its host chaperone proteins, and the residual tertiary or secondary structure removed by heating the naked DNA. This heating step also provides the mechanism by which the two DNA strands are separated (melted) into single strands ready for PCR amplification. As a further point, whereas *in vivo* DNA chromosomal replication involves the replication of many millions of nucleotides, PCR amplification products are generally designed to be shorter than 1,000 bp in length (a parameter largely imposed on the PCR amplification process by the type of heat-stable DNA dependent DNA polymerases used). However in extreme cases, the successful PCR amplification of regions of DNA over 35 kilobases in length have been reported using special heat-stable DNA dependent DNA polymerases combinations [Hu et al., 2002; Davies and Gray, 2002]. In principle, DNA from all types of viruses, cells (plant, animal and bacterial), or tissues (lung, brain, etc.) may be amplified using nucleic acid extraction techniques coupled to PCR amplification.

2.1.2 RNA

As defined in the "central dogma of molecular biology", the flow of genetic information *in vivo* proceeds from DNA to RNA (via transcription) and then finally to protein (via translation). However, since the central dogma was formulated, scientific discoveries have shown that the flow of genetic information may occur in the reverse direction, i.e. from RNA to DNA (reverse transcription). This process of reverse transcription occurs during the replication cycle of certain virus families (*Retroviridae*, *Hepadnaviridae* and *Caulimoviridae*), and the relevant enzymes (reverse transcriptases) have been isolated and utilized to generate DNA from RNA molecules in specially adapted PCR protocols. This special form of PCR is known as reverse transcriptase- or reverse transcription-PCR (often shortened to RT-PCR), and allows the detection and quantification of several types of RNA molecules, including messenger RNA (mRNA) from both prokaryotic and eukaryotic cells, as well as RNA-based genomes, e.g. Coronaviruses and Picornaviruses. Several non-PCR-based amplification methods have also been developed which do not utilise reverse transcriptase enzymes and hence do not make use of an intermediate DNA template. These techniques amplify RNA (and not double stranded DNA as is the case with PCR) using an initial RNA template.

2.2 Target DNA Strand Separation and Primer Annealing

During *in vivo* DNA replication, cell associated proteins (including DNAa, DNAb, DNAc, single stranded binding proteins, helicases and gyrases, ligase and a variety of polymerase subunit proteins), all act in concert to uncoil the stable α-helical DNA structure, break the hydrogen bonds between the purine and pyrimidine bases, and expose a DNA replication origin [Nasheuer et al., 2002]. In contrast, DNA required for PCR amplification is separated from its chaperone proteins using chemical and/or enzymatic extraction methods, with the DNA then being separated into single strands via thermal disassociation, i.e. via incubation at approximately 94°C for 30 seconds to 5 minutes, which causes breaking of the hydrogen bonds between the complementary base pairs present on opposite strands. This method is not available to living cells as proteins are rapidly denatured at 94°C and cellular integrity irretrievably breaks down. In fact, the melting temperature (T_m) of a DNA molecule is defined as the temperature at which half of it's constitutive bases are no longer paired to their complementary partner on the opposite strand. For most natural species of DNA, the melting temperature lies somewhere between 70°C and 100°C, with the actual T_m being dependent on both the length of the DNA molecule to be melted and on the base composition of the strands. Moreover, there exists a linear relationship between the percentage of guanosine and cytosine bases present in the DNA strands (referred to as the "GC content") and the melting temperature of that particular DNA helix, with GC-rich DNA melting at a higher temperature than adenosine and thymidine rich (AT-rich) DNA. Because of the fact that DNA melting curves are relatively steep, temperatures of T_m+5°C and T_m−5°C above or below the calculated T_m value will result in the double helix being completely denatured or completely intact, respectively.

Several other factors may also facilitate DNA helix destabilization, all of them operating by destabilizing the interaction between the various complementary DNA base moieties. Extreme pH values, denaturants such as formamide or urea and the overall salt concentration are important parameters in this respect (Rauch et al., 2000). Essentially, the addition of these reagents shifts the DNA melting curve to the left, with the result that the absolute T_m for the double stranded DNA value decreases. This decrease in T_m may be tens of degrees centigrade dependent on the factor included in the reaction mix. Conversely, several compounds actually stabilize the DNA double helix, including magnesium ions and elevated salt concentrations, which help neutralize the triple-negative phosphate charges on the opposing DNA strands in the duplex (therefore limiting electrostatic repulsion), as well as (partially) neutralizing the effect that the water dipole has on DNA hydrogen bonding.

The heating procedure used to melt DNA during the PCR process is a simple and reliable method for ensuring that DNA strand separation occurs and that DNA binding sites for specifically designed PCR primers are exposed. However, at DNA melting temperatures, PCR primers also remain disassociated from the target DNA, and effective binding of the PCR primers to the target DNA can only take place at a reduced temperature when the thermal energy is low enough to allow complementary base pairing. In most applications (random amplification of polymorphic DNA or RAPD excepted), PCR primers are designed to specifically bind to known sequences of target DNA and to anneal to the melted target DNA at a temperature of between 45°C and 65°C (the "annealing" temperature). Of course, at this temperature the target DNA also re-anneals to its complementary strand. However, the excess concentration of primers added to PCR mixes ensures that binding of at least some PCR primers will occur during each thermocycle (dependent of course on the presence of primer-complementary sequences in the target DNA).

2.3 DNA Dependent DNA Polymerase and Oligonucleotide Primers

DNA dependent DNA polymerase is an essential component of both *in vivo* DNA replication and the PCR process (Fig. 2.2), though the DNA dependent DNA polymerases found in the vast majority of organisms are heat sensitive (thermolabile), one of the major stumbling blocks hindering the initial success of PCR. However, the discovery of thermophilic organisms and the subsequent isolation of thermostable DNA dependent DNA polymerases from these organisms heralded a new chapter in PCR amplification technology [Saiki et al., 1988], such that successive cycles of heating and cooling (necessary for target DNA melting and primer annealing during PCR thermocycling) no longer resulted in the concomitant denaturation of the thermolabile DNA dependent DNA polymerase enzyme. The discovery of these thermostable enzymes allowed the whole PCR process to become far more convenient, less time consuming and more user-friendly for laboratory personnel. The first thermostable DNA dependent DNA polymerase enzyme to be widely used in PCR (Taq polymerase) was derived from the hot spring-dwelling bacterium *Thermus aquaticus*, though many other commercially available non-Taq thermostable DNA dependent DNA polymerase enzymes are now available on the market, e.g. Pfu from *Pyrococcus furiosus*, Vent from *Thermococcus litoralis*, etc. The advantages and disadvantages of some of these different thermostable polymerases with respect to PCR thermocycling are described more fully in Chapter 7.

DNA dependent DNA polymerases *per se* are actually incapable of performing DNA synthesis from a purely single stranded piece of DNA and require an additional shorter "priming" oligonucleotide to initiate DNA replication. In *in vivo* DNA replication, an RNA oligonucleotide (generated by a DNA dependent RNA polymerase called primase) acts as the primer for DNA replication. This primase enzyme synthesizes a short RNA molecule (of approximately ten nucleotides),

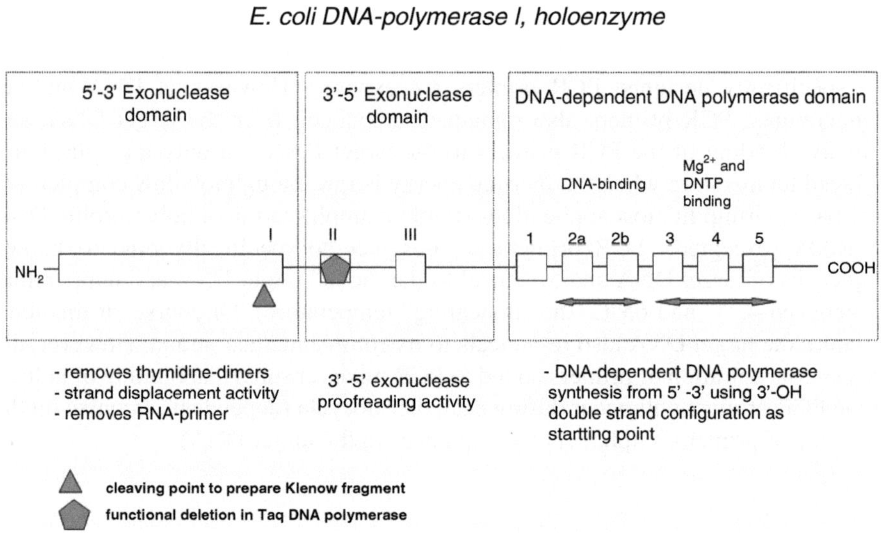

Fig. 2.2 Schematic outline of the structure of the DNA dependent DNA polymerase I holoenzyme from *Escherichia coli.* The amino terminal domain has exonuclease activity, which is able to remove RNA primers and thymidine dimers and to induce strand displacement. The deletion of this domain generates an enzyme that is deficient in 5′- to 3′- exonuclease activity (Klenow fragment). In contrast, the central domain exhibits 3′- to 5′- exonuclease activity, and the carboxy terminal domain is involved in the binding of DNA and co-factors such as magnesium ions and dNTPs. *Taq* polymerase essentially has the same structural organisation

which is complementary to the DNA region to be replicated, and once hybridized acts as a "primer" for DNA replication by cellular DNA dependent DNA polymerase. The primase enzyme itself does not require a primer in order to generate this short RNA oligonucleotide. In contrast, a typical PCR uses two specifically designed and synthetically synthesized 15–25 base pair long oligonucleotides to act as primers for the DNA dependent DNA polymerase enzyme (see Fig. 1.1, Chapter 1). PCR primers are designed such that one primer is complementary to the sense (coding) template strand and one primer complementary to the antisense (non-coding) template strand, and as such are the major components, in determining PCR specificity. Further, the two PCR primers determine the length of the DNA region to be amplified and facilitate the exponential (two-fold or doubling) of amplification products during each PCR cycle.

2.4 Deoxyribonucleotides and Additional Factors

In vivo DNA replication and PCR not only require specific primers (RNA for *in vivo* replication, DNA for PCR amplification) and a (thermostable) DNA polymerase, but also several other factors. The most essential of these factors are the building blocks or free deoxynucleotide triphosphate molecules that are incorporated into the growing DNA chain by the DNA dependent DNA polymerase. These compounds can be acquired *in vivo* as nutrients, but are generally synthesized via complex pathways involving the reduction of ribonucleotide diphosphates and addition of a further phosphate group by a kinase enzyme. The four major deoxynucleotides required for DNA synthesis (adenosine-, guanosine-, thymidine- and cytosine-triphosphate) contain an energy-rich triphosphate moiety, which is utilized by the DNA polymerase to catalyze a phosphodiester link between the 3′-hydroxy terminus of the primer (or previously added deoxynucleotide triphosphate) and the newly acquired deoxynucleotide triphosphate on the growing DNA strand. Hydrogen bonding between complementary nucleotides on adjacent strands then completes the double stranded primary structure of the DNA molecule. The mechanism by which deoxyribonucleotide triphosphates (dNTPs) are added to the growing DNA chain is identical for both *in vivo* replication and PCR amplification processes, though the concentration of dNTPs in *in vitro* amplification reactions may be easily manipulated to artificially high levels. As well as deoxyribonucleotides, successful *in vivo* DNA synthesis and amplification requires a large number of chemical components such as (1) the bivalent metal ions magnesium (a cofactor for the DNA polymerase enzyme) and manganese and (2) simple chemicals such as sodium chloride, etc. which help to maintain the correct pH of the reaction or allow for the synthesis of new DNA with a complex secondary structure (see Chapters 4–7 and Wilson et al., [2002] for a more detailed discussion).

References

Davies PA, Gray G. 2002. Long-range PCR. Methods Mol Biol 187:51–55.

Goren A, Cedar H. 2003. Replicating by the clock. Nat Rev Mol Cell Biol 4:25–32.

Hu M, Chilton NB, Gasser RB. 2002. Long PCR based amplification of the entire mitochondrial genome from single parasitic nematodes. Mol Cell Probes 16:261–267.

Kelman Z. 2000. DNA replication in the third domain of life. Curr Protein Pept Sci 1:139–154.

Nasheuer HP, Smith R, Bauerschmidt C, Grosse F, Weisshart K. 2002. Initiation of eukaryotic DNA replication: regulation and mechanisms. Prog Nucleic Acid Res Mol Biol 72:41–94.

Nasmyth K. 2002. Segregating sister genomes: the molecular biology of chromosome separation. Science 297:559–565.

Rauch J, Wolf D, Hausmann M, Cremer C. 2000. The influence of formamide on thermal denaturation profiles of DNA and metaphase chromosomes in suspension. Z Naturforsch 55:737–746.

Saiki RK, Gelfand DH, Stoffel S, Scharf SJ, Higuchi R, Horn GT, Mullis KB, Erlich HA. 1988. Primer directed enzymatic amplification of DNA with a thermostable DNA polymerase. Science 239:487–491.

Shcherbakova PV, Bebenel K, Kunkel TA. 2003. Functions of eukaryotic DNA polymerases. Sci Aging Knowl Environ 26:RE3.

Stewart RD. 1997. A theoretical investigation of cell cycle effects and interspecies radiosensitivities. Ph.D. thesis.

Wilson T, Carson J, Bowman J. 2002. Optimisation of one-tube PCR-ELISA to detect femtogram amounts of genomic DNA. J Microbiol Methods 51:163–170.

Chapter 3

The PCR in Practice

3.1 Brief Overview of PCR Requirements

3.1.1 The PCR Reaction Mix

A typical PCR reaction mix includes target DNA (Chapter 4), specifically designed oligonucleotide primers (Chapter 5), deoxyribonucleotide triphosphates/magnesium ions/a buffer component (Chapter 6), a thermostable DNA polymerase (Chapter 7) and water. The range of quantities for each ingredient of a typical PCR mix is indicated below:

- 10 fg–10 µg of chemically extracted DNA or reverse transcribed RNA in 1–10 µl of rehydration buffer or water
- 0.1–1.0 µM each of oligonucleotide primers, one complementary to the positive (sense) strand and the other complementary to the downstream (antisense) strand
- Approximately 200 µM of each of the four deoxyribonucleotide triphosphate DNA building blocks, i.e. adenosine-, cytosine-, guanosine- and thymidine triphosphate
- 0.2–2.0 units of Taq polymerase (other commercially available thermostable DNA polymerases may also be used but bear in mind differences in proof-reading ability and processivity)
- Tris-buffer (pH 8.3), 50 mM KCl (specific buffers are usually supplied free by the manufacturer of the thermostable enzyme used)
- 500–1,000 µg gelatin or Bovine Serum Albumin (BSA) per ml (optional).
- 0.5–5.0 mM of magnesium chloride salt solution ($MgCl_2$ is a DNA polymerase cofactor)

All prepared in a total typical PCR reaction volume of 50 µl (5–100 µl = the usual range).

Note that the composition of the PCR reaction mix may vary dependant on the nature (and quality) of the heat-resistant thermostable DNA dependent DNA polymerase used. Low concentrations of detergents, such as Triton X-100, Tween-20, betain or dimethylsulphoxide (DMSO) may also be included in the PCR mix to help increase the specificity of primer binding. The same compounds may assist in overcoming problems caused by secondary structure or possibly even inhibitory

E. Van Pelt-Verkuil et al., *Principles and Technical Aspects of PCR Amplification,*

compounds. The composition of the PCR mix should ideally be optimized for every new PCR protocol developed, as any change in the PCR methodology (including changes in primer design, deoxyribonucleotide composition, template nucleic acid and type of thermostable DNA polymerase used) may influence the specificity of amplification. All PCR reaction ingredients should be stored in a freezer in a dedicated "clean" room where strict guidelines are enforced to help prevent possible contamination of reaction mixes and ingredients (Chapter 10).

In order to decrease the likelihood of non-specific primer binding occurring during the initial (i.e. pre-first cycle) stage of the PCR thermocycling program (when the temperature is low enough to allow partial binding of the primer), one of the reaction components (usually the DNA polymerase or magnesium salt) may be omitted from the reaction mix and added once the initial target DNA melting temperature of 90–95°C is reached (PCR amplification will not proceed in the absence of magnesium ions). Unfortunately however, the simple act of opening and closing a PCR reaction tube greatly increases the chance of contamination of that tube. One elegant alternative to this problem is to physically separate the reaction components into two separate compartments, e.g. via a thin layer of wax. Magnesium chloride solution may then be added to the upper surface of the wax bead. The wax bead then melts at a temperature that is too high for non-specific binding of the PCR primers to the DNA template to occur. Some suppliers provide wax beads already impregnated with magnesium salt. Other elegant alternatives devised to decrease the likelihood of non-specific primer binding during the initial stage of the PCR thermocycling program, include heat-activatable thermostable DNA polymerases, which require approximately 10 minutes of heat activation at around 90°C before becoming activate. Other methods utilise polymerases coated with antibodies, which inhibit the action of the polymerase until the temperature is high enough to denature and remove the antibodies. All of these approaches are collectively named "hot-start" procedures (Chapter 7).

When performing multiple PCRs in a single "batch" (i.e. at the same time), it is advisable to prepare a "master mix" of reaction ingredients (comprising a multiple number of individual reaction mixes prepared in a single tube), and then pipette out the individual volumes into individual reaction tubes, rather than pipetting each separate component into every individual reaction tube one after the other. This process significantly reduces the number of pipetting actions that need to be performed and as such greatly contributes to the reproducibility and reliability of the PCR test protocol. The volume of a particular reagent required in a master mix is dependent on the number of PCRs to be performed, and may be calculated by multiplying the volume of the ingredient required for a single reaction, by the number of tests to be performed, including duplicates and positive and negative control samples. Additionally, for every master mix reaction prepared, an extra aliquot of each ingredient should be added (in order to account for volume errors inherent in any repeat pipetting action). Preferably, once prepared, the master mix should be immediately dispensed or stored on ice for short periods of time. Mastermixes and/ or individual PCR reaction mix aliquots may be stored frozen for a prolonged period of time and immediately thawed prior to use. In this case, it is preferable to prepare master mixes without thermostable DNA polymerase and then add the required volume of enzyme after thawing (immediately prior to PCR thermocycling). The time period during which frozen mastermix aliquots may be stored without loss of PCR amplification sensitivity should ideally be determined for each particular mix to be prepared and be routinely quality controlled [Kofler and Klausegger, 1999]. From the authors' own experience, PCR master mix aliquots without thermostable DNA polymerase may be stored for 2–3 months at −20°C without loss of PCR sensitivity.

3.1.2 The PCR Thermocycling Regime

Figure 3.1 shows the temperature profile of a typical single round (single cycle) of PCR amplification. Each cycle includes a series of three repeating "amplification steps", of which the primer annealing temperature step is the most variable. In fact, the temperatures and times of each step in the PCR cycle may have to be altered in order to optimise each PCR protocol used. However, by far the most important factors to consider in this respect are the primer annealing temperature (step 2) and the number of PCR cycles to be completed per PCR run (Chapter 8).

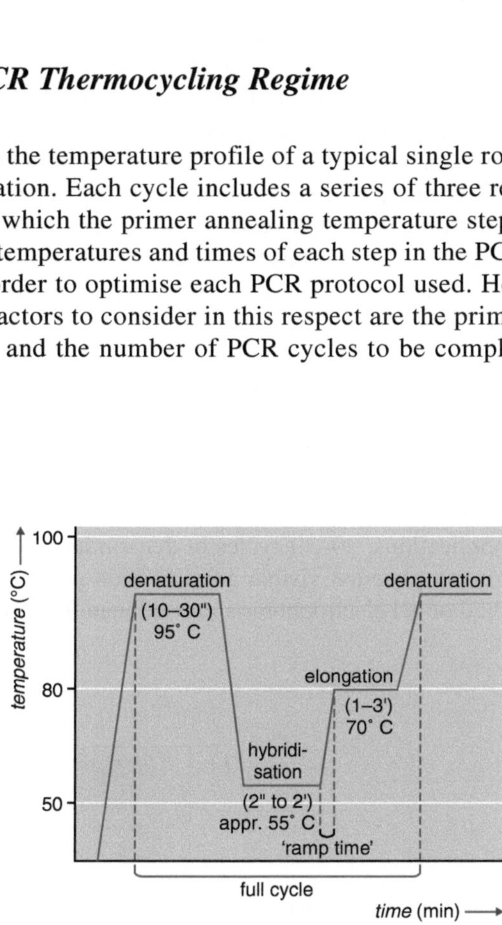

Fig. 3.1 Temperature profile of a typical three step PCR cycle. Initially, DNA is melted by an incubation step at 95–98°C (denaturation). Primers are then allowed to bind to the target DNA by lowering the temperature to the calculated annealing temperature of the primer pair used (usually between 45–65°C) (annealing). Subsequently, DNA extension takes place by increasing the temperature to the optimal temperature of the thermostable DNA dependent DNA polymerase, (usually around 70°C) (extension or replication). The following cycle of the PCR is then initiated by returning to the melting temperature of 95–98°C. Incubation times for each of these temperature steps may vary between 30 seconds and 2 minutes in an average PCR protocol

The different temperature cycling conditions required by PCR were originally achieved by transferring the reaction tubes between separate heated water baths. This methodology is nowadays redundant due to the development of dedicated automated PCR processors or thermocyclers, which precisely control the pre-programmed reaction temperatures, incubation times and number of cycles of the PCR. Many different brands of machines are currently available, using a range of principles for temperature control, e.g. metal Peltier elements (heating or cooling being achieved by passing an electric current through two conductors, and by far the most popular method), as well as hot water, and hot air. In the vast majority of these machines, sample heat exchange occurs via tight contact between the reaction tube and a metal (aluminium or even gold plated) heating block or via direct contact with heated fluids or air. Peltier element thermocyclers may be purchased in many shapes and sizes, ranging from those containing a single heating block, to those containing two or three individually controlled heating blocks per machine [Pray, 2004]. Moreover, these heating blocks are often interchangeable, allowing the same machine to be used for a range of PCR reaction tube sizes, as well as PCR applications, e.g. *in situ* PCR. Peltier-based machines can be obtained from a wide variety of companies (Applied Biosystems, Bio-Rad, Brinkman/Eppendorf, CLP, MatriCal, MJ Research, MWG Biotech, Stratagene, Techne, Thermo Electron and Whatman Biometra). Machines based on principles such as circulating air (Roche Applied Sciences, St. John Associates, Brooks Automation, Corbett Research, Idaho Technology), heating blocks (Stratagene), waterbaths (Abgene), electrically conducted polymers (Biogene), microfluidics (BioTrove) and ovens (Cepheid) are available as well, and more detailed information can be retrieved from company websites [Pray, 2004] (Fig. 3.2).

For many PCR applications, 20–30 cycles of denaturation, annealing and chain extension is sufficient to generate visible amplification products using the traditional detection method of gel electrophoresis in combination with ethidium bromide

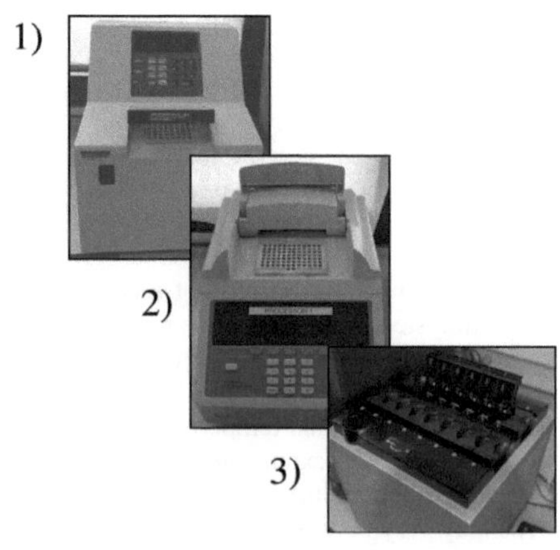

Fig. 3.2 Photographs showing 3 different types of PCR machines. (1) Applied Biosystems 480 PCR machine – one of the earliest thermocyclers, takes 48 × 0.5 ml tubes, requires oil overlay for each reaction, (2) Applied Biosystems 9,600 PCR machine – takes 96 × 0.2 ml tubes, has heated-lid, and (3) Cepheid SmartCycler II – real-time PCR machine taking 16 × 25 µl or 100 µl capillaries each with independent programmable thermocycling routines

staining (Chapter 9). If a higher sensitivity is required, for instance due to an apparent limit on the quantity of target DNA, then the number of programmed cycles may be increased from 30 cycles to a generally accepted maximum of 50 cycles. The amplification factor required for the determination of visible amplification product from biologically available quantities of starting DNA material using ethidium bromide staining lies in the order of 10^5–10^6.

The vast majority of PCR protocols currently use a ramping temperature/time of approximately 1°C/second, which allows for 20–30 cycles of amplification in approximately 2–3 hours. Less frequently used PCR protocols may considerably increase the average time for 20–30 cycles of amplification, e.g. >16 hours for PCRs intended to amplify multiple-kilobasepair sized fragments. Further, more recent technological developments have also facilitated the implementation of high-speed PCR tests. By using thin glass capillaries, the volume of the PCR reaction mix may be reduced and heat transfer rates (ramp rates) vastly increased from 1°C per second to 20°C per second. PCR processing times are also therefore greatly reduced from several hours to less than 30 minutes, though the number of samples that can be processed per PCR "run" are somewhat reduced in capillary PCR machines.

The duration of a particular PCR program is not only determined by the incubation time required for the individual steps and the total number of cycles programmed for that PCR protocol, but also includes; (a) the time required for the initial melting of the target DNA prior to PCR thermocycling *per se* (usually a "soak" time of between 2 and 10 minutes at 90–95°C used to ensure absolute melting of the target

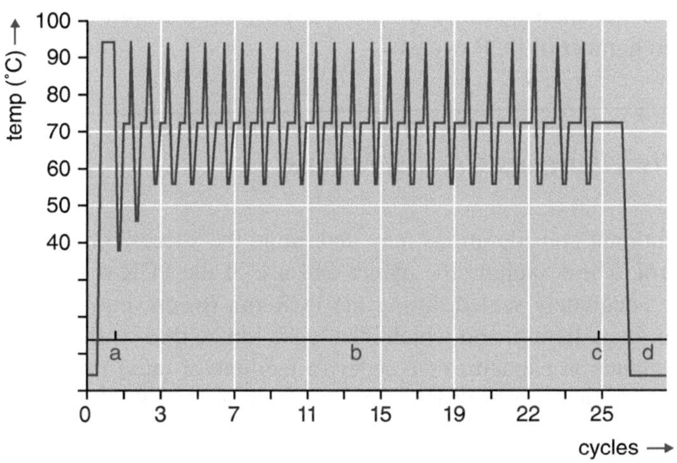

Fig. 3.3 A complete PCR temperature profile for a program consisting of 25 individual cycles. Different stages in the PCR process are indicated: (a) the time required for the initial melting of the target DNA prior to PCR thermocycling *per se*, (b) the time required for 25 PCR thermocycles, including the time required for temperature changes to be achieved between the three steps of each thermocycle, (c) a final "soak" time of approximately 10 minutes at 72°C in order to ensure that the many millions of amplified copies of DNA present are completely amplified, (d) PCR samples may be stored for several hours after thermocycling at 4°C until further processing is required

DNA and/or for activation of "hot-start" DNA polymerase enzymes), (b) the time required to change the temperature between the three steps of each PCR cycle, and (c) a final "soak" time of approximately 10 minutes at 72°C (added after the final PCR thermocycle is complete) in order to ensure that the many millions of amplified copies of DNA present at this stage of the PCR process are completely amplified. When the total PCR run is completed, the samples are usually cooled down to 4°C for an indefinite period, a task which may be performed by most modern PCR machines (some are even capable of freezing the samples), thereby providing stable very short-term storage conditions (Fig. 3.3).

3.1.3 Analysis of PCR Amplification Products

The visualisation of PCR amplification products is traditionally facilitated by gel electrophoresis and ethidium bromide staining, a chemical which inserts or "intercalates" between the two DNA strands and is detected by fluorescence during exposure to ultraviolet (UV) light (Chapter 9). Though this is still the method most widely used to detect PCR products, other fluorescent dyes are now available which are extremely useful in PCR applications, e.g. in the PCR sequencing of DNA. Recent developments have also facilitated the direct visualisation of amplified PCR products during actual PCR thermocycling *per se* (e.g. SYBR Green). This technology has allowed the development of "real-time" PCR and "semi-quantitative" PCR testing (where an estimate of the initial quantity of template DNA may be made based on calculations derived from the final quantity of specific post-PCR amplification after a pre-programmed number of PCR cycles).

3.1.4 Miscellaneous Considerations

Several additional considerations may determine the success or failure of a given PCR protocol. These include the nature and use of the PCR reaction tube, which needs to be adequately sealed during the PCR run (to prevent evaporation of the reaction mix ingredients), and which ideally should be thin-walled and have excellent heat conduction capacities (in order to facilitate a rapid rate of heat transfer between the external heating/cooling system and internal PCR mix environment). The most widely used PCR reaction tube formats include 0.2 ml Eppendorf tubes (during the early years of PCR, 0.5 ml sized Eppendorf tubes were used), sealable glass capillaries or microtitre plates. The current generation of reaction vials are designed for use in PCR machines with a heated lid, which ensures rapid thermal ramping rates and negates the need to add an evaporation-preventing layer of oil to the surface of the PCR mix. Further progress in PCR reaction vial design has led to the 96- or 384-well ELISA plate format, which allows for high-throughput (automated) PCR applications with a requirement for only limited hands-on time.

In diagnostic research, where results could affect patient treatment, the adequate quality control of the PCR mix ingredients (e.g. monitoring the "use-by" dates of enzymes, dNTPs, etc.) and strict adherence to the reaction mix composition, are a necessity. Changes in the quality of enzymes, primers and other ingredients may drastically affect the sensitivity and specificity of the PCR reaction. Adequate controls are particularly important, with known positive (high and low titre DNA standards) and negative (water) controls necessarily included in each batch of PCR reactions (Chapter 10). As part of the quality control procedure, it may also be prudent to design and develop written records, documenting that the necessary quality control requirements for a particular PCR batch have been met (Chapter 11). This "paper trail" procedure is particularly important in legal situations (e.g. forensic applications), as part of medical or drug studies, or when PCR testing is being compared to more traditional "gold standard" testing methods (e.g. [Zambon et al., 2001]).

The primary goal of PCR testing is the amplification and detection of small amounts of a certain sequence motif in nucleic acid extracts. This does not allow for assessment of cellular localisation or tissue distribution of the sequence motif. However, the *in situ* PCR technique may be helpful in such circumstances [Nuovo et al., 1991, 1993]. There are two distinct variants of *in situ* PCR amplification, differentiated by the mode of visualisation of the PCR amplified targets. During *in situ* (IS) PCR, labelled nucleotides are directly incorporated into the amplimers. These labelled DNA molecules can be visualised directly post-PCR. On the other hand, PCR-*in situ* hybridisation (PCR ISH) depends on the post-PCR detection of the amplimers by *in situ* hybridisation with a labelled synthetic probe. *In situ* PCR protocols are highly specific, but require modified heating blocks to accommodate pathological microscopy slides, and incubation times are typically longer than those for non-IS PCR protocols. Some *in situ* PCR protocols have recently found their way into the clinical laboratory (Chapter 13).

References

Kofler B, Klausegger A. 1999. Simplified PCR set-up using a frozen preformulated mix for the detection of cytomegalovirus. Diagn Microbiol Infect Dis 34:33–35.

Nuovo GJ, Gallery F, Hom R, MacConnell P, Bloch W. 1993. Importance of different variables for enhancing *in situ* detection of PCR-amplified material. PCR Methods Appl 2:305–312.

Nuovo GJ, Gallery F, MacConnell P, Becker J, Bloch W. 1991. An improved technique for the in situ detection of DNA after polymerase chain reaction amplification. Am J Pathol 139:1239–1244.

Pray LA. 2004. Consider the cycler. Scientist 18:34–37.

Zambon MC, Hays JP, Webster A, Newman R, Keene O. 2001. Relationship of clinical diagnosis to confirmed virological, serologic, or molecular detection of influenza. Arch Intern Med. 161:2116–2122.

Chapter 4

The Different Types and Varieties of Nucleic Acid Target Molecules

4.1 General Features

The amount of nucleic acid present within a particular cell differs greatly between the various classes of organisms and between tissue types, with microorganisms harbouring smaller genomes than the more complex nucleated cells of eukaryotes. This is partially due to the presence of various families of repeat DNA sequences present in eukaryotic DNA that are essentially absent in prokaryotes. These sequences do not seem to participate in gene coding and their precise function is still largely unknown. Viruses, which are dependent on host cell replication machinery, generally have the smallest recorded genomes, with certain plant viruses containing a genome not larger than a few kilobasepairs (kbp), for example the tobacco necrosis virus genome is only 3.8 kbp in length. However, viruses are not self-replicating and the smallest genome for a self-replicating organism belongs to the bacterium *Mycoplasma genitalium* at 5.8×10^5 bp [Fraser et al., 1995]. That said, the symbiotic mini-bacterium *Carsonella rudii* lives within the cells of the hackberry petiole gall psyllid *Pachypsylla venustai* (an insect) and has a genome of only 159,662 basepairs, or less than 300 genes. This mini-bacterium is dependant on its host for nutrients [Nakabachi et al., 2006]. In contrast, amphibians and flowering plants possess the largest genomes (4×10^8 to 1×10^{11} bp), with human cells possessing approximately 3×10^9 bp on 2×23 linear chromosomes. The overall number of nucleotides in human DNA is in the order of one billion, of which 3.5% are predicted to encode proteins approximating to 30,000 genes (if we assume an average gene length of 1,000 nucleotides). An optimised PCR amplification protocol is able to selectively amplify one individual gene (or fragment) from within all of these other genes, exons and accompanying nucleic acids, e.g. extrachromosomal plasmids, episomal viruses, ribosomal RNA, messenger RNA, etc.

In order for PCR amplification to be successful, the nucleic acid target molecule to be PCR amplified must be readily accessible to primers and DNA polymerase and free from inhibitory concentrations of contaminating proteins, lipids, carbohydrates and salts. This necessitates that the DNA or RNA target molecule is first liberated and (partially) purified from both the cellular environment in which it resides and the chaperone proteins which may accompany it. Many commercially

E. Van Pelt-Verkuil et al., *Principles and Technical Aspects of PCR Amplification,*

available kits for DNA and/or RNA purification are currently available, employing enzymatic, chemical or mechanical extraction techniques. Most of these use some form of affinity purification, where disruption of the cell or nucleus is followed by nucleic acid binding to a solid support. In this way, the nucleic acid target may be isolated and then purified by washing in order to remove contaminating proteins or unwanted nucleic acid. The most frequently used matrices for nucleic acid binding comprise silica-based particles, silica filters, (silica coated) magnetic beads [Boom et al., 1990; Bruno et al., 1996; Ito et al., 1992], and nitrocellulose or nylon filters. Due to their ease of use and minimal reliance on large quantities of hazardous chemicals, commercial extraction kits have largely replaced the older, more labour intensive methods of nucleic acid extraction which relied on phenol/chloroform/isoamylalcohol extraction, cesium chloride density gradient centrifugation or guanidinium isothiocyanate lysis followed by phenol-chloroform extraction. Alongside the introduction of commercially available nucleic acid extraction kits, appeared new extraction formats, which facilitate the high-throughput isolation of DNA and/or RNA for rapid diagnostic testing, e.g. the Magnapure apparatus (Roche Molecular Systems, USA). Finally, advances in PCR amplification technology *per se* have resulted in a new generation of methodologies that may be used to identify target nucleic acid molecules even though these molecules are still present *in situ* within their natural cellular or nuclear environments. Using such *in situ* PCR methods, negates the need for nucleic acid extraction, though the sensitivity of such *in situ* PCR methodologies may be reduced compared to *in vitro* PCR methodologies that utilise nucleic acid extraction techniques.

4.2 A Brief Description of *In Vivo* DNA and RNA Targets

Both DNA and RNA occur in the intracellular compartment and, with the exception of viruses, most nucleic acid species occur in well-defined cell types. This renders their localisation and their interaction with protein molecules highly specific. Mitochondria, chloroplasts and cilia possess circular, non-nucleated, maternally inherited DNA molecules which show structural resemblance to bacterial DNA. This organelle-specific DNA mainly serves as a template for several organelle specific proteins (e.g. the cytochrome C oxidase III enzyme of mitochondria), though the nucleus of the cell may also code for other organelle specific proteins (e.g. mitochondrial transcription factor A). However, one common feature of nearly all cellular and sub-cellular genomes is that the genome comprises double stranded DNA, the exceptions being a number of distinct viral families.

Viruses are packages of nucleic acid and obligate intracellular parasites, which require host cell transcription and translation machinery in order to self-replicate [Acheampong et al., 2003]. Viruses show a great diversity with regard to: (1) their genome size (ranging from approximately 3 kbp encoding 5 genes to approximately 200 kbp encoding 500 genes), (2) host and tissue specificity and (3) their pathogenic effect on the host (cell death, oncogenicity, etc.). In order to protect the nucleic acid, viral genomes are encapsulated by virus-specified proteins and sometimes a host-derived envelope (which also helps facilitate virus entry into the host cell). Once within the host cell, viral genomes exist as free (episomal) forms or integrate into the host genome (provirus), the actual site of virus integration being virus dependent. Viruses may be classified on the basis of the nature of their genetic material into:

Class I: Double stranded DNA viruses. This group contains relatively large viruses such as adeno- (eye, respiratory infections), vaccinia- and variola-viruses (skin infections) as well as smaller viruses such as the human papillomaviruses (warts, cervical cancer). They replicate in the host cell cytoplasm.

Class II: Single stranded DNA viruses which may be either + or – sense stranded DNA. Positive sense stranded means that the DNA or RNA has the same polarity as mRNA, whilst negative sense stranded DNA has a sequence complementary to mRNA. Their DNA may be encapsulated in virus-encoded proteins. One example is Parvovirus B19 (the causative agent of slapped cheek syndrome/erythema infectiosum/fifth disease).

Class III: Double stranded RNA viruses that show highly diverse host specificities. The genome is usually fragmented (often into 8–12 different double stranded RNA molecules), e.g. rotaviruses (infant diarrhoea).

Class IV: Positive sense single stranded RNA viruses, whose genomic material resembles messenger RNA. After penetration of the host cell, viral encoded proteins are immediately translated by the host cell ribosomes. During virus replication, a complete negative sense stranded RNA genomic copy is synthesized which serves as the template for generating new copies of the positive sense viral genome strand ready for packaging. Poliomyelitis virus is a well-known member of this family.

Class V: Negative sense single stranded RNA viruses including a characteristic virus-specific RNA-dependent RNA polymerase since the negative sense genomic RNA must be transcribed into positive sense RNA before translation can begin. Influenza viruses belong to this family.

Class VI: Retroviruses, which synthesize cDNA from their positive single stranded RNA genome using a unique "reverse transcriptase" enzyme. After reverse transcription, the double stranded cDNA can integrate into the host cell genome. These viruses frequently have transforming capabilities (the ability to prolong the life of the cell indefinitely), e.g. human T-cell lymphotropic virus I (adult T-cell leukemia), human immunodeficiency virus or HIV (the virus causing the acquired immune-deficiency syndrome or AIDS).

The differing nature of the genomic material carried by viruses in these six viral classes means that different approaches may be required for their detection, as the PCR technique *per se* utilizes DNA and not RNA as starting template material. Hence, the PCR amplification of RNA viruses must involve an initial "reverse transcription" step, converting the RNA into copy DNA (cDNA) prior to PCR processing (Section 4.5). Another point to note, is that the nucleic acid extraction methodology may need to be optimized prior to PCR amplification dependant on (i) the tissue in which the virus is distributed (faecal samples, serum sample, biopsy material, etc.) and (ii) the viral load (faeces contain approximately 10^7–10^9 virions/ml, whilst HIV may be found in concentrations of 1 copy per 10,000 CD4+ T lymphocytes up to 1,000 copies/per single CD4+ T lymphocyte [Wood et al., 1993]).

Bacteria are self-replicating microorganisms that require nutrients and the correct physical environmental factors (e.g. temperature, pH) to survive. They may be distinguished from eukaryotic cells by the fact that they do not contain a defined nucleus and that their genetic material comprises a circular and not a linear chromosome. The higher order structure of the bacterial chromosome is "loosely" defined within most bacterial species, with the circular bacterial chromosome ranging in size from 500 to 5,000 kbp. Frequently, smaller, multi-copy, extra-chromosomal circular DNA (plasmids) is also present within the bacterial cell. These plasmids range in size from a few kilobasepairs to hundreds of kilobasepairs in size and may be present in different copy numbers within the bacterial cell. Both chromosome and plasmid molecules contain functional genes, usually present as single copies, although transfer RNA (tRNA) and ribosomal RNA (rRNA) genes for instance may occur in multiple copies on the chromosome. Repeat sequences of DNA are usually present in bacterial chromosomes, though in smaller numbers than found in eukaryotic genomes. Current advances in the comparative genomics of bacteria allow degenerate PCR primers (Section 5.1.1) for the amplification of pathogenic genes from different species to be more easily designed, allow gene "knockout" studies to be designed and performed using PCR amplified DNA, as well as changing our insights into genomic flexibility [Koonin et al., 2000; Schoolnik, 2002]. Currently, the complete genome sequence of over 50 different eubacterial species is available.

Eukaryotes possess the largest genome sizes of all, usually condensed into complex, multi-structured chromosomes, with the degree of structure (or folding) being dependent on the transcriptional status of the DNA. In regions where few genes are found or where most genes are transcriptionally inactive, i.e. the centromeres and telomeres, the DNA is tightly packed into heterochromatin (Fig. 2.1). In contrast, regions that contain many genes are often more loosely packed into 30 nm fibres (or loops) of euchromatin and separated from the heterochromatin by an "insulator", which in vertebrates interacts with a "CCTC binding factor" protein. In mitotic nuclei, the degree of DNA condensation is at its maximum, with many meters of linear DNA being folded into an overall size of only a few micrometers. The number of chromosomes present within the cells of an individual varies between species, with changes in chromosome number or chromosome composition being implicated in the development of disease [Thomas and Cann, 2003]. Such changes in chromosome copy number are described using a specific terminology. In polyploid form, cells contain more than one haploid set of parental chromosomes, which is common in angiosperm plants, e.g. the peanut or banana. Aneuploidy describes the state where there is one missing or one extra chromosome, e.g. three copies of chromosome number 21 in Down syndrome [Abbott and Benn, 2002]. Mosaicism involves different numbers of chromosomes being found within different cells of an individual. Some chromosomes show a sex-dependent distribution. In humans, for example, the Y-chromosome is male-specific and chromosomal abnormalities associated with the Y-chromosome show themselves only in males, e.g. the aneuploidy

Klinefelter's syndrome [Morel et al., 2003]. Chromosomal aberrations may also occur via translocations, where parts of individual chromosomes are swapped with other chromosomes or duplicated on the same chromosome. These translocations are often associated with the development of malignant disease in humans. The reciprocal translocation between chromosomes 22 and 9 for instance, the Philadelphia chromosome, is found in 95% of chronic myeloid leukemias [Silliman et al., 2003]. All genes of eukaryotes are normally present in duplicate, one being a maternal and the other a paternal copy. Sometimes additional, non-transcribed, pseudogenes may be present, which may represent remnants of previously active genes. These pseudogenes may arise due to duplication or retrotransposition [Zhang et al., 2002]. In mammals, methylation patterns of CpG-rich regions of DNA may determine whether or not both copies of a gene (the maternal and paternal) will be expressed equally efficiently, influencing messenger RNA (mRNA) concentrations. Inhibition of gene transcription may result from C5 methylation of pyrimidine bases present in key positions of gene promoters, possibly resulting in the development of clinical diseases such as hepatocellular cancer [Lu et al., 2003]. This methylation process *per se* is facilitated by specific methylase enzymes.

Present within almost every eukaryotic gene are regions of non-coding DNA called introns. These introns are initially transcribed into RNA along with the rest of the gene but are then cut (spliced) out of the RNA transcript via the action of protein complexes called spliceosomes. The essential pieces of the RNA are then rejoined before migrating out of the nucleus to be translated on ribosomes. As a general rule the GC-content of introns is usually below that of the corresponding protein-encoding exon sequences. Problems with intron splicing have been linked to, e.g. Lou Gehrig's disease, and intron sequence data may be utilized to design PCR primers for sequencing and typing any intervening exons, the portions of RNA that actually code for protein and are part of the mRNA translated by ribosomes.

Many regions of the eukaryotic genome comprise highly repeated elements of DNA associated with both coding sequences (e.g. tRNA, rRNA, histone gene sequences) and non-coding sequence elements. The non-coding repeat sequence elements are very diverse and may be grouped into short-, middle- and large-sized repeat units of approximately 2–30 bp, 300 bp and >2,000 bp, respectively. Examples of these repeat regions include; the short interspersed nuclear elements (SINEs) Alu and Mariner, the long interspaced nuclear elements containing internal RNA polymerase III promoters (LINES), mini- and micro-satellites, centromeres and telomeres, etc. These repeat regions are often polymorphic (i. e. comprise different numbers of repeat units) between individuals due to frequent replication errors as a consequence of the presence of tertiary structure in combination with DNA polymerase infidelity, a process known as slipped strand mis-pairing [Pfeuty et al., 2001]. Short tandem repeat elements (STRs) are particularly useful as targets for forensic investigations [Agrawal et al., 2002]. Variation within a given STR may be associated with a particular disease state [Berg et al., 2000; De Leeuw et al., 2000] and the American National Cancer Institute (NCI) recommends several STR PCRs for clinical diagnostics

[Antin et al., 2001]. Such tests for repeat regions of DNA may also be adapted for the analysis of paraffin-embedded archival tissue sections, as long as the PCR products amplified are relatively short [Sieben et al., 2000; Umetani et al., 2000]. For forensic purposes, eleven different STR loci are usually investigated, often in a single multiplex PCR, with only eight to twelve cells being necessary to generate an individuals' specific DNA fingerprint [Benecke, 1997].

Individual genetic loci can sometimes be discriminated on the basis of the presence of a single nucleotide polymorphism (SNP or SNiP analysis). These naturally occurring point mutations may be found in both coding and non-coding regions of DNA and follow Mendel's rules for genetic inheritance [Pusch et al., 2002]. Moreover, they are the most common form of polymorphism present in the human genome, being present on average once in every 1,000 bp. SNPs may be used in both diagnostic and large scale epidemiological studies. As such, SNPs can be valuable indicators of genetic disease, genetic predisposition towards metabolic disease, or the compatibility of certain medications or treatment regimens [Barton, 2002].

As well as DNA chromosomes, living cells also contain various different sets of RNA molecules, where ribonucleic acid is the sugar component instead of deoxyribonucleic acid. Some of these different sets of RNA molecules are shared by all cells whereas others are highly cell-type specific. Most of the large RNA molecules are in one way or another involved in protein biosynthesis.

Messenger (mRNA) represents approximately 3% of the total RNA content in a transcriptionally active cell. Most eukaryotic mRNA molecules contain an approximately 200 nucleotide long poly-adenine region at the 3′- (tail) end and a modified guanine (*cap*) at the 5′- end, both which are added during post-transcriptional maturation. The 5′- *cap* structure protects the mRNA against exoribonuclease and phosphatase mediated RNA digestion, as well as promoting mRNA translation by the eukaryotic translation apparatus. Both the 5′- and the 3′- ends of mRNA molecules usually contain non-translated domains (untranslated regions or UTRs), many of which play an important role in post-transcriptional processing such as ribosome binding and disengagement and the binding of repressors and activators. Many UTRs contain specific protein binding domains that have a highly complex structure [Alberts et al., 1994; Mitsuhashi, 1996b]. In eukaryotes, mRNA is generated from heterogeneous nuclear RNA (hnRNA), which is spliced into mature mRNA within a very short period of time. These primary hnRNA transcripts usually range in size from 2 to 20 kb in length and undergo several splicing and ligation reactions before the final mRNA is ready for translation (Fig. 4.1). Despite the presence of this excess material, eukaryotic mRNAs are usually monocistronic, i.e. code for one protein per molecule, whereas the mRNA of prokaryotes are frequently polycistronic (code for several proteins on a single mRNA molecule). Most mRNAs are synthesised in low copy numbers dependent on the protein(s) that is/are encoded, the tissue of origin, the stage of the cell in the growth/replication cycle and the health of the cell. mRNA has a half life varying from a few minutes to many hours.

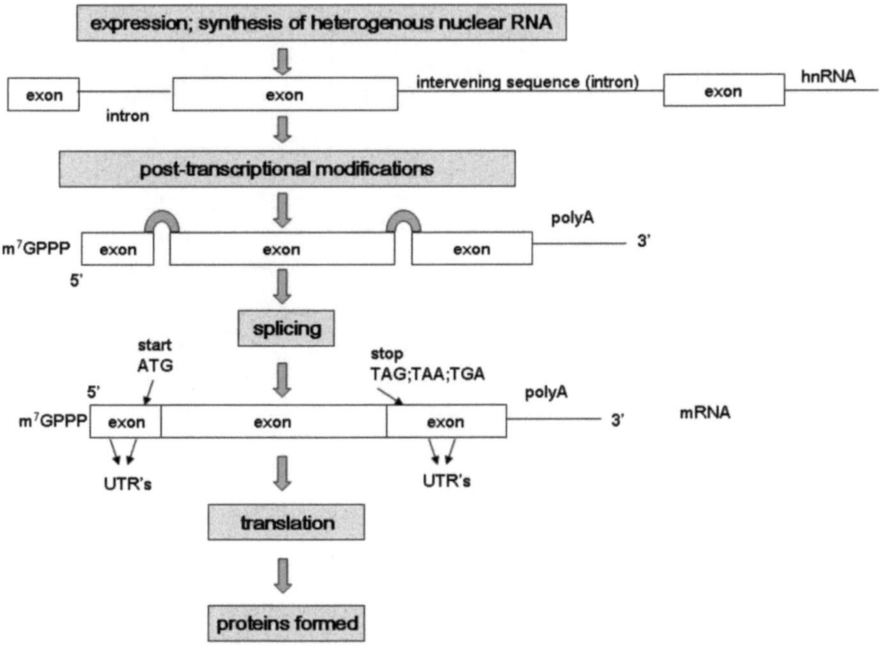

Fig. 4.1 Structural characteristics of eukaryotic messenger RNA and its splicing. Genes are built from blocks of exons (expressed at the protein level) and introns (intervening sequences). The latter are removed from the precursor heterogeneous nuclear (hn) RNA through splicing. The 5′-end of the mRNA molecule is equipped with a 7-methyl guanosine triphosphate moiety, whilst the mRNA 3′-end is usually enzymatically polyadenylated. The mature mRNA molecule can be translated into proteins

Transfer RNAs (tRNAs) are present in the cytoplasm and the nucleus of the cell and are involved in capturing and transporting individual amino acids as they are built into polypeptide chains and proteins on the ribosomal RNA. All tRNAs share the same three-dimensional structure, comprising a three base pair "anticodon" at one end (which binds to a complementary three basepair "codon" region of mRNA). The 3′-end of the molecule, where the amino acid is chemically linked, always ends in the nucleotide sequence CCA. Amino acids are enzymatically attached to the adenosine molecule by a range of tRNA aminoacyl synthetases [Tamura and Hasegawa, 1999]. Interestingly, mitochondria and chloroplasts contain tRNAs that are highly similar in structure to the tRNAs of bacteria.

Ribosomal (rRNA) is the most abundant group of RNA molecules within cells comprising over 90% of the total cellular RNA. rRNA molecules have a very complex secondary and tertiary structure [Ban et al., 2000] (Fig. 4.2), possessing regions with a high degree of inter-species conservation interrupted by regions with a high degree of sequence variability. These regions of high sequence variability

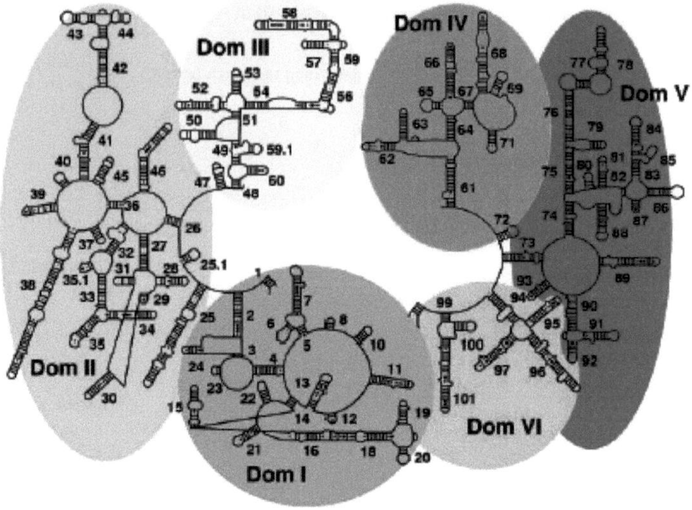

Fig. 4.2 Schematic secondary and tertiary structure diagram of the 23S small subunit (ssu) component of the large 50S prokaryotic ribosomal RNA molecule from *Haloarcula marismortui* (an Archean bacterium) and its colour coded domains. The complexity of the structure is obvious, containing several distinct RNA domains involved in different structure maintaining and/or translational activities (Reprinted from Ban et al., 2000. With permission from AAAS)

have great diagnostic value, especially in the field of microbiology [Van Burik et al., 1998; Hendolin et al., 2000]. Ribosomes, the cellular machinery by which mRNA is translated into polypeptides and proteins, contain copies of rRNA of various sizes (referred to as large and small subunits). Differences in the size of these subunits are apparent between prokaryotic and eukaryotic organisms. In prokaryotic organisms, the small ribosomal subunit (30S in size; where 30S relates to a sedimentation coefficient of 30 Svedburg units) comprises 16S RNA and a further 21 additional proteins, whilst the large ribosomal subunit (50S) comprises both 23S and 5S RNA and 34 other proteins. In contrast, in eukaryotes, the small ribosomal subunit (40S) comprises 18S RNA and a further 21 additional proteins, whilst the large ribosomal subunit (50S) comprises 28S, 7S and 5S rRNA. Ribosomal subunits are also associated with a large number of accessory proteins. Interestingly, the ribosomes found in chloroplasts and mitochondria of eukaryotes are more similar in structure to the 70S ribosomes of prokaryotes than the 80S ribosomes of eukaryotes.

There exists a broad range of nucleic acid molecules present within living cells which may potentially be utilised as targets for nucleic acid amplification and diagnostics. Molecular assays (including PCR) can be developed and adapted to utilise these targets so long as the chemical nature, structure, stability and quantity of the target material is taken into account during target isolation and PCR primer design.

4.3 DNA Samples

In principle, PCR can be used to amplify a single copy of a complete or partial gene sequence many millions of times. An optimised PCR assay, performed on "clean" DNA (i.e. DNA without associated histones and accessory proteins), should be able to detect a single copy of (for example) a viral genome against a background of 50,000 eukaryotic genomes. However, it should be noted that the more copies of the target sequence are present at the start of the amplification, the more robust is the PCR. In practice, approximately 100–30,000 copies of target sequence are usually present at the start of PCR thermocycling. Correspondingly, the detection of a single gene in a eukaryotic cell requires target DNA from approximately 6,000 cells, which is approximately equivalent to 20 ng of human DNA. For multi-copy genes, the actual number of copies per genome influences the lower limit of template DNA required for successful amplification. Since bacterial and viral genomes are much smaller than eukaryotic genomes, there will be relatively more copies of a particular target bacterial or viral gene for a given quantity of extracted bacterial and viral DNA than for the same quantity of DNA obtained from a eukaryotic source.

Relatively impure DNA samples (e.g. washed cells from cell culture, bacterial colony suspensions) may be used for PCR and added directly to the PCR amplification mix prior to thermocycling. However, the amount of material (cells, bacteria, etc.) added to the mix should always be kept low in order to reduce the effect of inhibitory substances, e.g. mucopolysaccharides, DNA-coupled glycoproteins, heme-moieties and certain fatty acids, which may facilitate physical obstruction or enzymatic inhibition of the thermostable DNA polymerase. In fact, PCR using non-purified DNA is a rather "hit and miss" affair, with reproducibility being a particular problem. For diagnostic purposes, target DNA or RNA should always be purified prior to processing, with the nature of the DNA isolation protocol being determined by the complexity of the material from which the nucleic acids are to be isolated (Section 4.3.1). Essentially, isolated DNA target material should be free of nucleases (which could degrade template and amplified DNA), endo- or exoproteases (which may inactivate the thermostable DNA dependent DNA polymerase through digestion) and other DNA binding proteins (which could help stabilise the double helix structure of the DNA and hence interfere with DNA unwinding, primer annealing and ultimately PCR amplification). These contaminating and PCR inhibiting proteins may be especially important in *in situ* PCR reactions, where "fixed" (e.g. paraffin embedded) material is used (Chapter 13), and where protein-DNA cross-linking interactions may completely block successful PCR amplification.

In rare instances, the secondary or tertiary structure of the target nucleic acid may limit the successful outcome of the PCR. Stable hairpins result in incomplete denaturation or melting of the DNA, leading to shortened PCR amplification products which cannot take part in further rounds of amplification, whilst guanosine–cytosine rich regions may cause the DNA dependant DNA polymerase to "stutter", increasing the probability of the polymerase "falling off" of the template DNA.

The addition of DNA destabilising agents such as DMSO or betain to the PCR reaction mix may alleviate the problems associated with target secondary or tertiary structure, though different types of thermostable DNA dependent DNA polymerase may exhibit different degrees of inhibition in different PCR mixes. The inclusion of modified, partially base-pairing deficient nucleotide homologues is another possible solution to overcoming PCR amplification inhibition due to the presence of PCR inhibitors [Ohtsuki et al., 2001].

4.3.1 DNA Isolation Procedures

Successful nucleic acid isolation protocols have been published for nearly all biological materials (Fig. 4.3). These protocols generally involve the physical and chemical processes of tissue homogenisation (to increase the number of cells or the surface area available for lysis), cell permeabilisation (using non-ionic detergents, enzymes, etc.), cell lysis (using hypotonic buffers), protein degradation and removal of nucleases (e.g. using alkaline protease), protein precipitation (heat, chaotropic salts), solubilisation of nucleic acids and finally various washing steps. Tissue homogenisation (e.g. using a blender) is particularly useful with respect to extracting the nucleic acid from whole or partial organ(s). Cell permeabilisation may be achieved with the help of non-ionic (non DNA-binding) detergents such as Tween, SDS, Nonidet, Laureth and Triton. Nonidet P40 is particularly interesting as it lyses eukaryotic cell membranes but does not degrade the nuclear membrane. Plant material is particularly resistant to nucleic acid extraction. The cell wall and various starch containing organelles contain complex polysaccharide protein conjugates and polyphenols that often co-purify alongside the DNA. Freeze drying or immersion in liquid nitrogen of leaf sections followed by physical destruction with a pestle and mortar in combination with an aggressive lysis buffer (e.g. cetyl trimethylammonium bromide (CTAB) or N-laurylsarcosine [Sarkosyl]/proteinase K mix) are often used in plant tissue extraction protocols.

Addition of a protease enzyme efficiently degrades released nuclease enzymes after cell lysis, whilst also helping to remove chaperone nucleoproteins from their nucleic acid host. After treatment, some proteinases can be removed by extraction or simple heat inactivation. For paraffin embedded material, prolonged incubation with proteases may be necessary for up to 5 days [Jackson et al., 1990]. A simple heating step will also inactivate several of the cellular nucleases and proteases released upon cell lysis. In the case of paraffin-embedded tissue sections, this heating process will also serve to melt the paraffin, thereby facilitating separation of the hydrophobic paraffin from the aqueous phase in which the nucleic acid is suspended, though several hydrophobic solvents, e.g. xylene and toluene may be used to achieve the same effect [Fredericks and Relman, 1999]. After protease treatment, the nucleic acid may be precipitated and concentrated into a smaller volume. Well-known nucleic acid precipitants include isopropanol, ethanol and polyethylene glycol (PEG). Figure 4.4 shows an example of the nucleic acid precipitating agent

N,N′-Bis[3,3′-(dimethylamino)-propylamine]-3,4,9,10-perylene-tetracarboxylic diimide (DAPER). This compound binds in stoichiometric proportions to any nucleic acid molecule over 70 bp in length. Aternatively, CTAB is a positively charged detergent with similar capacities to DAPER. At an elevated salt concentration, CTAB complexes to nucleic acid molecules and renders them insoluble. Chelex is another nucleic acid precipitant which binds most specifically at low ionic strength and at elevated temperatures. Moreover, Chelex also indirectly protects nucleic acids from hydrolysis by sequestering the metal ions that are required for nuclease activity and is often used in forensic science to remove several common PCR inhibitors [Burkhart et al., 2002]. However, it should be noted that the use of Chelex significantly reduces RNA yield in nucleic acid extractions.

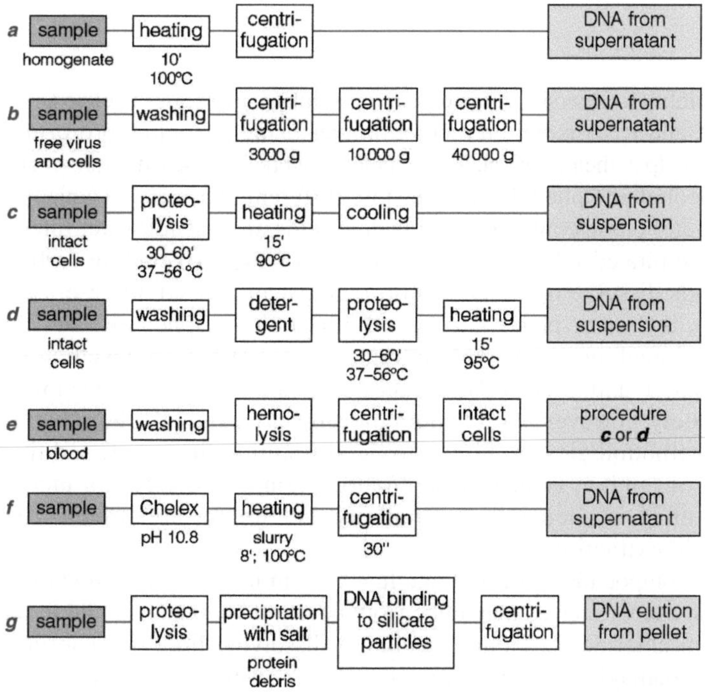

Fig. 4.3 Boxed scheme outlining a variety of different DNA purification protocols. The general protocol consists of a limited number of steps, including: (1) the dissolution of the material in which the nucleic acid is packaged (i.e. tissues, microorganisms, etc.), (2) the denaturation and washing away of contaminating proteins, lipids, polysaccharides, etc. and (3) nucleic acid precipitation and subsequent washing. From left to right several stepwise procedures are outlined, always starting from the (clinical) material and ending with the storage of (partially) purified DNA

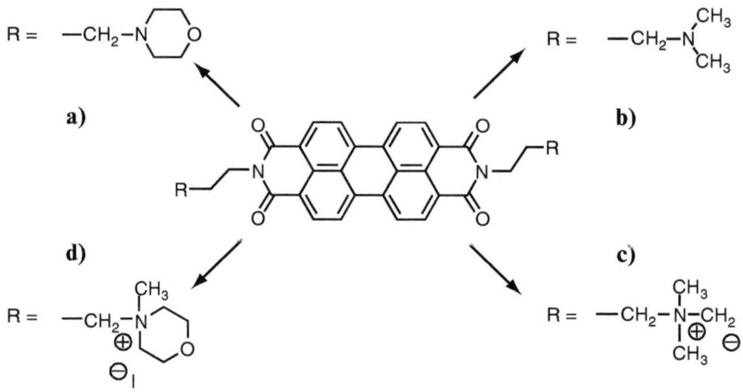

Fig. 4.4 DAPER (N,N″-Bis[3,3′-(dimethylamino)-propylamine]-3,4,9,10-perylene-tetracarboxylic diimide; $C_{34}H_{32}N_4O_4$; molecular weight 560) is a chemical capable of precipitating DNA. The alternatively substituted compounds are known as (a) Tel03, (b) DAPER, (c) DAPER01 and (d) Tel11. All compounds have DNA intercalating capacities

If PCR inhibitors persist in the extraction solution, even after nucleic acid isolation, then DNA and RNA isolation methodologies using phenol/chloroform/isoamylalcohol may help in their removal. These methods work because a mixture of chloroform, isoamylalcohol and phenol separates into different phases in an aqueous solution, with the water phase containing the nucleic acids, the phenol/water interphase collecting precipitated inhibitory proteins and polymers (including carbohydrates), whilst lipids dissolve in the chloroform and isoamylalcohol phase. After phenol extraction, the water phase is pipetted into a clean Eppendorf and the nucleic acid precipitated by adding 3 M sodium acetate pH 5.2, followed by washing in 100% ethanol and then 70% ethanol to remove residual phenol and salts. Unfortunately however, the phenol/chloroform/isoamylalcohol chemicals used in these methodologies are highly toxic, difficult to dipose of, and the methods are time consuming. Hence, phenol/chloroform/isoamylalcohol methodologies have lost much of their initial popularity and are not now frequently used for nucleic acid extraction.

Several compounds serve multiple functions in nucleic acid extraction protocols. Chaotropic agents such as guanidine isothiocyanate (GuSCN), sodium iodide (NaI) and lithium chloride (LiCl) are capable of dissolving lipid-rich capsules and cell envelopes (damaging the integrity of cells), as well as denaturing proteins and inactivating nucleases. GuSCN at high molarity (>4 M) precipitates high molecular weight DNA and RNA and can also be used to precipitate mRNA molecules at even higher molarities. Within this chaotropic salt-rich environment, silica compounds are capable of specifically binding DNA and RNA molecules, whilst contaminating lipids and proteins have only a moderate affinity for these compounds and may be readily washed away. Alternatively, by chemically changing the nature of the silica matrix, the matrix may be transformed to specifically bind contaminating

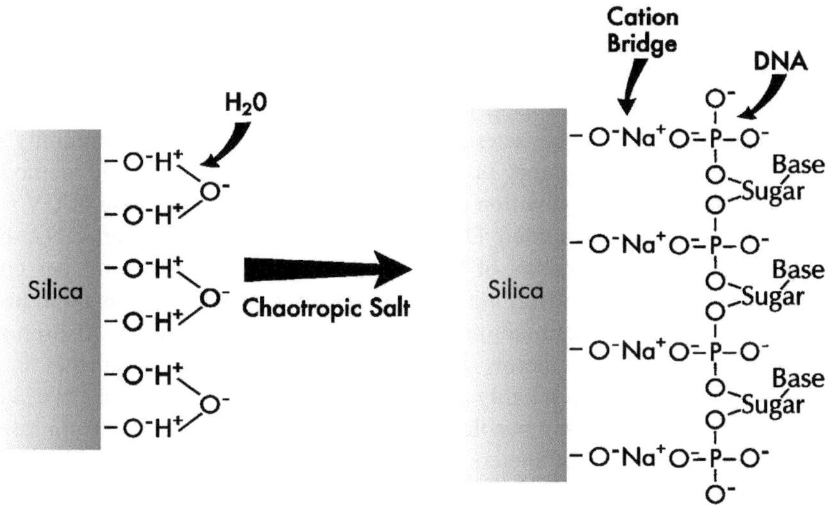

Fig. 4.5 Silica provides a universal medium for the immobilisation of nucleic acid molecules. In the presence of water and chaotropic salts the molecules interact ionically through sodium ions and the nucleic acid phosphate groups. After binding, contaminants may be removed by washing and centrifugation (From the Geneclean Guide to Protocols and Procedures, Bio101. Courtesy of MP Biomedicals LLC)

protein or vegetable polysaccharides instead of nucleic acids. Nucleic acid adsorption to silica matrices is actually enhanced in acid pH and high salt concentrations, with the consequence that a buffer solution containing an elevated pH and low salt conditions may be used to elute bound nucleic acid from the silica matrix [Boom et al., 1990; Tian et al., 2000). This silica method of DNA isolation is by far the most common DNA isolation method utilised today, and silica-based matrices have been introduced in most commercially available extraction kits (Fig. 4.5). The silica matrices within these kits may be found in many forms, including filters (e.g. Wizard spin columns, Promega, USA; Nucleospin columns, MN, ITK, The Netherlands), gels (e.g. Glassmilk, Geneclean, Amersham, UK; Silicagel, Hybaid, UK), suspensions (e.g. Celite, Acros Chemicals, Belgium; plain-coarse silicate, Boehringer-Mannheim, Germany), or even coating magnetic particles (e.g. Dynabeads, Dynal Biotech UK, Bromborough, UK). Filtration, centrifugation, or the use of a magnet allows separation of silica-bound nucleic acids from contaminating lipids, carbohydrates and proteins via convenient multiple washing steps. The use of such technology greatly facilitates the extraction of nucleic acids from complex specimens such as faeces, semen and plasma and allows for automation of the entire process. This has already led to the development of dedicated DNA extraction machines such as the MagnaPure manufactured by Roche [Van Doornum et al., 2003]. Alternative methods to silica-based matrices include the capture of specific DNA or RNA fragments (including poly-A tailed mRNA) via "hybridisation capture" (HC) techniques, which utilise specifically designed "probes" attached to solid matrices or magnetic beads [Millar et al., 1995; Marsh et al.,

2000]. One disadvantage of such capture techniques however is that during the binding and elution steps, nucleic acid shearing may lead to DNA fragmentation and loss of target.

As a final step in nucleic acid isolation, the yield and purity of the extracted nucleic acid may need to be determined, though this task is not commonly performed in the clinical laboratory prior to PCR amplification. A simple method for determining nucleic acid extraction efficiency and purity is via the use of UV spectrophotomctry. Single stranded DNA, double stranded DNA and RNA have specific absorption coefficients of 0.027, 0.020 and 0.025 μg per ml per cm at 260 nm, respectively. Moreover, the absorption ratio of 260/280 nm is an indicator for protein or phenol contamination since proteins have a maximum adsorption at 280 nm. When the ratio is below 1.8 a significant amount of impurities is still present within the sample. Highly purified samples of DNA have a 260/280 nm ratio of 1.8–1.9 whilst highly purified samples of RNA have a 260/280 nm ratio of 1.9–2.0. Many spectrophotometers will automatically calculate the 260/280 nm ratio and quantity of nucleic acid for you. Phenol/urea contamination may be assessed at 230 nm. Always calibrate the spectrophotometer with a blank prior to measuring nucleic acid concentrations. This blank should comprise the solution in which the nucleic acid is resuspended (e.g. nuclease free water, Tris EDTA buffer) only. Use disposable or cleaned quartz cuvettes for each new measurement. If using plastic disposable cuvettes, first ensure that they are suitable for the measurement of absorbance using ultraviolet light. Another method for assessing the efficiency of nucleic acid extraction involves the electrophoresis of a small sample of the extract on an agarose gel along with a molecular weight marker (ladder) and/or known quantities of nucleic acid. Indeed, it is now possible to buy molecular weight markers with bands comprising a known quantity of DNA (e.g. Smartladder, Eurogentec S.A., Seraing, Belgium). The size and quantity of the isolated nucleic acids may then be estimated by incorporating a DNA intercalating fluorochrome into the gel, either prior to, or after, electrophoresis. Several DNA intercalating fluorochromes may be used to stain nucleic acids in gels, including ethidium bromide, Hoechst 33258 (which almost exclusively binds DNA, preferentially binding to AT-rich regions), PicoGreen, YO-PRO-1, YO-YO-1, etc. Gel electrophoresis measurement tends to be much more sensitive than spectrophotometry, though in general, reliable measurements using both methods require a quantity of nucleic acid greater than 50 ng per loading.

After extraction, nucleic acids may be stored in buffers compatible with the PCR process. For example, TE buffer is often used to store DNA but inhibits the PCR as the EDTA present in the buffer chelates magnesium ions, and magnesium ions are essential for the proper working of the thermostable DNA polymerase component of the PCR mix. Sterile distilled water may also be used to store extracted DNA, though it should be noted that DNA in solution tends to form a weak acid, which may eventually lead to auto-degradation of the DNA. Phosphate buffers may also interfere with PCR efficiency since phosphate interferes with the nucleic acid structure. Tris has no inhibitory effect on Taq polymerase activity *per se*, though the success and reproducibility of a particular PCR is often dependent on the pH value of the reaction mix (hence the addition of a pH stabilising buffer to PCR reaction mixes). Some pathology fixatives are highly acidic, and will eventually facilitate the depurination and chemical destruction of stored nucleic acids. Conversely, alkaline conditions lead to the preferential destruction of RNA molecules. Under these conditions, the storage of extracted DNA is feasible for many weeks or months at −20°C and even several years at −80°C. However, storage of RNA is more problematic due to the large number and ubiquitous quantity of RNases present in the (laboratory) environment (e.g. on human skin, laboratory work benches, etc.). Hence, the storage of RNA should ideally be undertaken using RNase free buffers and storage tubes (Section 4.4). Certain commercial RNA storage solutions are commercially available, e.g. The RNA Storage Solution (Ambion, Inc, Texas, USA). In general, it is also advisable to keep the number of freeze/thawing events per sample to a minimum, possibly by storing multiple frozen aliquots of the extracted nucleic acid, and discarding any that have been thawed for use.

The increasing use of molecular techniques in the fields of diagnostics (human, animal and plant health, etc.), as well as in fundamental biological research (disease epidemiology, genomics, etc.), has led to much commercial attention being paid to the development and sale of easy to use nucleic acid extraction kits (the first step in PCR and other molecular protocols). Currently, a great many commercial kits are available on the market, the vast majority of which utilize the principles mentioned above for nucleic acid isolation and purification. Some of these kits are designed to facilitate both DNA and RNA isolation/purification using the same materials, whilst others are dedicated towards the isolation of for example, genomic DNA from blood or plant tissues, plasmid DNA from bacteria, total RNA or mRNA alone. These kits have become invaluable for the laboratory PCR user, and are becoming essential in the development of rapid nucleic acid extraction procedures for use with automated high-throughput PCR machines.

4.3.2 Comments on Nucleic Acids in Specific Sample Types

Cytological specimens and cultured cells: These specimens may be obtained through swabbing (exfoliative methods) or biopsy (punctates) and are usually rich in cells. Buccal cells, for instance, can be easily harvested and provide a perfect starting material for various forms of human genetic testing. Tissue samples may be converted into cellular suspensions by mild proteolysis followed by centrifugation (to pellet the cells), washing and immediate lysis. An aliquot of these preparations generally suffices for molecular diagnostics. Swabs are frequently used to diagnose epithelial infections with for instance human papillomaviruses or the bacterial species *Chlamydia trachomatis*, the agent of pelvic inflammatory disease (PID). They are also frequently used for forensic diagnosis and paternity suits. Various cellular secretion products such as mucous or other extra-cellular polysaccharides can interfere with nucleic acid purification. These contaminating products may be present in greater quantities than the DNA and/or RNA to be extracted and are present in most clinical materials, e.g. as secreted products from tissue cultures. Various, non-specific enzymes have been described for removing such contaminants including the enzyme hyaluronidase which digests persistent polysaccharides. Cells which adhere to the surfaces of culture vessels are best removed prior to nucleic acid extraction, e.g. via trypsin or versene treatment. These treatments should not adversely affect nucleic acid extraction at this initial stage.

Faeces: Faeces are frequently used in diagnostic and molecular microbiology as the preferred clinical material for detecting microbial pathogens of the gastrointestinal tract. However, faeces contain numerous PCR inhibitors at high concentrations, including food remnants and bile salts. Indeed, a 0.4% vol/vol suspension of untreated faeces is enough to completely inhibit PCR amplification. Several PCR "facilitators" have been described, including the addition of bovine serum albumin, betain or ps32 [Al-Soud and Radstrom, 2001], which may help reduce any PCR inhibitory effects associated with faeces. However, for best results regarding sensitivity and reproducibility, it is recommended that the nucleic acid be first extracted from faeces prior to PCR processing, rather than attempting PCR directly on a quantity of faeces.

Urine: Urine contains many factors that can inhibit PCR. The elevated level of salts present in urine means that dilution of the specimen prior to PCR is advisable. Since this is not always possible, nucleic acid extraction is frequently required. Mucoid secretions, which derive from the urethral epithelium and which may occasionally be present in significant quantities in urine, may be removed by hyaluronidase treatment. In general, it is better to extract the nucleic acid from urine prior to PCR processing, rather than attempting to PCR urine directly.

Blood: Blood is a frequently used material in clinical diagnosis. Indeed, 100 µl of blood yields approximately 3 µg of DNA, a huge amount for PCR. Therefore, an average sized blood spot may generate enough DNA to perform more than 20 different PCR tests. Heparinized or citrated blood, serum samples (where the cells have been removed, e.g. by centrifugation) and dried bloodspots may be used in PCR amplification. It is possible to enrich for a particular sub-population of cells within the blood sample prior to nucleic acid isolation using Ficoll-Hypaque gradient centrifugation. Flow cytometry provides a modern alternative to this Ficoll-Hypaque/centrifugation technique. Unfortunately, many potent amplification inhibitors (e.g. heparin, haemoglobin, porphyrin, lactoferrin, bilirubin and trivalent iron ions) are present in blood [Satsangi et al., 1994; Miyachi et al., 1998], and a concentration as low as 0.004% vol/vol of blood can completely inhibit a PCR reaction. The native immunoglobulin G (IgG) in serum is one of the most powerful inhibitors of PCR known, since it has an exceptional affinity for single stranded DNA. The inhibitory effect of IgG may in part be prevented by pre-incubation with non-specific DNA such as that isolated from bacteriophage λ [Al-Soud and Radstrom, 1998; Al-Soud et al., 2000; Al-Soud and Radstorm, 2001]. The inhibitory effect of heparin may be adequately eliminated via the use of specific paramagnetic beads [Deggerdal and Larsen, 1997]. Commercial kits or protocols specifically designed for the isolation of nucleic acids from blood are available. These kits are especially designed to be performed with (a) a minimum hands-on time and (b) a rapid turn around time, mainly due to the fact that the molecular testing of blood samples, e.g. for blood bank screening, is often a high-throughput process. Some commercially available kits take less than 8 minutes hands-on time per sample.

Virus containing fluids and virus transport medium: These samples should be treated in the same manner as cytological specimens. Episomal (not integrated into the host chromosome) viral nucleic acid is a more natural template for PCR than integrated viral DNA in that the DNA is not associated with host chaperone proteins. Optimized PCRs generally require between 10 and 100 viral genome copies for successful amplification.

Microorganisms: Nucleic acid isolation from microorganisms may require specific enzymatic treatments in order to remove the rigid cell walls of these organisms. This is especially true in the case of fungi, with their complex and tough outer cell wall. Clinical specimens often comprise microorganisms in the presence of other tissues or body fluids, and a combination of microorganism-specific and sample-specific steps may need to be performed in order to extract the relevant nucleic acid. For example, the detection of a bacterium from a cervical swab (e.g. *Chlamydia trachomatis*) needs a separate approach than the detection of say a fungus or yeast from blood. Particularly problematic is the detection of microorganisms in foodstuffs [Perelle et al., 2004], as (1) the actual volume of foodstuff may far outweigh the number of microorganisms present (sampling volume difficulties), and (2) it is very difficult to concentrate large volumes of foodstuffs without also concentrating the ubiquitous PCR inhibitors also present. In this respect, special adaptations may be required for the extraction and detection of nucleic acid from microorganisms contaminating foodstuffs [Parham et al., 2003; Kurdziel et al., 2001].

Semen and embryonic cells: These cells are similar to cytological material. Studies relating to (in)fertility, inherited diseases and forensic applications may be undertaken using these cells. It has been shown that a single cell from an eight-celled embryo can provide enough DNA for genetic identification using PCR amplification methods. Single ova and sperm cells may also be analyzed in detail by PCR [Wan et al., 2003].

Frozen specimens: When performing nucleic acid extraction on frozen specimens, the specimen should first be thawed, or else ground (for tough materials, e.g. plant tissues) in a mortar and pestle whilst still frozen. Nucleic acids frozen in tissues, etc. are generally well conserved, though multiple cycles of freeze/thawing may have deleterious effects on the nucleic acid with regard to the activity of DNases and RNases and with respect to physical damage of the nucleic acid (e.g. breaking of DNA duplex strands). Hence, it is advisable to freeze several portions/aliquots of the target material and thaw only one portion at a time for use followed by disposal. Even mitochondrial DNA from mammoths, frozen for centuries in the North Siberian tundra has been effectively amplified [Noro et al., 1998].

Histological preparations: Paraffin fixed biopsies are an interesting and valuable source of DNA for retrospective studies, as it is possible to extract and amplify DNA from such specimens even after 40 years. A single 0.1 mm^2 section generally contains approximately 1,000 nuclei, sufficient for detailed PCR analyses.

In general, the yield of DNA from fixed and embedded tissue is not significantly different from that of fresh material. Xylene or toluene may be used to extract paraffin, though direct extraction at 56°C or the application of microwave technology may also be used [Frank et al., 1996]. Commercial preparations such as Histoclear, may however be preferred because they are less toxic and the tissue pellets remain more clearly visible during extraction [Gugliotta et al., 1992]. Commercially available kits have been demonstrated to be very useful for these specimens [Wickham et al., 2000]. For processing pathological samples, treatment with detergents is deemed unnecessary as most of the lipids are removed during the tissue processing. More problematic are those samples that have been extensively fixed with fixatives that permit strong cross-linking [An and Fleming, 1991], as under such circumstances, chromosomal proteins tend to become highly resistant to proteolysis. Optimisation of the proteolysis step is therefore important with such preparations. The use of PCR cycling protocols incorporating extension times of several minutes and/or increasing the number of cycles in the PCR run may help in solving the problems caused by fixatives. In some cases, (1) a nested PCR protocol may need to be performed in order to attain sufficient amplification sensitivity [Howe et al., 1997; Camilleri-Broet et al., 2000), and (2) a touch-down PCR approach to inhibit the amplification of non-specific PCR products may also be required. Tissue processing in general and chemical fixation in particular, can severely damage nucleic acids. Notorious in this respect are the agents of picric acid, sublimate, formic acid, hydrochloric acid or non-buffered formalin, and various alternative procedures have recently been suggested omitting these agents [Wickham et al., 2000]. A simple alternative is to concentrate on amplifying only small DNA fragments of at most a few hundred basepairs or even shorter [Kleter et al., 1999] in the PCR. When multiple small fragments are amplified in an overlapping fashion, the net result can be a relatively long, composite PCR product. This approach allows for the sequencing of extensive regions of DNA for phylogenetic research or even the complete sequencing of mitochondrial DNA and is very suited for use with the types of degraded DNA often found in fixed tissue samples. The same holds true for the multiplex PCR approach, where several small DNA regions are targeted at the same time, facilitating the eventual amplification of larger fragments by the integration of the small fragments in a single large amplimer [Wera et al., 1991]. A new and promising technology is micro-dissection, where a scalpel or laser irradiation is used to isolate small sections of a microscope section, thereby allowing the removal of potentially amplification-interfering regions from the specimen and the targeting of cells that require particular attention, e.g. cancerous regions [Rubin, 2001] (Fig. 4.6). Micro-dissection can be combined with hematoxicillin-eosin (HE) staining [Burton et al., 1998] or specific immuno-histochemical approaches to highlight target regions with an even greater specificity, indeed single cell excision is among the latest developments in this field [Alcock et al., 1999].

Dessicated material: Materials such as hair (mitochondrial DNA), herbarium specimens, seeds, dried blood spots and even mummified remains contain useful amounts of well-conserved DNA. For these desiccated materials it is important to ensure complete (re)hydration of the specimen in order to allow for adequate enzymatic treatments.

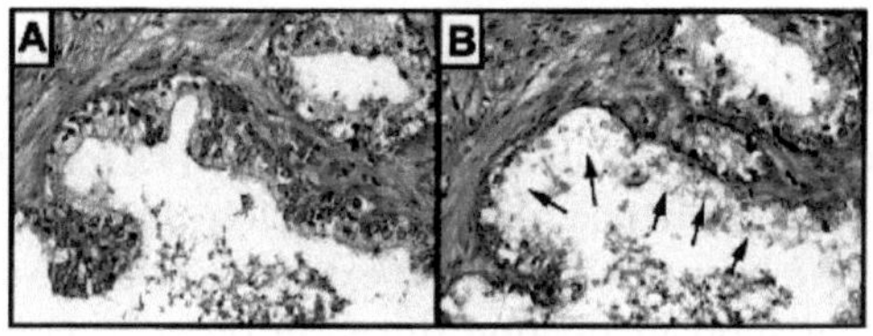

Fig. 4.6 Laser capture microdissection facilitates the physical separation of certain groups of cells from their neighbours. This picture shows a prostate tissue section before and after the removal of a group of cells (Reproduced from Rubin, 2001. With permission from Wiley)

Fresh plant specimens: For some plant species, the grinding-up of a leaf and inclusion of a diluted sample of this material in a PCR mix suffices for PCR amplification [Guidet, 1994]. This particular procedure has now been automated and made compatible with the processing of 192 samples at a time by automation. For some other plant materials, lyophilisation and extraction in the presence of a detergent or an organic solvent is all that is required. Inhibitory substances often include polysaccharides and polyphenols. Xylene for instance completely blocks PCR amplification at a concentration of 0.0025% (vol/vol). The inhibitory effects of these contaminants may be (partially) neutralised by simple dilution of the extracts, the addition of polyvinylpyrrolidone (PVP) to the PCR mix or CTAB extraction.

Insects: Archived insect material, most often stored in ethylacetate or dry, may be used for nucleic acid extraction. Fresh material first needs to be frozen at −20°C or −80°C. After a specific enzymatic treatment by, for example chitinase or β-N-acetylglucosaminidase, DNA that is usually quite degraded can be extracted by conventional methods [Reiss et al., 1977].

"*Forensic*" *samples*: Nucleic acids may need to be extracted from some very unusual "environmental" materials, particularly with regard to the science of forensics. Samples such as cigarette butts with desiccated saliva, hairs, saliva *per se*, skin scrapings, remains of bone, urine, dirt from underneath nails, dentine, chewing gum and many other materials require special attention. It is often the case that such samples are partially degraded, e.g. by DNA (auto)lysis or contaminated by "extraneous" nucleic acids. In some of these cases alkaline treatment may be an option, whereas template reconstruction using specific primer overhang-sequences may also be helpful [Golenberg et al., 1996].

DNA electrophoresed in agarose or acrylamide gels: Post-PCR analysis of amplification products (amplimers) frequently involves electrophoresis of samples in agarose gels. This provides information on the quantity and quality of the PCR products. If subsequent manipulations are necessary with respect to these products (cloning, sequencing, hybridisation, etc.), then it may be necessary to cut out and purify the required amplimers from the agarose or acrylamide gel. Various commercial kits are available for purifying DNA from contaminating agarose and acrylamide, most of which also effectively remove fluorochromes (e.g. ethidium bromide) present in the gel. In the vast majority of assays, the DNA is released from the agarose matrix using either a sodium iodide buffer solution and heating to 56°C for 10 minutes, chemical disintegration of the agarose (sodium periodate), or enzymatic cleavage (agarase enzyme). Chaotropic salts are also essential ingredients of many kits, as they facilitate the coupling of released DNA to a silica matrix, e.g. 96-wells ELISA plates (Sigma), spherical silica particles (BioRad) or spin columns containing a silica filter (Jetsorb, GenElute, SNAP). From the authors' own experience, the yield of DNA obtained may vary widely. Similar approaches have been developed for extracting DNA from acrylamide gels. It has been shown that short incubation times, in order to prevent leakage of inhibitory compounds from the acrylamide matrix, may be necessary [Etokebe and Spurkland, 2000]. Using the technique of agarose/acrylamide gel electrophoresis coupled to DNA extraction, fragments ranging in size from 20 to 45,000 nucleotides may be isolated in a pure form, i.e. free from non-specific amplimers, primer dimers, etc.

Direct PCR product purification: After amplification, the amplimer present in the final PCR mixture may be purified using commercially available PCR product clean-up kits in order to remove surplus primer, deoxyribonucleotide triphosphates, salts and polymerase enzyme. The most widely used PCR clean-up kits rely on the use of spin columns containing silica filters. The addition of chaotropic salts to the final PCR product mix facilitates the binding of amplified DNA to the silica filter, allowing elution of contaminants by washing in ethanol-based solutions, and leaving behind the required amplimer which may then be eluted using water. Examples of dedicated PCR clean-up kits include the Qiaquick PCR clean-up kit (Qiagen), the Zymoclean kit (Zymo technologies), the Magnesil approach (Promega) or the Dye Terminator Removal Kit (AbGene). Very small PCR fragments may be purified by performing silicate affinity chromatography at elevated temperatures [Smith et al., 1995].

4.4 RNA Samples

mRNA, rRNA, tRNA, hnRNA and viral RNA are nucleic acids that may also be analyzed by PCR. In general, PCR mediated RNA amplification is primarily used for the detection (absence/presence), quantifying (how many copies are present) and monitoring (how these RNA quantities vary over time) of growth-dependent gene expression in eukaryotes, bacteria and viruses. In addition, abundant rRNA molecules provide an interesting and important diagnostic target for the detection and identification of bacteria. Of special importance to note is that RNA is not an appropriate target for the thermostable DNA dependent DNA polymerases routinely used in laboratory PCR assays, meaning that the RNA must first be copied into single stranded copy DNA (cDNA) prior to PCR processing, a process achieved using special reverse transcriptase enzymes, and referred to as reverse transcriptase PCR, reverse transcription PCR or RT-PCR (Section 4.5).

Genes coding for mRNA in eukaryotes generally contain non-expressed intron sequences. In order to distinguish between genomic sequences and mature mRNA sequences, PCR primers may be designed in such a way that the primers span an intron splice site. After PCR amplification, the size of the amplimer will reveal whether it is derived from genomic DNA or from spliced mRNA. Of course, if PCR is performed instead of RT-PCR, then amplimers will only be generated from the longer genomic DNA, as thermostable DNA dependent DNA polymerases do not (usually) amplify RNA molecules. In contrast to eukaryotes, bacterial and viral genes do not contain introns, which means that no direct distinction can be made between the size of PCR products derived from reverse transcribed cDNA or genomic DNA, though DNase or RNase treatment prior to PCR may provide clues as to the initial origin of RT-PCR amplification products in these organisms. The presence of reverse transcriptase enzymes in tissues infected with retroviruses (the original source of reverse transcriptase enzymes) means that a different approach other than PCR may be necessary to detect these viruses, and several methods have been developed which do not rely on the amplification of nucleic acids using thermostable DNA dependent DNA polymerase, e.g. NASBA.

The synthesis of a full-length cDNA copy of the mRNA molecule by reverse transcriptase during the RT step of RT-PCR is important, as short fragments of reverse transcribed cDNA may not contain the complementary sequences against which the specific PCR primers have been designed. Fortunately, since most mRNA molecules are shorter than 3 kbp, full-length reverse transcription copies are usually synthesized. However, reverse transcriptase enzymes suffer from a distinct disadvantage in that they frequently mis-incorporate nucleotides into the growing cDNA chain, and these errors are not corrected as the enzymes lack proofreading ability (unlike several DNA dependant DNA polymerases, e.g. Pfu, Vent). RT-PCRs involving the amplification of fragments larger than approximately 5–10 kb are rarely successful, largely due to: (1) the presence of secondary structure within RNA molecules, (2) the ubiquity of RNA degrading RNase enzymes in the environment, and (3) the RNase H activity of reverse transcriptase (the degradation and removal of the RNA strand from RNA/cDNA hybrid molecules). The presence of ubiquitous and difficult to inactivate RNA degrading enzymes (RNases) is a problem that requires particular attention in the PCR laboratory if high quality, reliable and consistent RT-PCR is to be achieved.

4.4.1 Working Free of RNase Contamination

RNases are exceptionally stable, some even resist autoclaving at 120°C for 15 minutes in super-heated steam (though all RNases may be destroyed by heating at 180–200°C in dry heat for 6 hours or longer). The presence of ubiquitous and difficult to inactivate RNases in the working environment (on hands, work surfaces, etc.) means that working with RNA is especially problematic. In effect, contamination of work solutions, work surfaces, disposables, consumables, etc. by RNases needs to be avoided right from the beginning of RT-PCR processing, i.e. from the initial RNA isolation stage. As a rough guide to avoiding RNase contamination: (1) gloves should be worn during all RNA extraction manipulations as bare hands are a well-known source of RNases, (2) the use of disposable RNase free plastic consumables, pipette tips, Eppendorf tubes, etc., is recommended, and (3) attention should be paid to ensuring that pipetting instruments and work surfaces are also RNase free. RNases present on PCR reaction tubes, etc. may be inactivated by chemical treatment with diethyl pyrocarbonate (DEPC) or commercially available systems such as RNase AWAY (Molecular BIO Products), RNase ERASE (ICN Biomedicals), RNasin Ribonuclease inhibitor (Promega), RNA Clean (Hybaid), etc., though most companies now provide certification verifying that their pipette tips and Eppendorf tubes are already RNase free when purchased.

The use of DEPC treatment for materials to be used in RNA extraction protocols used to be particularly widespread, with a 0.1% solution (vol/vol) of DEPC in water being used to remove RNase from essentially all consumable articles. DEPC acts by blocking the active centre of RNase and when autoclaved, hydrolyses into harmless carbon dioxide and ethanol. Unfortunately however, the half-life of DEPC is low in aqueous environments and the presence of nucleophiles such as tris or borate accelerates disintegration of the compound. Water can be made RNase free by addition of 10 ml of a ten times diluted (in ethanol) solution of DEPC to 990 ml of water. This should be kept at room temperature for at least 16 hours, after which the water should be autoclaved. Other solutions can be treated similarly, although DEPC treatment of solutions containing nucleophiles is not advised for the reasons already mentioned above. Glassware should be rinsed with 0.1% DEPC in water and left to dry. It should be borne in mind that DEPC also hydrolyses the imidazole ring of purines and destroys nucleic acids. For this reason, reaction tubes, etc. treated with DEPC should always be autoclaved or heated at 180°C to inactivate the DEPC prior to use. DEPC should not be used on instruments constructed of polystyrene or polycarbonate materials (e.g. gel electrophoresis tanks, etc.) as it degrades these compounds. In these cases, a solution containing 10% hydrogen peroxide (incubated for at least 10 minutes at room temperature) should be used to remove RNase contamination, with thorough washing in RNase free water after treatment. It should be borne in mind that DEPC is a highly toxic carcinogen that requires appropriate safety precautions to be in force. Alternatives to DEPC include RNase AWAYand RNase Erase which can be sprayed onto work surfaces to eliminate RNase without any further follow-up treatment. The fact that RNases are so difficult to remove from the environment means that further steps may be necessary in order to achieve reproducibly high yields of RNA for RT-PCR.

In general, RNase proteins may be exogenous (see above) or endogenous in nature, i.e. present within actual biological tissues and aqueous solutions *per se*. The activity of endogenous RNases within the nucleic acid extract may be inhibited by the addition of various commercially available protein preparations. RNase OUT (Invitrogen) is a non-competitive, recombinant inhibitor of RNases which inactivates RNase A, B and C. RNase enzymes bind to this solid matrix and may be easily eliminated by centrifugation or filtration. RNasin (Promega Corporation, Texas, USA) is a 50 kD protein consisting of 460 amino acids isolated from the human placenta [Blackburn and Gavilanes, 1982]. RNasin binds in a 1:1 ratio with great affinity to RNase molecules ($K_m > 10^{14}$!). RNasin can be used in the temperature range of between 4°C and 37°C. The protein only looses activity when it is fully denatured [Saxena et al., 1991]. It can be conveniently used in reaction mixes incorporating reverse transcriptase enzymes and does not inhibit most of the available RNA polymerases used in *in vitro* transcription assays. The RNasin gene has been cloned into *E. coli* and a recombinant version of this RNasin is also for sale. Various other companies (Gibco, Ambion, Pharmacia, etc.) also market RNase inhibitors that work in a similar manner to RNasin. Some of these RNase inhibitors have even been impregnated into wet tissues and can be used to wipe down RNase contaminated work surfaces.

4.4.2 RNA Isolation for RT-PCR

The methods adopted for cell lysis during RNA isolation are essentially similar to those described previously for DNA (Section 4.3.1), with ideally, the final RNA preparation being free of denaturing agents such as sodium dodecyl sulphate (SDS), polysaccharides, EDTA, chloride ions, salt and last but not least, active nucleases (Section 4.4.1). RNA isolation can be conveniently performed when the buffer system includes components such as guanidine isothiocyanate, sodium nitrate, NP-40, DTT, sarcosyl, β-mercapto-ethanol and proteinase K. Alternatively, phenol/chloroform/isoamylalcohol may also be used for RNA extraction. By using the cationic surfactant Catrimox-14, the RNA is encapsulated in inverse micelles from which the RNA may be subsequently extracted in a relatively pure fashion. Alternatively, RNA sediments more rapidly than DNA during ultra-centrifugation in a caesium chloride gradient, facilitating subsequent physical separation of the RNA containing layer. Precipitation with lithium chloride is an alternative to lengthy ultra-centrifugation procedures and the RNA liberated only needs to be thoroughly washed using 70% ethanol. Mechanical processes, such as needle passage under significant pressure, have the advantage of breaking open cells as well as degrading the long strands of genomic DNA, a feature that often limits co-purification of DNA during RNA protocols. In all cases, the initial cell lysis step should be fast and efficient in order to ensure that no unlysed cells remain during subsequent steps in the protocol, as unlysed cells could introduce RNase contamination at a later stage in the isolation procedure. Blocking of RNase activity in crude samples may be achieved by including high concentrations of guanidine isothiocyanate (up to 14 M). This chemical agent does not display the reverse transcriptase inhibitory activity observed with methods using high concentrations of chloride ions. However, guanidine isothiocyanate cannot be used as the single lyzing agent for RNA extraction from formalin fixed specimens [Masuda et al., 1999]. In this case, a combination of guanidine isothiocyanate, phenol and low pH is very efficient [Chomczynski and Sacchi, 1987], the hydrophobic compounds dissolve in the phenolic phase, high molecular weight DNA and several protein species accumulate near the interphase, whilst the RNA remains soluble in the aqueous phase. Using this protocol, all types of RNA may be purified within a time period of 4 hours. Many rapid and efficient RNA isolation kits are currently commercially available, and have the advantage of using only small quantities of hazardous chemicals. Such commercially available RNA isolation kits are generally based on guanidine isothiocyanate chemistry in combination with the denaturing activity of β-mercapto-ethanol. Many of these tests also rely on the adhesive characteristics of silica, glass beads, latex beads or even celluloid for nucleic acids. Some nucleic acid isolation kits allow for the simultaneous isolation of both RNA and DNA [Otto, 1998], whilst several kits focus on nucleic acid extraction from specific specimens, such as blood, urine or plant material, taking into account the specific characteristics of these materials. Some kits even manage to remove the ubiquitous neurotoxic ribonuclease, an enzyme that is not usually removed using standard RNA isolation protocols.

Whether using "in house" or kit-based RNA isolation protocols, approximately 5–75 μg of RNA may be purified from a million mammalian cells. Eaton et al. (1997) used magnetic beads to capture circulating metastatic tumour cells prior to RNA isolation and detection. The sensitivity of this approach was very high, with the ability to detect as few as five tumour cells. For all methods of RNA isolation, the viscosity of the material from which the RNA is to be isolated is an important factor, as tissues rich in nucleic acids (e.g. spleen, thymus) may yield high concentrations of extracted nucleic acids, resulting in inefficient phase separation during extraction processes. This results in a corresponding increase in DNA/protein contaminant co-precipitation. Dilution of the specimen prior to RNA extraction provides a simple answer to this problem, though this may reduce the final RT-PCR sensitivity. Though the majority of isolated RNA material comprises ribosomal RNA, the vast majority of RT-PCR applications are designed for the amplification of specific messenger RNA (mRNA) molecules. Moreover, this pool of mRNAs within cells comprises a very heterogeneous mix of different mRNA fragments, and a single specific mRNA species may only constitute a minute fraction of the total mRNA pool. That said however, a sample of one million cells with a single specific messenger RNA species per cell still yields (theoretically) a million RNA copies per sample extraction, more than enough for RT-PCR amplification. Therefore, the purification of a single mRNA species is not a prerequisite for successful mRNA target amplification by RT-PCR, however it is usual for the mRNA fraction to be crudely purified from the total pool of extracted RNAs prior to RT-PCR. This crude purification is usually achieved via some form of affinity chromatography using a solid matrix to which oligo-dT molecules have been attached as the capture phase. Oligo-dT affinity capturing has been adapted to a variety of formats, some based on magnetic beads or membranes, others based on the 96 well ELISA format. Replacement of oligo-dT DNA by oligo-dT peptide nucleic acid (PNA) may result in better yields and even better removal of eukaryotic ribosomal RNA [Efimov et al., 1999]. RNA species with shorter poly-A tails are also isolated. In general, a successful RT-PCR amplification protocol for a specific mRNA species requires a pool of between 1 and 100 ng of oligo-dT purified mRNA. For samples containing a few cells, it may be worthwhile adding an amount of carrier RNA during the extraction protocol in order to efficiently (co)precipitate the RNA pool. Suitable carriers are glycogen or tRNA. An alternative approach is to add 2.5 M ammonium acetate to the extraction mix, which will result in the precipitation of all types of RNA molecules.

For RT-PCR amplification protocols, it is generally advisable to include a DNase treatment step immediately after isolation of the RNA in order to try and remove any contaminating DNA that could generate a false positive RT-PCR result. This step is imperative when determining relative and absolute mRNA gene expression values by RT-PCR. The high sensitivity of PCR amplification means that one contaminating molecule of DNA could feasibly be amplified to many millions of copies, yielding false positive results and hence false gene expression data results. Some techniques have been developed which can amplify gene specific mRNA in the presence of a gene specific DNA background (e.g. the Nucleic Acid Sequence-Based Amplification reaction or NASBA), though these techniques tend to suffer from a lack of specificity due to the relatively low annealing and cycling temperatures used (a consequence of the inherent thermal instability of the reverse transcriptase and other enzymes used in these protocols). In any case, the presence of gene specific contaminating DNA in (DNase treated) RNA can be easily determined by simply adding an aliquot of the isolated RNA to a reverse transcription mix that does not contain the reverse transcriptase enzyme (mock RT reaction). Processing of this "DNA control" specimen is then performed according to the normal RT-PCR protocol, alongside the specimens to be tested *per se*. After reverse transcription and PCR amplification, there should be no amplification products in the DNA control (provided that no contaminating genomic DNA was present in the initial extracted RNA sample).

Once isolated, it is advisable to use RNA immediately for downstream processing, or if storage is required, RNA may be stored at –70°C in ethanol containing solutions. Alternatively, deionised formamide is an excellent storage medium, even for large transcripts (though the formamide needs to be removed by ethanol precipitation prior to further RT-PCR processing). However, the constant problem of RNase contamination means that the use of commercially available RNA storage solutions is to be recommended, e.g. THE RNA Storage Solution (Ambion). Some RNA storage solutions, e.g. RNAlater (Ambion, Qiagen)/RNeasyProtect (Qiagen) have been specifically designed to stabilize RNA whilst *in situ* in its native cellular (tissue) environment. Tissues that are perfused with these liquids may be stored for a week at room temperature, 1 month at 4°C, or indefinitely at –70°C. Moreover, the RNA within these samples is able to withstand repeated cycles of freezing and thawing. The chemical compositions of these RNA storage solutions are currently withheld by the respective companies.

As with DNA isolation, the yield and quality of RNA extractions may need to be established. The basic method of achieving this is by using agarose gel electrophoresis of an aliquot of the isolated RNA. This requires that the RNA be denatured prior to electrophoresis, which is usually achieved by heating for 5–10 minutes at 65°C in a sample buffer containing formamide and formalin followed by immediate cooling on ice. This process "freezes" the RNA in an open-coiled, single stranded conformation. The progress of electrophoresis may be visualised using bromophenol blue dye, and after the allotted running time, RNA may be visualised using ethidium bromide, CyberGreen II or another fluorochrome. The larger ribosomal RNA species (28S or 23S rRNA) provide useful qualitative and quantitative markers. Messenger RNA is represented by a background smear comprising the cellular mRNA pool. If the smear is evenly distributed then the mRNA is probably intact, however if the smear comprises only shorter fragments, then the mRNA has probably been degraded. In this case, RT-PCR may still be possible, dependant on the size of the amplification product expected and the region and degree of mRNA degradation. Fixed and paraffin embedded archival tissue provide a less generous source of RNA than fresh tissues, with formalin-, Carnoy- or Omnifix-treated samples showing a tenfold or more reduction in RNA quantity than their fresh counterparts [Foss et al., 1994]. As mentioned in Section 4.3.1, the concentration of RNA may also be spectrophotometry measured at 260 nm, where an optical density of 1 indicates an RNA concentration of approximately 40 μg/ml. The quality of the RNA preparation can be determined based on an absorption ratio of 260/280 nm where an absorbance ratio of between 1.9 and 2.0 indicates purified RNA, and a ratio of 2.1 indicates highly purified RNA. A relatively novel way of assessing RNA quality is via the RT-PCR process itself. For instance, the RNA Inspector kit (Sigma) contains all the ingredients necessary to perform a variety of RT-PCRs, and is designed to amplify various universally present mRNA molecules of different sizes. These molecules serve as internal quality controls and are the ultimate control parameter for RNA preparations.

4.5 Reverse Transcription and RT-PCR

In order to be successfully amplified by PCR, RNA needs to be to be first converted into a DNA copy (cDNA), a process most efficiently achieved using an RNA-dependent DNA polymerase enzyme (reverse transcriptase, RT). RT is not an enzyme normally found in eukaryotic or prokaryotic cells, but in a limited number of viruses, including retroviruses. Indeed, the two most widely used and commercially available RT enzymes were originally isolated from the oncogenic retroviruses Avian Myeloblastoma Virus (AMV) and Moloney Murine Leukemia Virus (MMLV). However, both of these RTs are low fidelity enzymes, generating cDNA containing randomly mis-incorporated nucleotides compared to the original RNA molecule sequence copied. In essence, this RT low fidelity is advantageous for retroviruses in the natural situation, as it facilitates the rapid development of mutant (e.g. antiviral drug resistant) progeny viruses. Physiologically, the AMV RT enzyme functions as a dimer, with significant reverse transcription activity at 42–52°C, meaning that this enzyme may be used at relatively elevated temperatures compared to MMLV (MMLV is less thermostable than AMV with an optimum reaction temperature of 37°C). AMV is therefore more suited for reverse transcribing full-length cDNA from GC-rich RNA templates than MMLV. Further, AMV RT has a higher processivity than MMLV RT since the AMV enzyme is faster at incorporating nucleotides into the growing cDNA chain. However, the mutation frequency of AMV RT is twice as high as that of its MMLV counterpart, with mutation frequencies of 1:30,000 and 1:17,000, respectively. Both AMV and MMLV enzymes lack 3′-5′- exonuclease proof-reading activity and do not need a double stranded primer region (e.g. provided by primer binding) for the initiation of cDNA synthesis. The presence of such a double stranded region simply acts to accelerate the reaction kinetics of reverse transcription. As a final comment, MMLV RT has less RNase H activity than AMV RT (RNase H degrades the RNA strand of DNA/RNA duplexes) and RNase H negative enzymes, such as SUPERSCRIPT (Invitrogen), MMLV RNase RT and THERMOSCRIPT (both Life Technologies), are clearly superior to the wild type enzymes with regard to thermostability (up to 70°C), ability in dealing with GC-rich RNA templates with stable secondary structure, and the reverse transcription of long RNAs >10 kbp in length (the average RT is capable of synthesizing transcripts up to 3 kbp in length, with larger mRNA molecules, e.g. viral RNA genomes up to 30 kbp in length, being very difficult to reverse transcribe). Enzyme cocktails, combining the best qualities of both MMLV and AMV RT enzymes in a single mix, e.g. Revers-It (Abgene), *C. therm* polymerase (Roche), Sensiscript and Omniscript (both Qiagen) are now commercially available.

Retroviral RTs are by far the most common RT enzymes used in RT-PCR protocols. However, several thermostable DNA polymerase enzymes (isolated from bacteria such as *Thermus thermophilus*, *T. flavus*, *T. litoralis*, *T. brokianus*, and *Carboxydithermus hydrogenoformans*) also show some (rather limited) reverse transcriptase activity. By altering the buffer constituents, the action of these enzymes may be subtly altered; to facilitate reverse transcription of RNA into cDNA, manganese ions are added. The manganese ions may (after a period of time allowed for reverse transcription to proceed) be later bound by specific chemical chelators and subsequently replaced by magnesium ions to change the RT activity of the enzyme into a DNA dependent DNA polymerase activity. This may however require the reaction vessel to be opened to add the chelator and magnesium ions, increasing the chance of PCR contamination (though single unique reaction buffers are now available) [Hu et al., 2003]. A further advantage of these "dual" function thermostable enzymes is that reverse transcription may be performed at elevated temperatures as compared to the optimum reaction temperatures of AMV and MMLV RTs, increasing the specificity of primer binding in the reverse transcription reaction. Also, stable secondary structures unfold at these temperatures increasing the likelihood of complete reverse transcription of the whole target RNA molecule even through regions of high GC-content. A disadvantage of using reverse transcriptase enzymes with elevated optimum temperatures is the fact that hybridisation of the DNA primer to the RNA template may be less efficient, especially if this primer is not one of the primer pair to be used in the subsequent PCR amplification. If oligo-dT or random hexanucleotide primers are used in the reverse transcription protocol (Section 4.5.1) along with thermostable DNA dependant enzymes, then an initial incubation at a low temperature may be performed (during which period of time the primers can hybridise and be extended to a size allowing for efficient hybridisation, before reverse transcription at a high temperature). For oligo-dT primers, a pre-incubation at 25°C

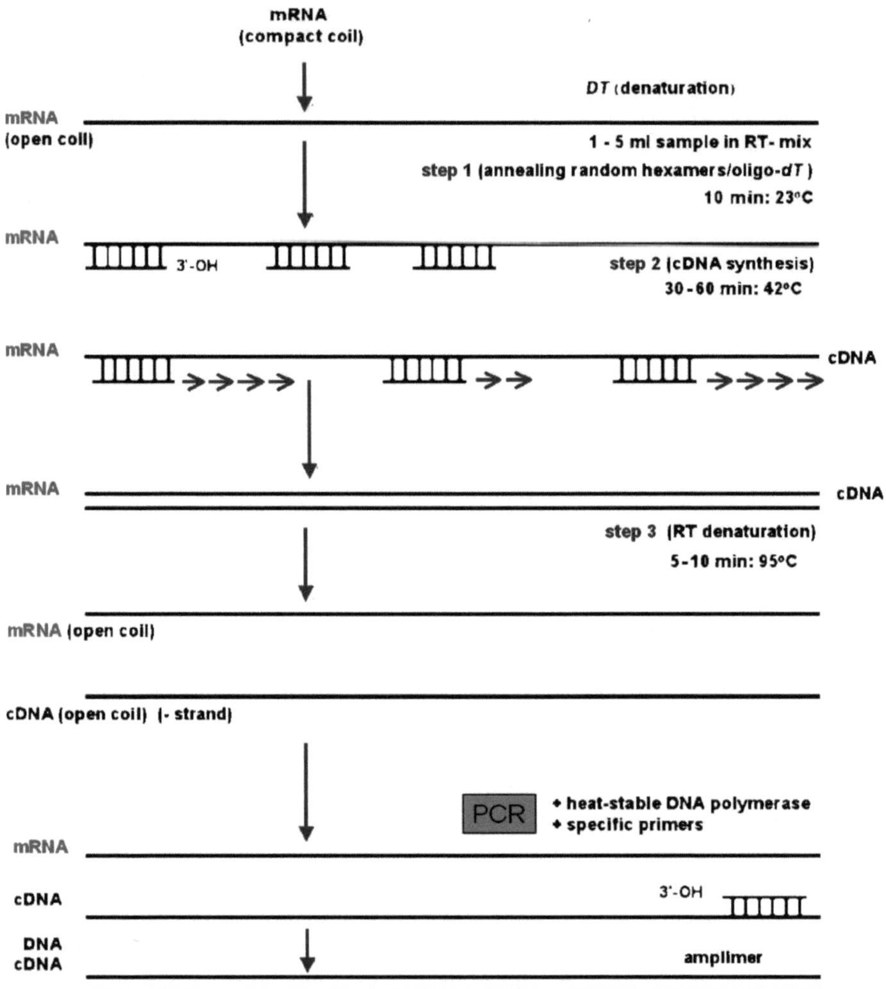

Fig. 4.7 Reverse transcriptase (RT-PCR) initially involves the extraction of RNA followed by the reverse transcription of RNA into a copy DNA (cDNA) molecule by the enzyme reverse transcriptase, and finally PCR thermocycling. After heat denaturation of the extracted RNA molecule (to remove any secondary structure which may prevent primer binding) hybridisation of reverse transcription primers, e.g. short random hexanucleotides or oligo-dT primers, to the RNA template is allowed to occur. RNA to cDNA reverse transcription is facilitated by incubating the mix 1 hour at 37°C or 42°C (dependant on the origin of reverse transcriptase used). At this stage, RNA/cDNA hybrids may be separated by heat or enzymatic action, or a portion of the reverse transcription mix simply added direct to a PCR mix ready for PCR thermocycling. In eukaryotes, non-coding intragenomic sequences (introns) may be present in the heterogenous RNA (hnRNA) which are spliced away out of the hnRNA to make the protein coding messenger RNA (mRNA). This means that there are often distinct differences in the sequences and lengths of genomic DNA compared to its corresponding mRNA transcript. These differences may be exploited to discriminate between a particular mRNA and its corresponding genomic DNA

for 10–30 minutes prior to high temperature reverse transcription is usually sufficient Fig. 4.7.

As a general solution to RNA secondary structure problems, alternative methods for melting stable secondary structures (not relying on the application of high temperatures and thermostable DNA dependent enzymes) include the addition of 1 M betain, 20% glycerol or 2–10% DMSO during cDNA synthesis [Frackman, 1998]. Sodium pyrophosphate may also be included (in concentrations up to 4 mM), as it reduces the rate of cDNA synthesis and thereby reduces the probability of the RT dissociating from the RNA template due to the presence of secondary structure elements. Inclusion of some of these compounds in the RT mix may also enhance the processivity of the RT though paradoxically, also their susceptibility toward inhibitory compounds.

4.5.1 cDNA Synthesis

The most popular method for reverse transcription of RNA into cDNA, initially involves making two different reverse transcription reaction mixes. The first mix includes the isolated RNA template and the primer to be used to initiate reverse transcription, whilst the second mix includes dNTPs, RNase inhibitor, RT enzyme, RT buffer (including the appropriate concentration of magnesium ions), dithiothreitol and nuclease free water. The first mix (containing RNA and RT primer) is incubated at 80°C for 2 minutes and then immediately placed on ice, which helps remove and "freeze open" secondary structure within the RNA, increasing the specificity of primer binding. After 5 minutes on ice, the second RT mix is added to the first mix and the new complete RT mix is incubated at 37°C for 1 hour. After 1 hour, the resultant reverse transcription products may be directly added to the appropriate PCR mix and PCR cycling performed (Fig. 4.7).

Various fundamentally different cDNA syntheses schemes have been developed. For example, it is possible to use a PCR primer in the RT reaction mix rather than a specifically designed RT primer, however this primer has to be complementary to the RNA molecule to be transcribed or else it will not bind to the single stranded RNA molecule. Alternatively, the whole of the extracted RNA pool (including the tRNA and rRNA if crude RNA preparations are used) may be reversed transcribed into cDNA using commercially available non-specific random hexanucleotide or decanucleotide primers (Invitrogen, Ambion). This methodology generates a library of cDNA molecules within a single reaction mix, allowing the same reaction mix to be used in several different PCR protocols, all using different PCR primer pairs. However, using random hexanucleotides results in a reduced yield of specific cDNA for the PCR, as thousands of fragments from all RNAs present in the RNA pool are reverse transcribed. An important issue to consider in this respect is the ratio of the random hexanucleotides versus the template concentration, as an excess of hexanucleotides leads to the reverse transcription of many short cDNA molecules. Oligo-dT primers are fragments of DNA containing multiple thymidine residues only (e.g.

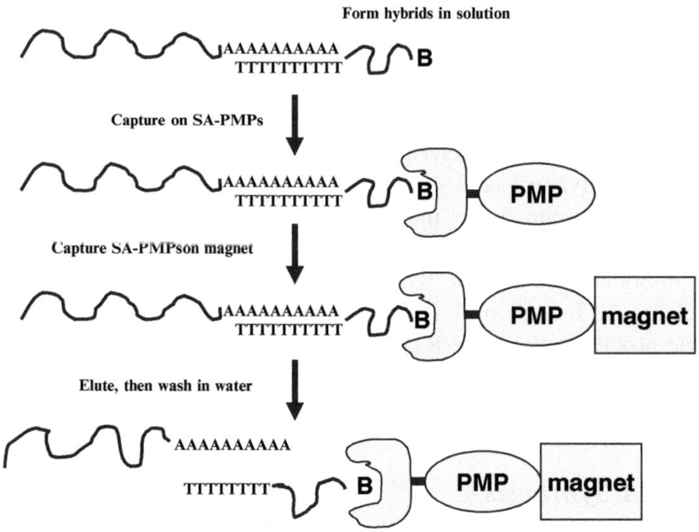

Fig. 4.8 Immobilised oligo-dT can be used to capture polyadenylated mRNA onto a solid support. After hybridisation the interaction between biotin and streptavidin attached to pyro-magnetic particles (PMP) allows mRNA to be extracted (from www.bioscience.hu). B = biotin

12–18 deoxythymidine residues (Invitrogen); 25 oligo (dT)-cellulose (New England Biolabs). Oligo-dT primers may be used to reverse transcribe the pool of poly-A tailed mRNA within the total RNA population, increasing the yield of reverse transcribed mRNA molecules. When these oligo-dT molecules are fixed to a solid support, affinity capture of poly-adenylated mRNA enhances the specificity of cDNA synthesis as well (Fig. 4.8). If these oligo-dT primers are designed to contain one or two extra nucleotides at their 3′- termini, then these 3′-end nucleotides will add extra specificity to the primer hybridisation reaction. The most specific means of reverse transcribing mRNA from the gene of interest is to use a gene specific primer. This primer is usually one of the primers utilised in the subsequent PCR amplification protocol and should be complementary to the (m)RNA molecule to be reverse transcribed. However, if it is not known which strand of the DNA is transcribed into mRNA, then both PCR primers may be used in the reverse transcription mix. Ideally, in order to facilitate efficient annealing, the reverse transcription primer should bind to a region of RNA that does not possess a high degree of secondary structure. The use of such gene specific reverse transcription primers (instead of random hexanucleotides, decanucleotides and oligo-dT primer, etc.) facilitates the use of increased incubation temperatures during reverse transcription. However, dependant on the reaction kinetics of the reverse transcription and the relative abundance of the required RNA molecule, random cDNA synthesis could prove to be more sensitive than the single, specific primer approach.

Some RT-PCR protocols include an RNase H treatment step immediately after cDNA synthesis. RNase H digests the original RNA template strand in RNA/cDNA reverse transcribed hybrid molecules, theoretically making the cDNA molecule more accessible to PCR primers during the initial cycles of PCR amplification. However, many RT enzymes inherently possess RNase H activity and the high DNA duplex "melting" temperatures used in PCR cycling effectively ensure the separation of RNA and cDNA strands in RNA/cDNA hybrids prior to PCR amplification. As further thought, RNase H negative reverse transcriptase enzymes do possess some major advantages over RNase H positive enzymes.

4.5.2 cDNA Synthesis Using RACE

The synthesis of full-length cDNA is often problematic for longer mRNA molecules, as the 5′- end of the reverse transcribed cDNA molecule may be absent. Rapid amplification of cDNA ends (RACE) can provide a solution to this problem in some instances [Gujar and Michalak, 2006; Zhang and Frohman, 1997; Frohman et al., 1988], and also allows the amplification of mRNA from a known internal sequence region (through an unknown sequence) to either the 3′-or 5′-end of the mRNA target. The RACE technique (also known as "one-sided" PCR or "anchored" PCR) requires prior knowledge of the sequence of the central region of the specific mRNA to be reverse transcribed, as well as 5′- and 3′- terminal sequence data and primers designed to contain specific restriction sites. If full-length mRNA sequence is not required, then either a 3′ or 5′ RACE protocol may be performed. In the 3′ RACE protocol, oligo-dT is initially used as a primer to reverse transcribe part of the mRNA target molecule in the 5′ to 3′ direction towards the centre of the target mRNA molecule. Next, a primer which hybridizes to a known exon sequence on the newly synthesized cDNA strand is added to the mix and a PCR performed. This protocol allows amplification from a known exon sequence to the poly-A tail of the target mRNA (through regions of unknown sequence) to be achieved. In a typical 5′ RACE protocol, a primer is designed which binds to an internal sequence of the target mRNA so that first strand synthesis (reverse transcription) is performed towards the 5′-end of the target mRNA molecule. Next, the cDNA molecule is tailed with dATP at its 3′- end using terminal adenylyl transferase and PCR amplification performed using a poly-dT primer and a second mRNA specific primer as PCR primers. This protocol allows amplification from a known exon sequence to the 5′-end of the target mRNA (through regions of unknown sequence) to be achieved. The 3′ and 5′ RACE products may be directly sequenced or used to generate probes. Moreover, by including known restriction sites in the RACE and PCR primers, RACE products may be cloned or ligated together to generate full-length cDNA copies of the original target mRNA molecule. The RACE principle is explained in more detail in Fig. 4.9.

The RACE technique relies very much on the presence of a complete target mRNA molecule, as degraded mRNAs give rise to artificial (prematurely terminated) ends. Some RACE kits (e.g. the Gene Racer RLM-RACE kit, Invitrogen Life Technologies) ensure that only full length capped target mRNA is amplified and not degraded target mRNA products by making use of the fact that truncated

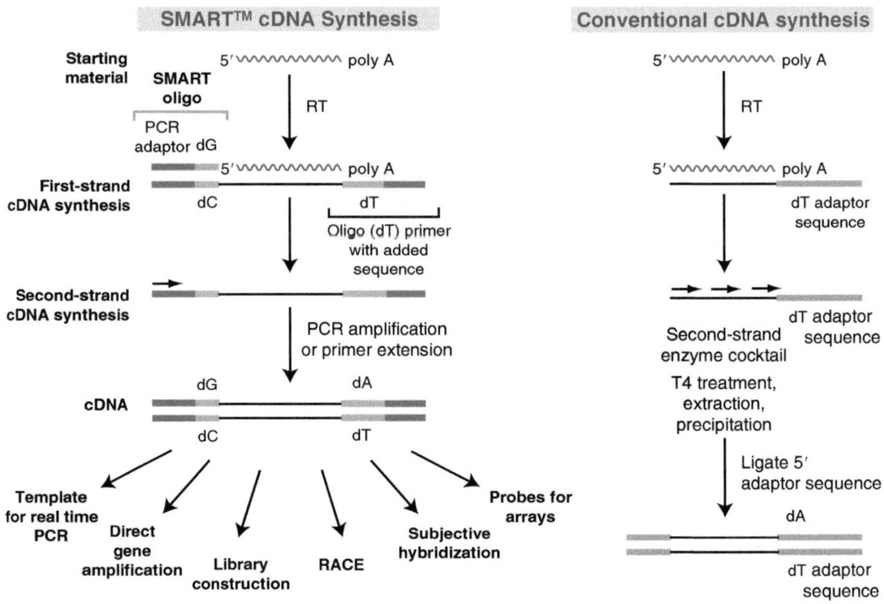

Fig. 4.9 RACE, a method to generate large numbers of cDNA molecules from a target mRNA molecule for which complete sequence information is missing. The RACE protocol approaches 5′- and 3′- end amplification of the mRNA molecule differently. The 3′ RACE protocol transcribes mRNA into cDNA via the poly-A tract, followed by addition of a central primer designed from known target mRNA exon sequence which facilitates PCR amplification. The 5′- RACE protocol transcribes mRNA into cDNA via a central primer designed from known target mRNA exon sequence. This cDNA molecule is then tailed with dATP, after which PCR amplification is performed using an oligo-dT primer and another central primer designed from known target mRNA exon sequence. The addition of restriction site sequences to the primers allow cloning, etc. of the amplified RACE products. The illustration shows specialised oligo-dT primed selection of mRNA from the total RNA sample. The primer contains a built-in sequence for PCR and restriction site generation (1). Next, reversed transcriptase reaches the end of the mRNA molecule and adds a stretch of cytosines due to its intrinsic terminal transferase activity (2). Then a specialised oligonucleotide hybridises to the cytosine stretch in order to create an extended template for the reverse transcriptase. The enzyme continues cDNA synthesis until the end of the specialised oligonucleotide and, consequently, both cDNA ends now carry specific adaptor sequences (3). The second strand cDNA synthesis is synthesised by primer extension or long distance PCR amplification using primers corresponding to the adaptor sequences (4). The double strand DNA can now serve as template for real time PCR, direct gene amplification, library construction, array analysis or subtractive hybridisation (Reproduced from www.clontech.com. With permission from Clontech Laboratories, Inc.)

(degraded) mRNA may be dephosphorylated by calf intestine phosphatase, whilst full-length mRNA is not dephosphorylated due to the presence of the cap structure. This dephosphorylation prevents the subsequent ligation of a specially designed primer to degraded mRNA. The specially designed primer may then be ligated to the full-length target mRNA by first removing the cap structure (using tobacco acid pyrophosphatase) and ligating the primer to the de-capped target mRNA molecule (using T4 RNA ligase). As well as ensuring the amplification of full length mRNA, this technique also generates a novel, 5′- end specific cDNA priming region.

4.5.3 RNA Extraction and cDNA Synthesis Controls

The monitoring of extracted RNA quality and the use of internal RT-PCR controls are important quality control factors in RT-PCR applications and are an absolute necessity for quantitative gene expression studies. For example, it is important to be sure that the absence of RT-PCR amplification product is not due to RNA degradation. This can be determined by RT-PCR amplification of mRNA from constitutively expressed "housekeeping" genes from the same sample of extracted RNA, or even within the same RT-PCR mix that is to be used for RT-PCR amplification. Well-known examples of such "housekeeping" genes include glyceraldehyde-3-phosphate, β-actin, hypoxanthine phophoribosyltransferase, tubulin, 18S rRNA and 28S rRNA [Suzuki et al., 2000; Radovic et al., 2004]. Although the concentration of the mRNA species may vary depending on the growth status of the cells from which the mRNA has been purified, the presence of mRNA from these "housekeeping" genes is a guarantee of correct RNA isolation procedures. In order to detect possible inhibition during RNA isolation and RT-PCR amplification, commercially available or self-made RNA species may be added in order to "spike" the nucleic acid isolation protocol or RT reaction. For controlling the presence/absence of DNA contamination, internal or external control RT-PCR reactions using primers specific for pseudogenes should generate no specific RT-PCR amplification products. Alternatively, a separate RT-PCR may be run using the same primers as the normal RT-PCR protocol, but omitting the RT enzyme. If only RNA is present then no PCR amplification product should be observed after RT-PCR, if DNA is present then it will act as a target for PCR amplification. If DNA contamination is suspected, then an additional DNase I treatment may alleviate the problem.

References

Abbott MA, Benn P. 2002. Prenatal genetic diagnosis of Down's syndrome. Expert Rev Mol Diagn 2:605–615.

Acheampong E, Rosario-Otero M, Dornburg R, Pomerantz RJ. 2003. Replication of lentiviruses. Front Biosci 8:156–174.

Agrawal S, Muller B, Bharadwaj U, Bhatnagar S, Khan F. 2002. Evaluation of short tandem repeat loci in forensics: north Indian populations. J Forensic Sci 47:686–689.

Alberts B, Johnson A, Lewis J, Raff M, Roberts K, Walter P. 1994. Molecular biology of the cell. Third Edition. Garland, New York.

Alcock HE, Stephenson TJ, Royds JA, Hammond DW. 1999. A simple method for PCR based analyses of immunohistochemically stained, microdissected, formalin fixed, paraffin wax embedded material. Mol Pathol 52: 160–163

Al-Soud WA, Radstrom P. 1998. Capacity of nine thermostable DNA polymerases to mediate DNA- amplification in the presence of PCR- inhibiting samples. Appl Environ Microbiol 64:3748–3753.

Al-Soud WA, Johnson LJ, Radstrom P. 2000. Indentification and characterization of Immunoglobulin G in blood as a major inhibitor of diagnostic PCR. J Clin Microbiol 38:345–350

Al-Soud WA, Radstrom P. 2001. Purification and characterization of PCR- inhibitory components in blood cells. J Clin Microbiol 39:485–493.

An SF, Fleming KA. 1991. Removal of inhibitor(s) of the polymerase chain reaction from formalin fixed, paraffin wax embedded tissues. J Clin Pathol 44:924–927.

Antin JH, Childs R, Filipovich AH, Giralt S, MacKinnon S, Spitzer T, Weisdorf D. 2001. Establishment of complete and mixed donor chimerism after allogeneic lymphohematopoietic transplantation: recommendations from a workshop at the 2001 Tandem Meetings of the International Bone Marow Transplant Registry and the American Society of Blood and Marrow Transplantation. Biol Blood Marrow Transplant 7:473–485.

Ban N, Nissen P, Hansen J, Moore PB, Steitz TA. 2000. The complete atomic structure of the large ribosomal subunit at 2.4 A resolution. Science 289:905–920.

Barton JH. 2002. Patents, genomics, research and diagnostics. Acad Med 77:1339–1347.

Berg KD, Glaser CL, Thompson RE, Hamilton SR, Griffin CA, Eshleman JR. 2000. Detection of microsatellite instalbility by fluorescence multiplex polymerase assay. J Mol Diag 2:20–28.

Blackburn P, Gavilanes JG. 1982. Identification of lysine residues in the binding domain of ribonuclease A for the RNase inhibitor from human placenta. J Biol Chem 257:316–321.

Boom R, Sol CJ, Salimans MM, Jansen CL, Wertheim-van Dillen PM, van der Noordaa J. 1990. Rapid and simple method for the purification of nucleic acids. J Clin Microbiol 28:495–503.

Bruno JG, Yu H, Kilian JP, Moore AA. 1996. Development of an immunomagnetic assay system for rapid detection of bacteria and leukocytes in body fluids. J Mol Recognit 9:474–479.

Burkhart CA, Norris MD, Haber M. 2002. A simple method for the isolation of genomic DNA from mouse tail free of real-time PCR inhibitors. J Biochem Biophys Methods 52:145–149.

Burton MP, Schneider BG, Brown R, Escamilla-Ponce N, Gulley ML. 1998. Comparison of histologic stains for use in PCR analyses of micro-dissected, paraffine-embedded tissues. Biotechniques 24:86–92.

Camilleri-Broet S, Devez F, Tissier F, Ducruit V, Le Tourneau A, Diebold J, Audouin J, Molina T. 2000. Quality control and sensitivity of Polymerase Chain Reaction Techniques for the assesment of immunoglobulin heavy chain rearrangements from fixed- and paraffine-embedded samples. Ann Diagn Pathol 4:71–76.

Chomczynski P, Sacchi N. 1987. Single-step method of RNA isolation by acid guanidinum thiocyanate-phenaol-chloroform extraction. Anal Biochem 162:156–159.

Deggerdal A, Larsen F. 1997. Rapid isolation of PCR-ready DNA from blood, bone marrow and cultured cells, based on paramagnetic beads. Biotechniques 22:554–557.

De Leeuw WJ, Dierssen J, Vasen HF, Wijnen JT, Kenter GG, Meijers Heijboer H, Brocker Vreinds A, Stormorken A, Moller P, Menko F, Cornelisse CJ, Morreau H. 2000. Prediction of a mismatch repair gene defect by microsatellite instability and immunohistochemical analysis in endometrial tumours from HNPPC patients. J Pathol 192:328–335.

Eaton MC, Hardingham JE, Kotasek D, Dobrovic A. 1997. Immunobead RT-PCR: a sensitive method for detection of circulating tumor cells. Biotechniques 22:100–105.

Efimov VA, Buryakova AA, Chakhmakhcheva OG. 1999. Synthesis of polyacrylamide N-substituted with PNA-like oligonucleotide mimics for molecular diagnostic applications. Nucleic Acids Res 27:4416–4426.

Etokebe GE, Spurkland A. 2000. Method for avoiding PCR-contamination when eluting DNA from polyacrylamide gels. Biotechniques 29:694–696.

Foss RD, Guha-Thakurta N, Conran RM, Gutman P. 1994. Effects of fixative and fixation time on the extraction and polymerase chain reaction amplification of RNA from paraffin-embedded tissue. Comparison of two housekeeping gene mRNA controls. Diagn Mol Pathol 3:148–155.

Frackman S. 1998. Betaine and DMSO: enhancing agents for PCR. Promega Notes 65:27–29.

Frank TS, Svoboda-Newman SM, Hsi ED. 1996. Comparison of methods for extracting DNA. from formalin-fixed paraffin sections for nonisotopic PCR. Diagn Mol Pathol 5:220–224.

Fraser CM, Gocayne JD, White O, Adams MD, Clayton RA, Fleischmann RD, Bult CJ, Kerlavage AR, Sutton G, Kelley JM, Fritchman JL, Weidman JF, Small KV, Sandusky M, Fuhrman J, Nguyen D, Ytterback TR, Saudek DM, Philips CA, Merrick JM, Tomb J, Dougherty BA, Bott KF, Hu P, Lucier TS, Peterson SN, Smith HO, Hutchison CA, Venter JC. 1995. The minimal gene complement of *Mycoplasma genitalium*. Science 270:397–403.

Fredericks DN, Relman DA. 1999. Paraffin removal from tissue sections for digestion and PCR analyses. Biotechniques 26:198–200.

Frohman MA, Dush MK, Martin GR. 1988. Rapid production of full-length cDNAs from rare transcripts: amplification using a single gene-specific oligonucleotide primer. Proc Natl Acad Sci USA 85:8998–9002.

Golenberg EM, Bickel A, Weihs P. 1996. Effect of highly fragmented DNA on PCR. Nucleic Acids Res 24:5026–5033.

Gugliotta P, Pacchioni D, Bussolati G. 1992. Staining for beta-galactosidase in immunocytochemistry and *in situ* hybridisation. Eur J Histochem 36:143–148.

Guidet F. 1994. A powerful new technique to quickly prepare hundreds of plant extracts for PCR and RAPD analyses. Nucleic Acids Res 22:1772–1773.

Gujar SA and Michalak TI. 2006 Characterization of bioactive recombinant woodchuck interleukin-2 amplified by RLM-RACE and produced in eukaryotic expression system. Vet Immunol Immunopathol 112:183–198.

Hendolin PH, Paulin L, Koukila-Kahkola P, Anttila VJ, Malmberg H, Richardson M, Ylikoski J. 2000.Panfungal PCR and multiple liquid hybridization for detection of fungi in tissue specimens. J Clin Microbiol 38:4186–4192.

Howe JR, Klimstra DS, Cordon-Cardo C. 1997. DNA extraction from paraffine-embedded tissues using a salting-out procedure: a reliable method for PCR amplification of archival material. Histol Histopathol 12:595–601.

Hu Y, Shahidi A, Park S, Guilfoyle D, Hirshfield I. 2003. Detection of extrahepatic hepatitis C virus replication by a novel, highly sensitive, single-tube nested polymerase chain reaction. Am J Clin Pathol 119(1):95–100.

Ito T, Smith CL, Cantor CR. 1992. Sequence specific DNA purification by triplex affinity capture. Proc Natl Acad Sci USA 89:495–498.

Jackson DP, Lewis FA, Taylor GR, Boylston AW, Quirke P. 1990. Tissue extraction of DNA and RNA and analyses by the polymerase chain reaction J Clin Pathol 43:499–504.

Kleter B, van Doorn LJ, Schrauwen L, Molijn A, Sastrowijoto S, ter Schegget J, Lindeman J, ter Harmsel B, Burger M, Quint W. 1999. Development and clinical evaluation of a highly sensitive PCR-reverse hybiridzation line probe assay for the detection and identification of anogenital human papillomavirus. J Clin Microbiol 37:2508–2517.

Koonin EV, Aravind L, Kondrashov AS. 2000. The impact of comparative genomics on our understanding of evolution. Cell 101:573–576.

Kurdziel AS, Wilkinson N, Langton S, Cook N. 2001. Survival of poliovirus on soft fruit and salad vegetables. J Food Prot 64:706–709.

Lu GL, Wen Jm, Xu JM, Zhang M, Xu RB, Tian BL. 2003. Relationship between TIMP-3 expression and promoter methylation of TIMP-3 gene in hepatocellular carcinoma. Zhonghua Bing Li Xue Za Zhi 32:230–233.

Marsh I, Whittington R, Millar D. 2000. Quality control and optimized procedure of hybridization capture PCR for the identification of *Mycobacterium avium* subsp. *Paratuberculosis* in faeces. Mol Cell Probes 14:219–232.

Masuda N, Ohnishi T, Kawamoto S, Monden M, Okubo K. 1999 Analysis of chemical modification of RNA from formalin-fixed samples and optimization of molecular biology application for such samples. Nucleic Acids Res 27:4436–4443.

Millar DS, Withey SJ, Tizard ML, Ford JG, Hermon-Taylor J. 1995. Solid phase hybridization capture of low-abundance target DNA sequences: application to the PCR detection of

Mycobacterium paratuberculosis and *Mycobacterium avium* subsp *silvaticum*. Anal Biochem 226:325–330.

Mitsuhashi M. 1996b. Technical report: Part 2: Basic requirements for designing optimal PCR primers. J Clin Lab Anal 10:285–293.

Miyachi H, Masukawa A, Ohshima T, Fusegawa H, Hirose T, Impraim C, Ando Y. 1998. Monitoring of inhibitors of enzymatic amplification in PCR and evaluation of efficacy of RNA extraction for the detection of hepatitis C virus using the internal control. Clin Chem Lab Med 36:571–575.

Morel F, Bernicot I, Herry A, Le Bris MJ, Amice V, De Braekeleer M. 2003. An increased incidence of autosomal aneuploidies in spermatozoa from a patient with Klinefelter's syndrome. Fertil Steril 79:1644–1646.

Nakabachi A, Yamashita A, Toh H, Ishikawa H, Dunbar HE, Moran NA, Hattori M. 2006. The 160-kilobase genome of the bacterial endosymbiont *Carsonella*. Science 314(5797):267.

Noro M, Masuda R, Dubrovo IA, Yoshida MC, Kato M. 1998. Molecular phylogenetic inference of the woolly mammoth *Mammuthus primigenius* based on the complete sequences of mitochondrial cytochrome b and 12S ribosomal RNA genes. J Mol Evol 46:314–326.

Ohtsuki T, Kimoto M, Ishikawa M, Mitsui T, Hirao I, Yokoyama. 2001. Unnatural base pairs for specific transcription. Proc Natl Acad Sci USA 98:4922–4925.

Otto P. 1998. Separate isolation of genomic DNA and total RNA from single samples using the SV total RNA Isolation System. Promega Notes 69:19–23.

Parham N, Spencer J, Taylor D, Ternent H, Innocent G, Mellor D, Roberts M, Williams A. 2003. An adapted immunomagnetic cell separation method for use in quantification of *Escherichia coli* O157:H7 from bovine faeces. J Microbiol Methods 53:1–9.

Perelle S, Josefsen, M, Hoorfar J, Dilasser F, Grout J, Fach P. 2004. A LightCycler real-time PCR hybridization probe assay for detecting food-borne thermophilic Campylobacter. Mol Cell Probes 18(5):321–327.

Pfeuty A, Gueride M. Lecellier G. 2001. Expansion/contraction of mammalian mitochondrial DNA repeats in *Escherichia coli* mimics the mitochondrial heteroplasmy. J Mol Biol 314:709–716.

Pusch CM, Sasiadek MM, Blin N. 2002. Hirschprung, RET-SOX and beyond: the challenge of examining non-mendelian traits. Int J Mol Med 10:367–370.

Radovic A, Thulke S, MacKay IM. 2004. Guidelines for reference gene selection for quantitative real-time PCR. Biophys Biochem Res Comm 313:856–862.

Reiss E. 1977. Serial enzymatic hydrolysis of cell walls of two serotypes of yeast from *Histoplasma capsulatum* with alpha and beta glucanase, pronase and chitinase. Infect Immun 16:181–188.

Rubin MA. 2001. Use of laser capture microdissection, cDNA microarrays and tissue microarrays in advancing our understanding of prostate cancer. J Pathol 195:80–86.

Satsangi J, Jewell DP, Welsh K, Bunce M, Bell JI. 1994. Effect of heparin on polymerase chain reaction. Lancet 343:1509–1510.

Saxena SK, Rybak SM, Winkler G, Meade HM, McGray P, Youle RJ, Ackerman EJ. 1991. Comparison of RNases and toxins upon injection into *Xenopus* oocytes. J Biol Chem 266:21208–21214.

Schoolnik GK. 2002. Functional and comparative genomics of pathogenic bacteria. Curr Opin Microbiol 5:20–26.

Sieben NL, Ter Haar NT, Cornelisse CJ, Fleuren GJ, Cleton Jansen AM. 2000. PCR artifacts in LOH and MSI analysis of microdissected tumor cells. Hum Pathol 31:1414–1419.

Silliman CC, Tyson RW, Wei Q, Karrer FG, Davies SM, Blake M, McGravan L. 2003. Acute Philadelphia chromosome positive leukemia in an adolescent boy after liver transplantation. J Pediatr Hematol Oncol 25:565–568.

Smith LS, Lewis TL, Matsui SM. 1995. Increased yield of small DNA fragment purified by silica binding. Biotechniques 18:970–975.

Suzuki T, Higgins PJ, Crawford DR. 2000. Control selection for RNA quantitation. Biotechniques 29:332–337.

Tamura K, Hasegawa T. 1999. Relationship of the CCA sequence of tRNA with the early evolutional aspect of aminoacyl tRNA synthestases. Nucleic Acids Symp Ser 42:211–212.

Thomas G, Cann H. 2003. Irruption of genomics in the search for disease related genes. Gut 52:1–5.

Tian H, Huhmer AF, Landers JP. 2000. Evaluation of silica resin for direct and efficient extraction of DNA from complex biological matrices in a miniaturized format. Anal Biochem 283:175–191.

Umetani N, Sasaki S, Watanabe T, Ishigami H, Ueda E, Nagawa H. 2000. Diagnostic primer sets for microsatellite instability optimized for a minimal amount of damaged DNA from colorectal tissue samples. Ann Surg Oncol 7:276–280.

Van Burik JA, Myerson D, Schreckhise RW, Bowden RA. 1998. Panfungal PCR assays for detection of fungal infection in human blood specimens. J Clin Microbiol 36:1169–1175.

Van Doornum G, Guldemeester J, Osterhaus AD, Niesters HG. 2003. Diagnosing herpesvirus infections by real-time amplification and rapid culture. J Clin Microbiol 41:576–580.

Wan QH, Qian KX, Fang SG. 2003. A simple DNA extraction and rapid specific identification technique for single cells and early embryos of two breeds of *Bos taurus*. Anim Reprod Sci 77:1–9.

Wera S, Bollen M, Stalmans W. 1991. Purification and characterization of the glycogen bound protein phosphatase from rat liver. J Biol Chem 266:339–345.

Wickham CL, Boyce M, Joyner MV, Sarsfield P, Wilkins BS, Jones DB, Ellard S. 2000. Amplification of PCR products in excess of 600 base pairs using DNA extracted from decalcified paraffin wax embedded bone marrow trephine biopsy. Mol Pathol 53:19–23.

Wood R, Dong H, Katzenstein DA, Merigan TC. 1993. Quantification and comparison of HIV-1 proviral load in peripheral blood mononuclear cells and isolated CD4+ T cells. J AIDS 6:237–240.

Zhang Y and Frohman MA. 1997. Using rapid amplification of cDNA ends (RACE) to obtain full-length cDNAs. Methods Mol Biol 69:61–87

Zhang Z, Harrison P, Gerstein M. 2002. Identification and analysis of over 2000 ribosomal protein pseudogenes in the human genome. Genome Res 12:1466–1482.

Chapter 5

PCR Primers

Chemically synthesised nucleotide primer molecules (also called oligonucleotides or oligos) are essential components to the success of PCR amplification. After (specific) hybridisation to a complementary region of the target DNA, the primers provide the 3′ hydroxyl ends by which DNA polymerase mediated DNA synthesis proceeds (Fig. 1.1). The position of the primers along the single stranded DNA strands defines the total length of the PCR amplification product generated. Primers designed to yield relatively short PCR amplification products (100–200 nucleotides), generate product (amplimers) more efficiently than primers designed to yield relatively long PCR amplification products (3,000–5,000 bp). This is especially true for DNA template molecules where significant secondary structure is present. Hence, PCR primers are usually designed to amplify a region of maximum 3,000–5,000 bp in length. However, using specifically adapted "long and accurate" PCR protocols, the amplification of fragments up to 40 kbp in length is feasible.

PCR primers are the main determinants of PCR specificity. In order to amplify a specific DNA sequence, the sequence of the primer annealing sites on the target DNA must be known. Further, in order to limit the chance of non-specific primer binding (where one or more bases are not complementary to the target sequence), or the chance of the primers binding to a similar sequence elsewhere on the target DNA, primers are usually designed to be in the order of 18–22 nucleotides in length (Section 5.1). A limited group of generally short primers has the capacity to hybridise to multiple sites on target nucleic acid, these include oligo-dT-based primers and random hexamer/decamer primers. Another category of PCR primers that have the potential to anneal to multiple sites are those that target genomic repeat sequences, which may be used to amplify genomically polymorphic repeat regions (i.e. regions varying in repeat sequence length between individual isolates) and generate "DNA fingerprints" for individual isolates [Van Belkum et al., 1998]. Also, primers designed to be complementary to conserved DNA sequences present in ribosomal genes are able to amplify specific ribosomal gene sequence fragments from all organisms, whether eukaryote or prokaryote.

E. Van Pelt-Verkuil et al., *Principles and Technical Aspects of PCR Amplification*,

5.1 PCR Primer Design and Quality Requirements

The base composition and length of primer are particularly important characteristics in facilitating the amplification of specific PCR products. In addition, the GC-content, secondary structure, the presence of poly-purine (adenosine or guanosine nucleotides) or poly-pyrimidine (cytosine or thymidine nucleotides) stretches and any possible homology of the fragment region to be amplified with other regions of DNA within the target DNA, are also important parameters influencing the design of PCR primers [Mitsuhashi, 1996]. Fortunately, various computer programs are available to help researchers in the design of specific PCR primer pairs, e.g. MEDUSA, Primer3, PrimerQuest, etc. Such software programs are free to use, freely downloadable, or may be purchased from commercial enterprises. The use of PCR primer design software is recommended when developing a new PCR protocol. It is however still possible to "manually" design PCR primers, based on a few simple rules. Indeed, from the authors' own experience, it is possible to manually design effective PCR primers with characteristics that would not be acceptable to primer design software programs (e.g. primers with low annealing temperatures, too many hairpins, formation of primer dimers, etc.). However, the main rules for effective primer design are discussed below.

To limit the chance of a PCR primer binding to a sequence of DNA that is not the intended binding site (non-specific primer hybridisation), PCR primers should be designed to be between 18 and 30 nucleotides (bp) in length. The "uniqueness" of a random 18-mer primer is almost guaranteed, as an 18 bp primer theoretically comprises $4^{18} = 6.9 \times 10^{10}$ different permutations. With the human genome being 3×10^9 nucleotides in size, the chances of a random sequence being found in extracted human DNA that is complementary to an 18–30 bp primer sequence is almost negligible. In extreme cases, functioning PCR primers have been designed with lengths as low as 8–10 bp [Afonina et al., 1997], although the efficiency of these PCRs depends on the presence of certain proteins that stabilise the binding between PCR primer and target DNA. Also, it should be noted that the length of a given PCR primer also influences the annealing temperature of that particular primer (Section 5.3).

When designing PCR primers, the sequence of an individual primer should be so designed that there are no internal complementary regions (e.g. multiple guanosine and cytosine repeats) present within that particular primer, as internal complementary regions can give rise to the formation of hairpin structures (loops), which may prevent complete hybridisation of the primer to the target sequence thereby hindering PCR amplification. The lower the "hairpin energy" calculation for a PCR primer is, the better the suitability of that primer for PCR [Mitsuhashi, 1996]. Interestingly, some PCR assays have been developed to actually make use of such internal hairpin structure, based on the fact that primers containing such structure hybridise inefficiently at low annealing temperatures. At higher annealing temperature, they loose their intrinsic structure and may function as efficient PCR primers. This type of PCR primer design has been utilised in certain nested PCR primer protocols, circumventing the need to open the PCR reaction vessel after completion of the first round of PCR amplification cycles (which increases the risk of second round PCR contamination) [Kaboev et al., 2000]. The reverse situation can also be useful as raising the temperature during PCR thermocycling is a basic feature of "touch-up" PCR using "TULIP" primers [Ailenberg and Silverman, 2000]. Additionally, PCR primers containing internal hairpin structure may be equipped with fluorescent and quencher (absorber) groups at each end of the primer (molecular beacons). At lower temperatures, hairpin formation results in the fluorescent and quencher groups being in close proximity to each other resulting in no or very little signal. At higher temperatures, the hairpin formation is disrupted resulting in the fluorescent and quencher groups being distant to each other, allowing a fluorescent signal to be detected. These types of primers (and probes) are very useful and widely used in real-time fluorescence PCR strategies (Sections 5.1.1 and 9.3.2).

Complementary regions between the different primers included in single PCR reaction mix (and especially complementary regions at their 3′ termini), may facilitate primer/primer hybridisation, giving rise to the amplification of "primer dimer" artefacts (short pieces of primer-based DNA of approximately 50–100 bp in length, Fig. 5.1). The PCR amplification of shorter primer dimer DNA proceeds much more favourably than the amplification of longer fragments of DNA (i.e. the intended target region), and removes from the PCR reaction mix a large fraction of the necessary primers and dNTPs that are required in later cycles. The chances of primer-dimer formation occurring are especially high in multiplex PCR protocols, where several different primers are included in an individual PCR mix in order to amplify several different target regions at the same time. If it is absolutely necessary to use PCR primers which yield primer dimer products, then the formation of such products may be partially inhibited by exchanging guanosine nucleotides with analogous 7-deaza-2′-deoxyguanosine nucleotides in the primer design. This guanosine-analogue still pairs with cytosine, but the energy content of the hybridisation is significantly lower than the usual guanosine/cytosine base pairing energy. Other factors that contribute to the formation of primer dimers include: an excessive number of PCR amplification cycles, an initial target concentration that is too low for efficient PCR amplification, and too high a concentration of thermostable DNA

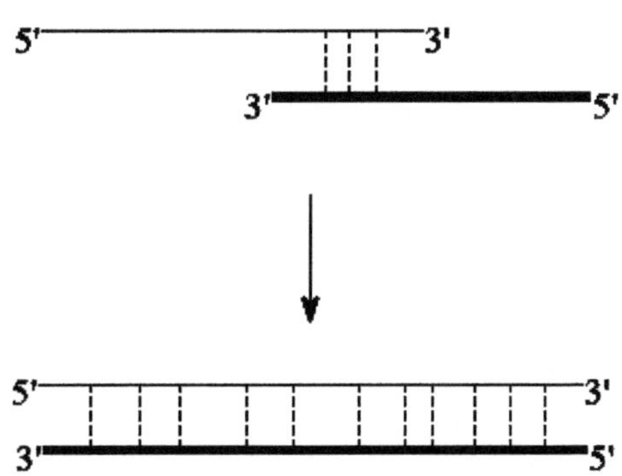

Fig. 5.1 If two or more PCR primers within a single PCR mix contain regions of complementarity, then there will be an increased risk of the preferential amplification of "primer dimer" artefacts instead of the intended target. This effect is especially likely when the regions of complementarity occur at the respective 3′ termini of the primers

polymerase/primer in the PCR reaction mix. Several thermostable DNA dependent polymerases are capable of adding single nucleotides to the 3′-end of a blunt-ended dsDNA molecule (a phenomenon utilised in "TA cloning reactions" [Zhou and Gomez-Sanchez, 2000]), which if complementary on differently amplified molecules, may increase the chance of primer-dimer formation.

Several authors have suggested that further criteria may need to be considered in the design of the 3′-end of a PCR primer. For example, a non-thymidine nucleotide at the 3′-end of PCR primers is preferred by some, whilst less efficient primer extension has been encountered with PCR primers containing a 3′-terminal adenosine residue. In contrast, other authors claim that the 3′-end of the primer is better designed to be A/T rich, which would theoretically help to reduce the chance of primer-dimer formation because of the low energy content of A/T base pairing (adenosine-thymidine base pairing is weaker than guanosine-cytosine base pairing). Further, three guanosine/cytosine basepairs at the 3′-end of a primer could possibly promote more efficient initiation of DNA synthesis, and it is documented that a penultimate (last but one) adenosine or thymidine residue may also be sub-optimal (for more detailed information see [Ayyadevara et al., 2000]).

As part of PCR primer design and sequence considerations, it is also necessary to take into account the guanosine and cytosine (GC) content of each primer sequence, as this has a large influence on primer/target hybrid stability. In general, a GC-content in the order of 50% is recommended, with increased GC-content values resulting in a higher annealing temperature and hence increased stringency of primer annealing. If higher annealing temperatures are required (i.e. nearing the PCR extension temperature of 72°C), then it is usually recommended that the length of the primer sequence and not the GC-content of the primer sequence, be increased. This is because: (1) for many organisms a GC content of greater than 50% in any given region of DNA is unusual, (2) long stretches of GC repeats in the primer sequence may lead to the formation of internal hairpin loops and (3) DNA dependent polymerases have great difficulty in amplifying stretches of DNA containing runs of guanosine or cytosine residues. If the region to be PCR amplified is relatively poor in GC-content, then it is advisable to design PCR primers that initiate amplification outside of the target region, where the GC-content is more suitable for PCR primer design. If problems persist with the amplification of specific product despite increasing the annealing temperature, then a nested amplification protocol may be a useful alternative.

It is advisable to select PCR primer sites that amplify "genetically stable" regions (i.e. regions where mutations are less likely to occur) of the target DNA. If a target region accumulates mutations at a high frequency then it is not well suited for PCR primer design, especially for diagnostic PCR amplification purposes. Conserved sequences on the other hand, are more stable and are therefore the preferred targets for PCR amplification. Whether or not a sequence is unique can be partially assessed via the use of specific computer search-programs. These programs can screen a whole database library for similarities between the designed PCR primer sequences and known sequence data deposited in the database, e.g. the National Centre for Biotechnology Information "GenBank" database (http://www.ncbi.nlm.nih.gov/).

Using the same approach, primer sequences that are conserved (or relatively conserved) for a particular gene may be used to design PCR primers that amplify the same region within different species or even species that have not yet been studied. In this case, the gene sequences do not have to be identical, as "degenerate" primers may be designed, where the majority of the primer sequence is conserved but where at a specific primer location(s) different nucleotides are introduced in order to reflect the sequence differences that may occur between individual species.

PCR primer pairs should be designed so that their annealing temperatures do not significantly differ (i.e. by ±5°C), as this may favour either the amplification of single stranded products or non-specific hybridisation by one of the PCR primers. With regard to balanced primer annealing temperatures, primer/target DNA sequence mismatches may lead to changes in the apparent annealing temperature of the primer/target DNA hybrid (Sections 5.3 and 5.6). Hence, the correct calculation of balanced primer annealing temperatures is most likely when primer sequences are 100% homologous to their intended target DNA sequences.

No matter how well-designed the PCR primers, attention still has to be paid to optimising each new PCR protocol *per se*, especially with regard to magnesium ion/primer concentration and primer annealing temperature. For the vast majority of PCR protocols, the PCR primer pair concentration should be equimolar in order to facilitate a "balanced" amplification reaction. However, some PCR protocols utilise uneven quantities of primers resulting in "a-symmetric" PCR amplification, and leading to the accumulation of single stranded DNA formed from the excess primer. This strategy was often applied for generating templates for DNA sequencing (Section 9.4; [Mazars and Theillet, 1996]). Relative differences in individual primer concentrations for some assays (e.g. some Taqman assays) may vary in the order of four- to seven-fold concentration differences (Section 9.3.3; [Tucker et al., 1999]).

The use of the universally binding inosine nucleotide (Section 6.2 and Fig. 6.1) in primer design may allow for the efficient amplification of target DNA that is otherwise difficult to amplify, for example due to sequence heterogeneity in the target DNA or the presence of stable secondary structures. The inosine base is promiscuous in the sense that it can pair with all four classical bases present in DNA [Martin et al., 1985], a feature that allows modification of PCR stringency by substitution of the nucleotides in a PCR primer by inosine. The stability of an inosine-containing basepair is less than that of the classical Watson-Crick pairings. However, it should be noted that the inclusion of inosine in a PCR primer also enhances the tolerance of DNA polymerase enzymes for mismatches between primer and target DNA, thereby possibly reducing the specificity of PCR amplification. Finally, the use of inosine derivatives in sequencing reactions is considered very useful for GC-rich templates [Shen et al., 2003].

Mis-hybridization (non-specific hybridization) occurs when one or more of the PCR primers in a pair does not hybridize to the specific target sequence intended. The causes of mis-hybridization include; an annealing temperature that is too low (allowing partial hybridization of the primer and effectively reducing the primer length), incorrect primer design (e.g. to a non-conserved target sequence), and/or too high a concentration of thermostable DNA polymerase added to the PCR reaction mix. Primers greater than 18 nucleotides in length should in theory specifically hybridize to a single nucleic acid target sequence within the total number of sequences found in a eukaryotic or prokaryotic cell. Fortunately, non-specific hybridization of PCR primers usually results in smears of PCR products or the presence of products of the incorrect length upon gel electrophoresis. This particular problem is therefore relatively easy to detect. However, on some occasions, supplementary testing (e.g. DNA sequencing, Southern hybridization, etc.) may be required in order to verify the (non)specific nature of the amplimers generated. It should be noted that certain PCR applications actually rely on the ability of primers to hybridize to very similar target sequences (e.g. PCR consensus primers designed to detect either closely related or currently unidentified species or sub-species of microorganisms, or homologous genes [Riemersma et al., 2003]). For example, the human papillomavirus group includes at least 75 genotypes based on DNA sequence analysis, many of which share regions of significant sequence identity.

5.1.1 Different Primer Species

As well as the "traditional" type of PCR primers, which were simply designed to anneal to target DNA and act as primers for thermostable DNA dependent polymerases, a variety of modified primers have now been developed which greatly broaden the scope of PCR applications. These applications may be grouped into three distinct categories (Fig. 5.2).

The "5′-end modified" category of primers, are PCR primers which have been designed to be complementary to specific regions of target DNA (similar to "traditional" PCR protocols), but contain additional nucleotides at their 5′-ends that are not found in the target DNA. These additional 5′ sequences are an integral part of the PCR primer and are co-amplified along with the rest of the amplified PCR product. The use of these "5′-end modified" primers may sometimes require the use of lowered annealing temperatures during the first steps of the PCR cycling protocol. Examples of 5′-end modifications include:

(1) *Restriction sites*: The presence of a restriction site at the 5′-end of an amplification product facilitates cloning of the PCR product into a relevant vector. If different restriction sites are designed into the 5′-end of each of the PCR primer pair, then various PCR product cloning strategies become available. In order to enable efficient recognition of the site by its restriction enzyme, four to five "terminal nucleotides" need to be present immediately upstream of the 5′-end of the restriction site sequence. These nucleotides may be chosen at random, or adapted to fit in with the optimal design of the PCR primer sequence (e.g. 50% G/C-content; Section 5.1). Of course, 5′-end restriction sites should be chosen such that the restriction enzymes used do not cut within the amplified PCR product itself.

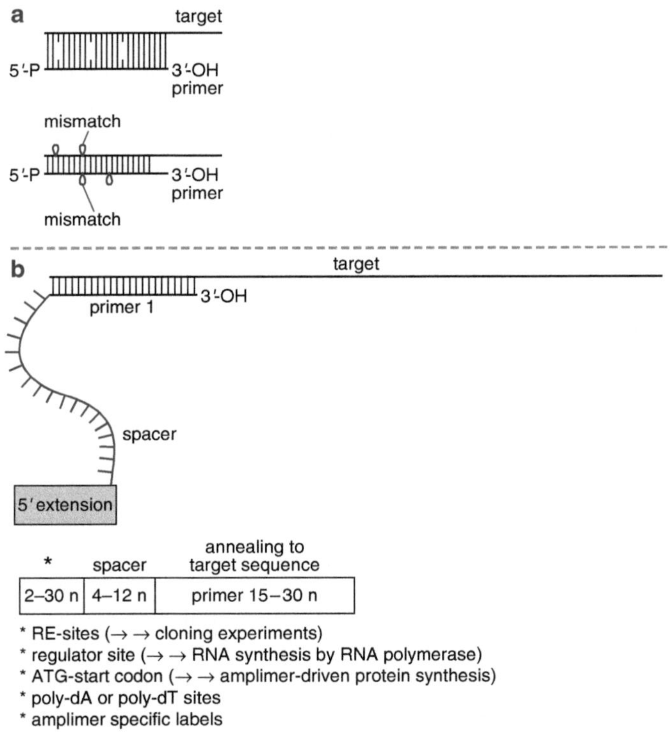

Fig. 5.2 A variety of modified PCR primers and their applications have been described. Example a: PCR primers may be designed to introduce point mutations (top), insertions or deletions (bottom), etc. into the amplified product. Example b: PCR primers designed with various 5′-end extensions including restriction enzyme recognition sequences (for cloning of the PCR product), or regulatory sequences (e.g. an RNA polymerase initiation site, ribosome binding site, and ATG start codon for translating the PCR product into protein), or fluorescent labels used to detect or visualise the presence of the PCR product after amplification

(2) *Regulatory sequences*: A 5′-end promoter sequence for DNA dependent RNA polymerase enzymes can be used to transcribe the PCR product into (m)RNA copies. This may allow a large amount of synthesised RNA to be generated and used for example as RNA probes in hybridisation studies, or for *in vitro* protein synthesis studies (see below). The most frequently used promoters used for *in vitro* RNA and protein synthesis are the bacteriophage T7 and SP6 promoters.

(3) *Promoters and ribosomal binding sites (RBS)*: A functional 5′-end promoter sequence designed in conjunction with a ribosomal binding site and an ATG start codon may be used for optimised *in vitro* coupled transcription and translation of PCR products. This allows mRNA and protein to be produced in a single "coupled" reaction system, e.g. the TNT T7 Quick for PCR DNA system from Promega.

(4) *Capture haptens and fluorophores*: Dependent on the intended use of a PCR product, various sequence elements can be added to the 5′-end of the primer.

Homonucleotide stretches like poly-dT or poly–dA can be used for purification of the PCR product (e.g. via poly-dT paramagnetic beads) or for RACE-like applications (Section 4.5.2). PCR primer sequences can also be labelled with various haptens such as biotin and digoxygenin as well as a wide variety of fluorophores, e.g. fluorescein. These labels enable convenient immobilisation of the PCR product on a solid support, e.g. through the interaction between PCR primer labelled biotin and streptavidin bound paramagnetic beads, or enable amplimer visualisation. Alternatively, the attachment of a specific monoclonal antibody to the fluorophores may be important in the direct detection of the amplified PCR products.

(5) *GC-clamp*: When the 5′-end of a primer is increased by the addition of 20 or more GC bases, then this GC sequence will dominate the melting behaviour and electrophoretic mobility of the final amplified PCR product. This effect is utilised in the denaturing gradient gel electrophoresis or DGGE protocol (Section 9.2.1.v).

The second category of primer species is that of the "degenerate-modification" primers, which are designed not to be fully complementary to their intended target sequence, but to introduce target mutations into a specific piece of amplified DNA. For this purpose, they may either lack one of the nucleotides of the target sequence or they may differ slightly in their primary sequence. However, it should be noted that this degeneracy may result in incomplete hybridisation at the target site, leading to a hybrid structure involving "bulging" or "looped-out" nucleotides. These bulges or loops have a destabilising influence on the hybrid structure, necessitating the use of lowered PCR primer annealing temperatures, especially during the first few PCR cycles. Up to 4 or 5 mutations can be tolerated in the 5′-half of a 20 nucleotide long primer, though it may also be necessary to extend the primer beyond its 5′-end to increase the complementary base pair matches between the degenerate PCR primer and the target template. Moreover, if these mutations are located at or near the 3′-end of the primer, PCR amplification may be completely disrupted, since 3′-end binding of a PCR primer to its target is crucial in facilitating DNA polymerase-mediated amplification. For this reason, degenerate sites should be located nearer to the 5′- rather than the 3′-end of the PCR primer. Alternatively, extending the primer sequence by one or two nucleotides at the 3′-end may solve this problem. If degenerate primers are used to amplify PCR products from target regions with a high frequency of mutation, then the nucleotide inosine (which universally binds to G, C, A and T nucleotides; Section 6.2), may be incorporated into the primer design instead of one of the other nucleotides. Inosine incorporating PCR primer protocols usually also require relaxed annealing conditions during the first few cycles of PCR (as long as the inosine mismatches do not adversely affect the stability of the primer/template hybrid of course). Degenerate primers have been applied in various adapted PCR protocols, including:

1. *Site directed mutagenesis*: If sequences in the 5′-half of the primer differ from those of the target DNA, then these sequence changes (mutations) will be found in the final amplified PCR product. Post-PCR manipulations, such as cloning/plasmid mediated transformation or transformation using the PCR product itself, may then allow the introduction and study of this mutated DNA *in vivo*, i.e. back into its original host or into another suitable host.

2. *DOP PCR*: Degenerate oligonucleotide primer (DOP) PCR can be used to increase the amount of target DNA in a non-specific manner [Aubele and Smida, 2003]. In this protocol, a large random mixture of primers is used in a PCR with as a net result the amplification of random genomic sequences. Although success is not guaranteed, this approach has been shown to be useful for the general amplification of for instance bacterial genomic DNA [Sayada et al., 1994], extremely small amounts of DNA [Cheung and Nelson, 1996], and DNA from archival anthropological samples [Buchanan et al., 2000]. DOP PCR can also be combined with downstream methods such as single nucleotide polymorphism (SNP) detection [Grant et al., 2002].

3. *Translating amino acid sequences into genomic sequences*: When part of a protein sequence is known (for instance after sequencing of a protein purified by classical chromatographic methods), the amino- and carboxyl-terminal protein sequences can be reverse translated into two "degenerate" PCR primers (degenerate primers because during mRNA translation into protein, various tRNA triplet codons can encode for the same amino acid, e.g. CGU, CGC, CGA, CGG, AGG and AGA triplets all code for the amino acid arginine). Sequencing of the PCR product generated using this degenerate primer pair will help identify the gene sequence from which the original protein was translated and transcribed. For successful PCR amplification, this protocol usually requires a pool of possible degenerate PCR primer combinations, all derived from the reverse translation of the protein sequence, which ensures that the PCR primer sequence that exactly matches the actual gene sequence encoding the protein is present. However, an oligopeptide sequence of five amino acids encodes a primer of 15 nucleotides in length and if each amino acid is coded for by four different triplet codons, then the total number of different primer combinations required to generate a single 15 base pair PCR primer that is exactly complementary to the actual DNA sequence is $4 \times 4 \times 4 \times 4 \times 4$ (or 4^5), equal to 1,024 possible combinations. To try and alleviate this problem, the nucleoside inosine may be incorporated into the degenerate primers (see above), or alternatively, specific computer search programs such as CODEHOP (COnsensus-DEgenerate Hybrid Oligonucleotide Primers) may help in the design of the most selective PCR primer sequences [Rose, 2005]. Also, some organisms possess a typical "codon usage" pattern, where particular triplet codons are preferred for a particular amino acid during protein biosynthesis [Ohno et al., 2001]. If the codon usage pattern for that particular organism is known, then an "educated guess" can be made regarding the most likely degenerate PCR primer sequences to be used. If pools of degenerate primers are to be used, then the first few cycles of the PCR protocol should utilise relatively low annealing temperatures, to ensure that all possible primer combinations bind to the target DNA. Finally, as several different degenerate primers may actually hybridise to the required target sequence and generate PCR amplification product, it should be noted that the final amplimer may contain sequences that are slightly different from that of the original target sequence.

4. *Universal primers* (*also known as broad-range or consensus primers*): These primers may be used to amplify similar target sequences present within a variety of

different (sub)species of organisms. Often, a limited degree of degeneracy needs to be designed into the PCR primers in order to account for all of the mutations found in the intended target sequence. For example, different human papillomaviruses contain conserved sequence motifs that are present across all of the currently identified genotypes. These elements can be targeted by a "consensus" PCR generating a PCR product no matter which HPV type is present. Subsequent restriction digestion analysis, probing or sequencing of the PCR product then allows the actual HPV genotype to be identified. This may be clinically important, because approximately 25 genital HPV types exist and a few of these are linked to oncogenicity in humans. An interesting advantage of this broad-range PCR approach is that specimens may be screened in a single PCR for all HPV types thereby saving on time and materials, also new genotypes of HPV may be detected without having to modify the PCR protocol. Further examples of the usefulness of such universal primer PCRs include; the PCR amplification of conserved sequences within the ribosomal RNA genes of bacteria [Hilali et al., 1997], the identification of immunoglobulin heavy-chain variability in B-cell malignancies [Korganow et al., 1994], and the assessment of minimal residual disease after treatment for T-cell leukaemia [Morgan and Pratt, 1998], which when optimised allows the detection of a single leukemic cell within a background of 100,000 healthy cells.

5. *Competitor primers*: These primers are designed to be added to the PCR mix and directly compete with one or more of the primers that have been designed to amplify the specific target DNA. Competitive primers usually share part of the sequence of the amplifying primer pair but have a dideoxynucleotide analogue, (e.g. dideoxythymidine in place of deoxythymidine) inserted into the 3′-end of the primer. The addition of this dideoxynucleotide analogue effectively inhibits DNA polymerase mediated replication of the complementary DNA strand to which the competing primer has annealed, though the relative abundance of competitor primer in the PCR reaction mix may have to be carefully balanced. Competitor primers may be used in several PCR protocols, and may be useful in preventing the amplification of non-specific products when PCR amplification is performed using extremely low amounts of target DNA [Atamas et al., 1998]. Conversely, inhibition of PCR amplification on specimens containing very abundant target sequences may also be important. For instance, in the case of RT-PCR, ubiquitous "control" mRNA species are often used as internal control targets (e.g. for quality control and quantitative purposes). If the genuine target cDNA is far less abundant than the "control" cDNA, it may be necessary to inhibit the amplification of the control cDNA to a level comparable to the abundance of the target cDNA. Competitor primers may also be used during the reverse transcription of mRNA to cDNA in quantitative RT-PCR protocols [Benavides et al., 1995] (Fig. 5.3). As with typical PCR primers, 5′-end primer modifications may also be added to competitor primers.

The third category of primer species includes a "miscellaneous" set of primer species.

1. *Repeat sequence primers*: The genomes of organisms are generally crowded with DNA repeat sequence regions. These can be organised into homopolymeric/heteropolymeric direct repeats, homopolymeric/heteropolymeric inverted repeats, composite heterogeneous repeats, or degenerate repeats, and may be scattered

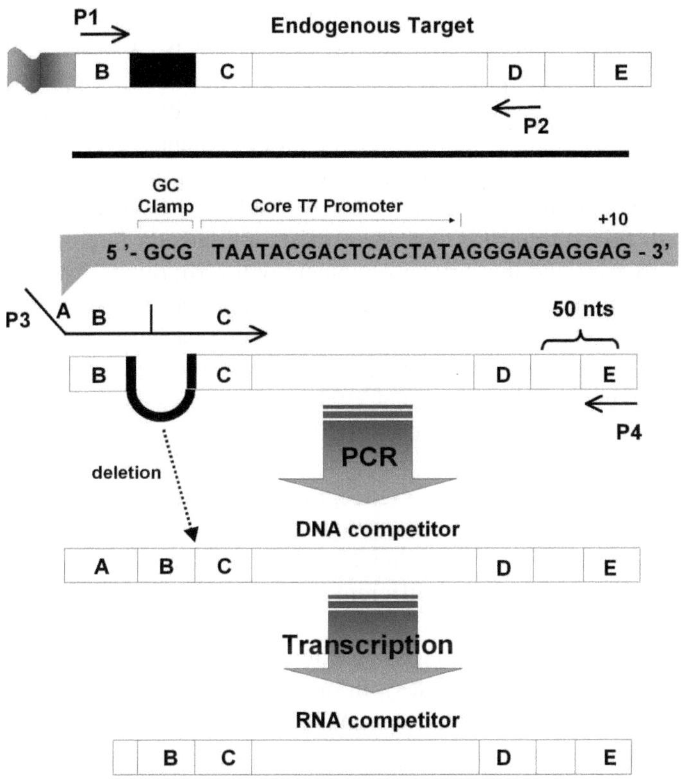

Fig. 5.3 Competitor design and use in competitive PCR. When two primers (P1 and P2) are being used for PCR amplification of a given target, two additional primers are used to generate a deletion construct for competitive RT PCR. Primer P3 includes a T7 promoter primer at its 5′ end, the P1 sequence in its middle and a downstream target specific sequence at its 3′ end. P3 creates the deletion and incorporates a transcription promoter for subsequent synthesis of an RNA competitor. P4 primes synthesis of from the target cDNA at a site approximately 50 nucleotides downstream of the P2 binding site. The extra 50 nucleotides are essential for RNA competitors as it assures that the competitor is reverse transcribed as efficiently as the endogenous target. For instance, when random sequence primers are used for reverse transcription, the extra nucleotides are required for efficient priming and synthesis through the entire amplification region. When sequence specific primers are used, the extra sequence allows RNA structure to form around the primer binding site of the competitor as it would for the endogenous target (From Ambion Technical Bulletin 185 (Dr. D. Brown), Ambion)

throughout the entire genome of an organism [Van Belkum et al., 1998]. DNA repeat sequence regions may be targeted using single repeat sequence primers or multiple repeat sequence primer pairs, where one primer of the pair is specific for one particular repeat sequence whilst the other primer of the pair is specific for a different repeat sequence. When sequence repeats vary with regard to genomic location and/or number between individual species and isolates, then the number of successfully amplified PCR product fragments may differ, resulting in the production of specific DNA fingerprints for each species.

2. *Concatemeric primers*: Used to generate a single large PCR amplification product by coupling together overlapping PCR products. A basic feature of this approach is that the PCR protocol is designed to generate overlapping amplimers, the individual strands of which can partially hybridise. When these amplimers are melted into single strands (via the application of heat) and then re-annealed, some of the amplimers will hybridise to each other and act as primers for DNA polymerase. This will lead to the generation of a new and longer DNA product. The major problem with this approach is that there exists a relatively large probability that mutations will arise and that primer oligomers will form due to frequent mispriming. Furthermore, there may occur preferential amplification of some of the smaller amplimers. This said however, the concatemer PCR approach has been very important in amplifying significantly large-sized fragments of DNA from, for instance, mummified specimens [Woodward et al., 1994], and samples that have been subjected to "violent" extraction protocols resulting in highly sheared and fragmented DNA [Marota et al., 2002].

3. *Multiplex PCR primers*: In essence, the concatemer protocol is an example of a multiplex PCR, though in a normal multiplex PCR, the primer pairs are designed not to anneal to each other (Section 12.9). Multiplex PCR protocols involve the addition of up to ten different PCR primer pairs within the same PCR reaction mix. In order to facilitate this, an efficient multiplex PCR normally contains primers which are somewhat extended (28 nucleotides instead of the more customary 18–22 nucleotides), in order to enhance primer specificity and help prevent unwanted primer/primer annealing interactions. The use of computer software is recommended when designing multiplex PCR primers, e.g. Vector NTI (Novel software) and Lasergene (DNAstar software), as it is often very difficult to balance the GC-content, annealing temperatures, possible primer to primer interactions etc of multiple primer pairs manually [Gorelenkov et al., 2001].

4. *Megaprimers*: Megaprimers comprise primers of between 450 and 800 bp long which may be used in mutagenesis protocols. After initial hybridisation at 37°C, a limited number of PCR cycles (n = 7) are performed whereupon the megaprimers are extended from the 3′ end outwards. After limited PCR cycling, a normal PCR protocol is performed using the amplification products from the megaprimer PCR and another pair of normal length, specifically designed PCR primers. These normal PCR primers may also be equipped with additional, site-directed mutations [Peuschel, 2000].

5. *Molecular beacons*: In PCR protocols utilising molecular beacons, a primer or probe contains a self-complementary sequence that adopts a hairpin-like secondary

Fig. 5.4 (continued) with a complementary sequence designed to occur downstream of its own, nascent strand. Inset shows a duplex scorpion that exchanges the stem – loop structure for a primer element terminally labelled with the fluorophore and a separate complementary oligonucleotide labelled with a quencher at the 5′ terminus (Reproduced from Huang et al., 2004. With permission from Oxford University Press)

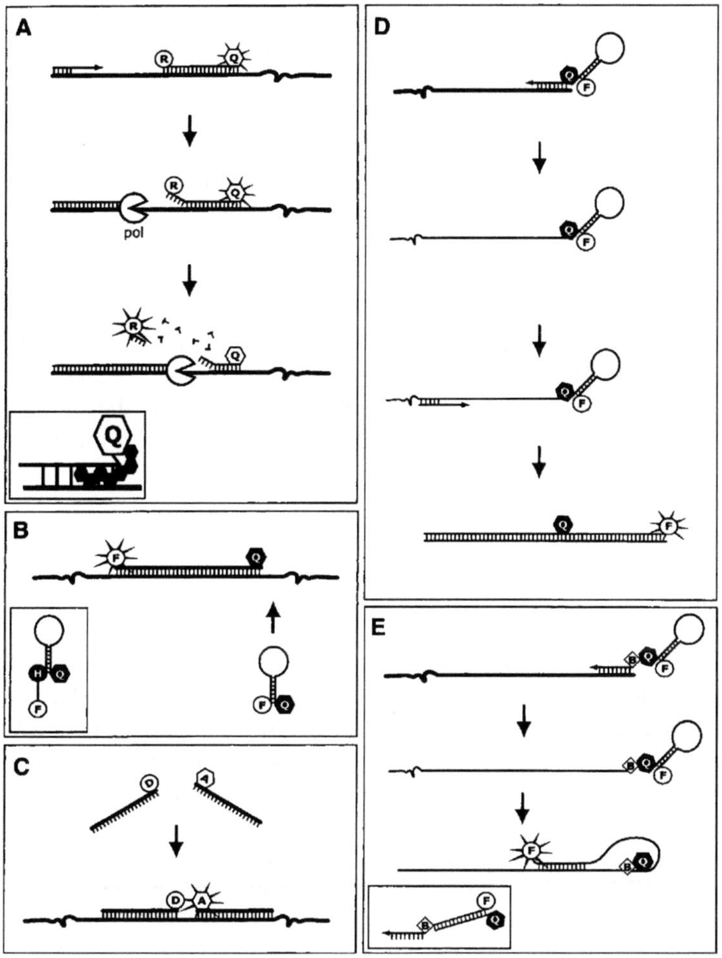

Fig. 5.4 Oligoprobe chemistries, (A) 5′ Nuclease oligoprobes. As the DNA polymerase progresses along the relevant strand it displaces and then hydrolyses the oligoprobe via its 5′-to-3′ endonuclease activity. Once the reporter is removed from the extinguishing influence of the quencher, it is able to release excitation energy at a wavelength that is being monitored by the instrument and different from the emissions of the quencher. The inset shows a pictorial diagram of a TaqManMGB nuclease oligoprobe, incorporating an improved 3′ non-fluorescent quencher that exhibits low background fluorescence (Q,) and a "minor groove binding" (MGB) moiety (filled hexagons), that stabilizes the hybridized probe such that the melting temperature (T_m) of the probe is increased. (B) Hairpin oligoprobes. Hybridisation of the oligoprobe to the target separates the fluorophore and non-fluorescent quencher sufficiently to allow emission from the excited fluorophore, which is monitored. Inset shows a wavelength-shifting hairpin oligoprobe incorporating a harvester molecule. (C) Adjacent oligoprobes. Adjacent hybridisation results in a FRET signal due to the interaction between the donor and acceptor fluorophores. (D) Sunrise primers. The opposite strand is duplicated so that the hairpin's structure can be disrupted. This separates the labels, eliminating the quenching in a similar manner to the hairpin probe. (E) Scorpion primers. The primer does not require extension of the complementary strand: in fact it blocks extension to ensure that the hairpin in the probe is only disrupted by specific hybridisation

structure. Folding of the hairpin, positions a fluorescent “reporter” molecule and a “quencher” (absorber) molecule closely together. In this conformation no fluorescence occurs (Chapter 9, Fig. 9.16). However, once the molecular beacon primer is incorporated into amplified DNA product during the first PCR cycle, the fluorescent group and its quencher become spatially separated, allowing fluorescent signal to be measured upon laser excitation. Alternatively, a molecular beacon probe (e.g. Taqman 5′- nuclease oligoprobes) may specifically bind to a PCR amplimer and be digested by the DNA polymerase upon replication, generating free reporter groups [Mackay et al., 2002]. Several different types of molecular beacon are shown in Fig. 5.4. When designing molecular beacons, the hairpin structure sequence should not anneal to any homologous sequence elsewhere on the target DNA. In most cases the self-complementary primer sequence should comprise at least ten nucleotides with a T_m between 30°C and 36°C. The melting temperature of the hairpin structure formed should be compatible with the annealing temperature of the intended PCR product [Kaboev et al., 2000].

5.2 Primer Hybridisation (Annealing)

Correct primer hybridisation to the target DNA is the most fundamentally important aspect of ensuring PCR amplification specificity. Fundamentally, dissociation of duplex DNA into the single stranded form allows specific complementary base pairing interactions and primer annealing to a target DNA sequence. Although PCR primers are relatively small (compared to Southern blot hybridisation probes, etc.), their presence in large excess during PCR cycling makes the primer/target annealing process very efficient, a process usually complete within a few seconds. In fact, most PCR cycling protocols allow on average 30–60 seconds for the primer to “find” its relevant complementary target and fully anneal. The exact annealing time programmed is dependent on the ramp rate and heating capacity of the PCR thermocycler used, the volume of the PCR mix, and the concentrations of the primer and target template. Also, several biophysical characteristics are key factors influencing the hybridisation kinetics of PCR primer to template DNA hybridisation, including the nature and complexity of the target (e.g. size, base composition, purity), the composition of the sample buffer, primer length, and GC-content (short yet GC-rich primers exhibit the fastest hybridisation rates). In general, if thin-walled reaction vessels are used and the PCR volume is kept below 50 µl, then 15 seconds is usually sufficient to achieve completion of primer binding to the template (i.e. to reach equilibrium). All of these factors are gathered under the term “stringency”, a concept which indicates the degree of PCR primer hybridisation to the target DNA. Another very important factor involved in PCR primer hybridisation to target DNA (and influencing the rate and accuracy of a PCR protocol) is the “annealing temperature”. The annealing temperature (T_m) of a particular PCR primer is that temperature at which the PCR primer will completely and accurately anneal to a complementary sequence of DNA. In general, the PCR annealing temperature used in a particular PCR protocol should be in the region of 5°C below the T_m, with the T_m for both PCR primers used being ideally "balanced" (i.e. within ±5°C of each other). Mismatches between primer and target DNA affect the T_m such that degenerate primer PCR protocols use a recommended T_m of 10–20°C below the T_m of the primer/target sequence (Section 5.1.1). For an accurate prediction of DNA hybridisation temperature, the formula of Baldino et al. [1989] can be used:

$$T_m = 16.6\log[Na^+] + 81.5 + 0.41[GC] - 0.72[\%formamide] - 500/ n\text{-}1.5 [\%mismatch]$$, where

$[Na^+]$ = the sodium ion concentration of the hybridisation mix, [GC] = the percentage GC nucleotides in the DNA strands analysed, and n = the length of the nucleic acid molecules. Nowadays, formamide is not usually incorporated into PCR mixes and the factor "500" differs within various literature citations. The other parameters are self-explanatory. However, this formula was especially developed for Southern hybridisation protocols where DNA probes ranging in size from 200 to 1,000 nucleotides are mostly used. This means that the relatively small size of PCR primers tends to render the term 500/n irrelevant, and also means that the addition or deletion of even a single nucleotide to a PCR primer has a disproportionately greater effect on the calculated T_m (a single nucleotide mismatch in a 20 nucleotide PCR primer yields 5% mismatch, but a single nucleotide mismatch in a 200 nucleotide Southern hybridisation probe yields only 0.5% mismatch). Also, formamide is an agent that is very effective in lowering the annealing temperature, but is rarely used in PCR reaction mixes. One final note, the formula takes note of the sodium ion concentration (Na^+) but not the magnesium ion concentration (Mg^{++}), magnesium ions being essential components of any PCR reaction mix for efficient DNA polymerase activity. Hence, the Baldino formula is not really useful for PCR.

When we apply the Baldino formula to PCR primer T_m calculations, the GC-content of the PCR primer appears to be the prime determinant of the T_m. Hence, a simpler alternative to the Baldino equation takes into account only the contribution of each basepair to the T_m [Thein et al., 1986]. This formula simply adds 2°C for every A or T basepair and 4°C for every G and C basepair present in the DNA sequence and reads: $T_m = 2\,(T+A) + 4\,(G+C)$, and allows manual calculations of T_m to be easily calculated. However, this formula only works well for DNA sequences up to approximately 14 nucleotides in length, and is less accurate for PCR primers 18–25 nucleotides in length. Moreover, due to its simplicity, this formula significantly underestimates the contribution of G- and C-rich regions to the overall thermodynamic stability of the DNA sequence. Taking into account primer composition, a more accurate equation for the calculation of PCR primer annealing temperature is:

$$T_m = 64.9 + 41\,([yG + zC - 16.4]/[xT + wA + zC])\ °C$$

where the letters y, z, x, and w identify the number of G, C, T and A nucleotides in the molecule, respectively.

5.3 Thermodynamic Approach of T_m Calculations

A major shortcoming of the equations stated above however, is that they do not take into account one important aspect of double helix stability, namely "base stacking". The aromatic residues of the individual bases tend to align in separate horizontal planes. This contributes in a positive sense to the stability of the double helix since hydrophobic groups are assembled in a "shield-forming" (i.e. protective) conformation. This nearest neighbour interaction lowers the free energy content (ΔG) of the complex, with GC-interactions contributing more significantly than AT-interactions to this base stacking effect (Table 5.1). If we consider the general thermodynamic formula: [ΔH] = [ΔG] + T[ΔS] where H = enthalpy, G = Gibbs' free energy, T = temperature and S = entropy, then it can be deduced that T = ([ΔH]-[ΔG])/[ΔS]. Further, by definition the term melting temperature (T_m) of a length of double stranded DNA is used when 50% of the basepairs present within the DNA molecule are unpaired and the Gibbs free energy when 50% of the basepairs are unpaired is actually zero. There will be equilibrium between the solid and fluid form of DNA and [ΔG] will be zero as well. Consequently, when T is considered in degrees Kelvin its absolute value will be T = [ΔH]/[ΔS]. However, the intermolecular interactions during PCR are highly complex and depend heavily upon a variety of molecular characteristics. This usually implies that the precise value for ΔH (in kcal/mole) and ΔS (in kcal/degree Kelvin × mole) cannot be calculated. Values for these parameters have been defined on the basis of well-controlled model experiments. These values should be considered rational estimates (Table 5.2). The individual contributions of many stacking interactions have actually been defined experimentally (Table 5.3 and Fig. 5.5). Sometimes major differences in thermodynamic approach are noted. It has to be emphasised that these approaches may be biased towards individual preferences and that different tables for entropy and enthalpy values are sometimes used. Mueller et al. [1999] uses a version of the formula where they deleted the ΔS factor.

The formula of Rychlyk [Rychlik et al., 1990; Freier et al., 1986], provides a reasonably accurate estimate of the T_m of a primer/template complex, by taking into account such base stacking effects:

$$T_m\,[^\circ C] = \{[\Delta H]/([\Delta S] + R\ln(C/4))\} - 273.15 + 16.6\ {}^{10}\log[Na^+]$$

T_m [°C]: calculated annealing temperature in degrees centigrade, where 273.15 is a correction factor to change degrees Kelvin into degrees centigrade (the value of ΔH and ΔS have been defined at 25°C)

ΔH: enthalpy of base stacking interactions corrected for helix initiation factors

ΔS: entropy of base stacking interactions corrected for helix initiation factors in 1 M NaCl

R: universal gas constant [1.987 (cal/°C × mol)]

C: oligomer concentration (which should ideally be corrected at each cycle for the varying primer concentrations during the PCR, though a value of 250 pM gives good agreement with experimental results)

[Na^+]: the salt concentration, if different from 1 molar (most PCRs are performed with a Na^+ concentration of 50 mM).

Rln(C/4): is a correction factor, which is included in this formula to compensate for the excess number of probe molecules (in this case the PCR primers) compared to the number of target molecules usually present in DNA hybridisation methodologies. This approach includes primer sequence information and introduces the energy

Table 5.1 Unified values for the free energy parameters at 37°C (kcal/mol) for all nearest neighbour combinations of nucleotides within an oligonucleotide lacking self-complementarity at pH 7 and 1 M NaCl [From SantaLucia, 1998]

First	Second nucleotide			
Nucleotide	A	C	G	T
A	−1.00	−1.44	−1.28	−0.88
C	−1,45	−1.42	−2.17	−1.28
G	−1,30	−2.24	−1.84	−1.44
T	−0,58	−1.30	−1.45	−1.00

Table 5.2 Experimental T_m values for oligonucleotides compared to theoretical values obtained by calculation

Sequence	T_m	T_m	T_m	Oligo	Potassium
5′ to 3′	exp	eq. 1	eq. 2	mmol/l	mmol/l
GATGAGTTCGTGTCCGTAC	55	58	46	0.66	50
ATTACCGGTAAT	27	28	5	12	9
CAAGCTACGAGC	51	55	30	100	70
CGCGAATTCGCG	63	36	36	0.65	100

T_m eq.1: T_m = [delta-H{delta-S + S-R.ln(C/4)}]−273.16 + 16.6log[K]

T_m eq.2: T_m = 81.5 + 16.6.log[K]−675/l + 0.41.(%GC)

T_m exp: T_m observed in model experiments

Table 5.3 Thermodynamic standard figures for the generation of basepairs between two oligonucleotides [From SantaLucia, 1998]

Sequence	Delta-H	Delta-S	Delta-G
	(kcal/mol)	(cal/k.mol)	(kcal/mol)
AA/TT	−7.9	−22.2	−1.00
AT/TA	−7.2	−20.4	−0.88
TA/AT	−7.2	−21.3	−0.58
CA/GT	−8.5	−22.7	−1.45
GT/CA	−8.4	−22.4	−1.44
CT/GA	−7.8	−21.0	−1.28
GA/CT	−8.2	−22.2	−1.30
CG/GC	−10.6	−27.2	−2.17
GC/CG	−9.8	−24.4	−2.24
GG/CC	−8.0	−19.9	−1.84

$$\Delta G_{total} = -(\Delta g_i + \Delta g_{sym}) + \Sigma_x \Delta g_x$$

3.1 1.9 1.9 3.1

↓ ↓ ↓ ↓

G-G-A-A-T-T-C-C

C-C-T-T-A-A-G-G

↑ ↑ ↑

1.6 1.5 1.6

$$\Delta G_{predicted} = -(5.0+0.4) + (2\times 3.1) + (2\times 1.6) + (2\times 1.9) + (2\times 1.6)$$

$$\Delta G_{predicted} = 9.3 \text{ kcal}$$
$$\Delta G_{observed} = 9.4 \text{ kcal}$$

Fig. 5.5 Calculation of the delta-G (change in Gibbs free energy) of a random sequence. Delta-G(total) indicates the energy required to calculate the energy contribution of the basepairs. Delta-g(sym) corrects for possible self-complementarity of the helix. Delta-g(i) indicates the helix initiation energy. For further explanation see Breslauser et al. [1986] and also Tables 5.1–5.3

gain provided by each and every stacking interaction [Breslauer et al., 1986; Kämpke et al., 2001].

The use of dedicated software (e.g. OLIGO [Rychlik and Rhoads, 1989]) allows a speedy and relatively precise T_m to be calculated using the above formula. The use of computational software in PCR primer design is recommended, as these software programs can search a sequence of DNA for user defined primer values allowing for example the T_m of the chosen PCR primer pair to be set to a given value. Some of these software packages even allow for "target walking" with a constant indication of the T_m value as the program moves along the target DNA sequence. Such programs are especially useful in recognising possible PCR primer design problems, e.g. primer complementarity, sub-optimal primer termini and secondary structure elements (hairpins and palindromes). Some programs even allow the design of PCR primers for TGGE or DGGE applications to be performed. The PrimerSelect (DNAStar), Interactive DoPrimer, OLIGO6 and Primer3 and the Internet-based oligo-Analyzer are examples of frequently used PCR primer design programs. Some programs are included as part of larger commercial nucleic acid analysis packages. Examples of such software packages include; Primer Premier 5 (PREMIER Biosoft, Palo Alto, Ca), RedaSoft (Toronto, Canada), Informax (Bethesda, Md), Expression (Genamix, Hamilton, New Zealand), iOligo (Portsmouth, NH) and Primer Designer 4 (Sci-Ed Software, Durham NC). These software packages are not only suited for PCR primer design, but can also be used for PCR optimisation, primary sequence analysis and comparison and hybridisation modelling. Some of these software packages even help with the interpretation of gel electrophoretic analysis of PCR products. It may sometimes be necessary to determine the "uniqueness" of any newly designed PCR primer, in which case sequence comparison using the GenBank sequence depository (http://www.NCBI.NIH.-gov/GenBank), in combination with the Basic Local Alignment Search Tool (BLAST) program, will allow searches of known DNA (or protein) sequences against DNA (or protein) sequence databases to be performed [Altschul et al., 1990].

5.4 Primer Synthesis

Once the design process is complete, it becomes necessary to synthesize the PCR primer. In the vast majority of cases, primer synthesis is performed by commercial companies rather than "in house", as such companies provide cost-effective and rigorously quality controlled primers. Most PCR primers are physically synthesised via non-enzymatic, organic chemistry reactions, with the individual oligonucleotides being linked together by special nucleic acid synthesising machines (e.g. the ABI 3400 DNA synthesizer, Applied Biosystems, USA). These machines can currently generate primers with a maximum length of approximately 70 nucleotides and are also able to introduce specific chemical modifications into the oligonucleotide chain. Options which may be selected by the customer include the scale of synthesis (from 0.05 to over 5 μM per oligonucleotide), the type of chemical approach (phosphoramidite, phosphothioate), the purity of the end-product (with or without HPLC mediated purification), type of labelling (biotin, FITC, etc.), special base moieties (inosine, uracil), inclusion of random sequences, inclusion of degenerate nucleotide positions or whether the oligonucleotide PCR primer is delivered in solution or as a lyophilised powder. One of the most frequently used synthesis schemes involves β-cyanoethyl phosphoamidite and the H-phosphonate cycle [Kraszewski and Norris, 1987].

If we consider the steps necessary to synthesise a PCR primer, then initially, all of the various reagents are stored in separate compartments of the DNA synthesiser with their reactive sites (i.e. the 3′- and 5′- ends of the sugars and the reactive atoms in the nucleotide bases themselves) "blocked", such that chemical reactions are only possible once the blocking groups have been removed by acid or alkaline treatment. To couple one nucleotide to another, a sequence of five synthesis steps is required which proceed in the 5′ to 3′ direction (Fig. 5.6). It is worth noting that the starting nucleotide is fixed with its 3′-end to the solid support in a concentration that will yield the required concentration of primer at the end of synthesis, and that as a final step in the synthesis protocol, the oligonucleotide is fully "de-protected" using ammonium hydroxide and uncoupled from the solid matrix via alkaline hydrolysis. After neutralisation and extraction of the solution to remove traces of organic compounds the oligonucleotide is essentially ready for use. Not all of the individual steps in PCR primer oligonucleotide synthesis proceed with 100% efficiency, which limits the final yield of product and prohibits synthesis of oligonucleotides much longer than 70 bases in length. The final oligonucleotide solution is usually ≥90% pure. This purity level suffices for most PCR applications. However, for high-stringency PCR and PCR fingerprinting protocols, the primers will need to be purified to almost 100% purity [Weirich et al., 1997]. This may be achieved using HPLC and is usually performed by the company delivering the oligonucleotides upon request. Various commercially available purification kits allow "do-it-yourself" purification, with spin columns based on silica affinity chromatography being the most common, though agarose gel electrophoresis followed by gel excision is another option. The purity of the cleaned oligonucleotide may be determined spectrophotometrically

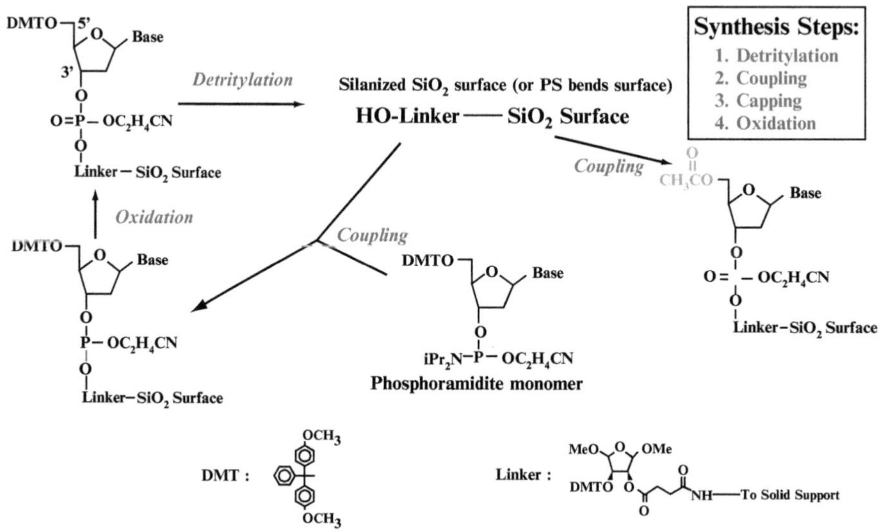

Fig. 5.6 Oligonucleotide synthesis using the β-cyanoethyl phosphoramidite method. Five distinct steps are involved. Step 1, the 5′ blocking chemical (dimethoxytrityl, DMT) is removed using dichloro-acetic acid, thereby exposing the 5′ hydroxyl group of an immobilsed nucleotide. Step 2, the following nucleotide is 3′ activated with tetrazole. Step 3., this activated nucleotide is attached via a condensation reaction to the 5′ hydroxyl group of the nucleotide that is already attached to the solid matrix. Step 4, requires iodide to oxidize the phosphite ester into a phosphate ester. Step 5, is a preventive step in that all of the 5′ hydroxyl groups that have not reacted are "capped" in order to prevent their activity in subsequent cycles. If this step were not included, a number of truncated oligonucleotides would be generated, which would contaminate the intended primer sequence [From Cheng et al., 2002]

using the A260/A280 ratio (Section 4.3.1). When the A260 has been determined, the oligonucleotide concentration can be determined using the formula:

$$\text{oligonucleotide concentration [mol/l]} = \text{(A260 value)/Em}$$

Em represents the overall extinction coefficient, which is calculated by adding together the individual molecular extinction coefficient values for each of the individual nucleotides; these are: A = 15,200 l/mol.cm, G = 12,010 l/mol.cm, T = 8,400 l/mol.cm and C = 7,050 l/mol.cm, valid for a 1 cm spectrophotometry cuvette at 260 nm.
Alternatively, the following equation may be used:

$$\text{oligonucleotide concentration (pmol/}\mu\text{l)} = \text{A260/(100/[1.5nA + 0.71nC + 1.2nG + 0.84nT])}$$

where n = the number of respective A, C, G and T nucleotides present within the oligonucleotide primer.

This formula also takes into account the individual absorption values of the various nucleotides as well as the actual length of the nucleic acid molecules. Synthesized PCR primers are chemically stable compounds that can be bought as a lyophilised powder or as a solution dissolved in a variety of buffers. Both formats may be conveniently stored as "stock" solutions at −20°C or −70°C. It is advisable to prepare "working" solutions from the stock solution when necessary and that working solutions are replaced regularly to prevent DNA and/or DNase contamination and to ensure that the primer concentration remains constant. Stock solutions may in general be freeze-thawed 5–20 times without apparent loss in quality, though this figure should be determined experimentally.

5.5 Non-radioactive Primer Labelling

Many PCR applications require the use of chemically labelled PCR primers. One such label is biotin (a water-soluble member of the B-complex group of vitamins), which binds to streptavidin (a tetrameric protein), with one of the largest free energies of association yet observed for non-covalent binding of a protein and small ligands in aqueous solution ($K\_assoc = 10^{14}$). Biotinylated amplimers may therefore bind with high affinity to streptavidin coated particles or surfaces. Moreover, if only one of the PCR primers in a pair is biotin labelled, then the non-labelled strand can be dissociated and washed away after streptavidin capture, yielding single stranded DNA which is then available for DNA/RNA hybridisation or sequencing studies. Other labels include a variety of fluorophores, with fluorophore-labelled primers or amplimers being especially useful in PCR sequencing protocols. Dual labelled oligonucleotides or molecular beacons (fluorescent reporter and quencher groups present on the same oligonucleotide) can be included in real-time PCR protocols, in *in situ* PCR, or in specific PCR product capture assays. These labels have largely replaced the use of ultra-sensitive radioactive labels in PCR product analysis, as fluorophore labels are less hazardous and more convenient to work with. Labels such as the fluorophores JOE, ROX, FAM, TAMRA, HEX, etc., or haptens, e.g. biotin – streptavidin, are now widely used to label PCR primers allowing post-PCR detection using for example spectroscopic and immunochemical methods [Hahn et al., 2001]. If PCR amplification product detection sensitivity is an absolute requirement in the PCR protocol, for example in some forensic DNA fingerprinting techniques, then the incorporation of a radioactive label to one of the PCR primer pair is still the method of choice.

The process of primer labelling *per se* may be facilitated by a variety of chemical and enzymatic procedures, with some labels being attached via the introduction of an amino-group to the 5′-end nucleotide of the PCR primer (Fig. 5.7). Alternatively, an organo-synthetic route may be used to insert different fluorophores during the final cycle of oligonucleotide primer synthesis, a procedure that requires fluorophore-derived phosphoramidites instead of AminoLink. Using this protocol, fluorophores may be introduced into the middle as well as the ends of the

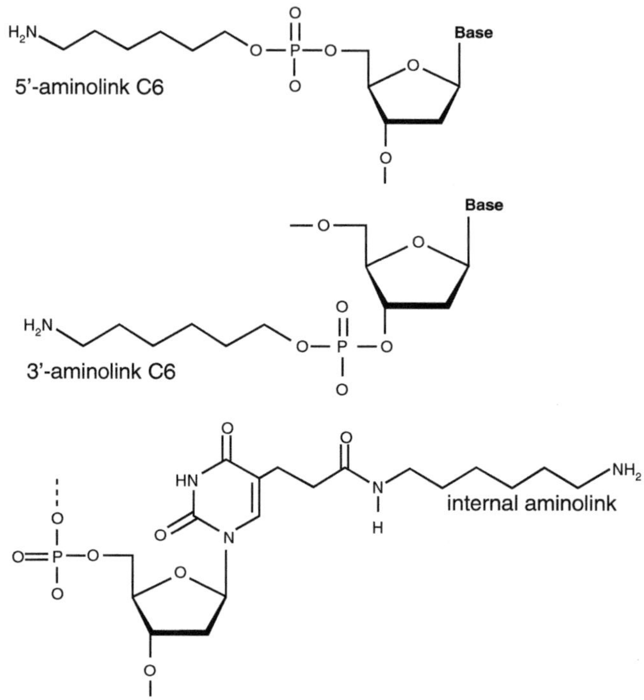

Fig. 5.7 The attachment of labels to the 5′-end of PCR primers frequently involves the fluorinated reagent AminoLink (Pierce Biotechnology). AminoLink can be directly applied to the oligonucleotide, whereupon it generates a reactive amino group. The AminoLink-treated oligonucleotide primer may then be equipped with hapten labels or fluorophores (From www.biomers.net)

synthesized oligonucleotide. The positioning of several fluorophore groups close to the 5′-end of a PCR primer will increase label density (increasing the sensitivity of detection of PCR products), without major implications for hybridisation efficiency. However, it should be noted that fluorophore groups that are not attached close to the 5′-terminal end of a PCR primer may interfere with PCR primer hybridisation, thereby reducing both specificity and/or sensitivity. As well as using chemical methods, the 5′-end labelling of PCR primers may also be achieved enzymatically in three simple steps (Fig. 5.8). In the first step, the enzyme T4 polynucleotide kinase transfers a thiophosphate group to the OH-group of the 5′-end terminal nucleotide. This 5′-thiol group may then be used to couple fluorochromes (e.g. via 5-iodoacetamidofluorescein), haptens etc to the oligonucleotide. Commercial kits are available which use this process, e.g. the 5′ EndTag labelling system (Vector Laboratories, CA, USA). Haptens (e.g.biotin and digoxigenin), and fluorophores (e.g. FITC or rhodamine), may also be incorporated into PCR amplification products during PCR cycling protocol *per se*, by including hapten- or

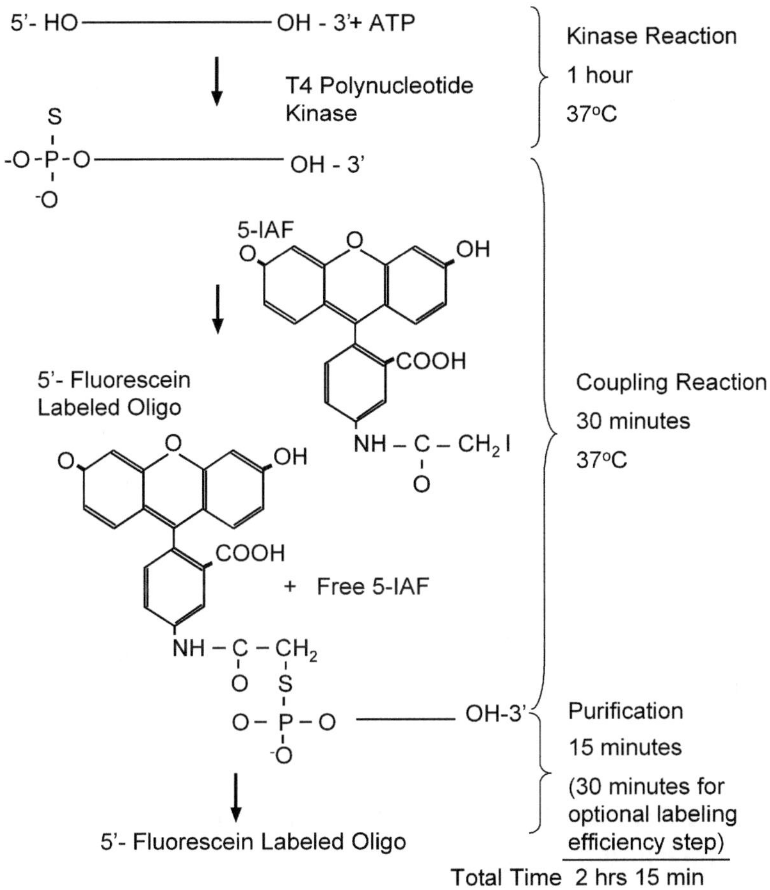

Fig. 5.8 5′-end primer labelling using T4 polynucleotide kinase and the poly-aromatic compound 5-IAF (5-iodoacetamidofluorescein), generates a fluorescein-labelled primer within 2 hours and in three simple steps. Essentially, a thiophosphate group is transferred to the 5′-end of the nucleotide, to which a fluorophore may be coupled via 5′-iodoacetamidofluorescein (From the Molecular Dynamics Users Guide, Amersham – Pharmacia Biotech. Copyright 1999 General Electric Company. With permission)

fluorophore-labelled nucleotides in the PCR mix. PCR amplification products may also be labelled using non-PCR enzymatic methods (e.g. the ARES DNA labeling kit, Molecular Probes, USA), with approximately 1 fluorochrome label per 12–20 nucleotides. Such labelling allows the use of PCR products in "downstream" applications such as fluorescence *in situ* hybridisation (FISH) and microarray hybridisation experiments. Table 5.4 shows a range of frequently used fluorochromes available for PCR primer/product labelling.

Table 5.4 Fluorochromes that may be coupled to oligonucleotide PCR primers or deoxynucleotide tri-phosphates

Fluorochrome	Lambda absorbance	Lambda emission	Manufacturer
BioDIPY/FI-XdUTP	505	513	Molecular Probes
FAM	488/495	518/535	PE Biosystems
Fluor X-dCTP	494	520	Biological Detection Systems
Fluoresceine-11-dUTP	490	520	Amersham
Atto 520	488/514	547	Fluka
SYBR Green	488/494	521	Molecular Probes
Fluoresceine-12-dUTP	494	533	Molecular Probes
TET	488	538	PE Biosystems
Eosin	519	542	Dupont NEN
VIC	488	552	PE Biosystems
Atto 520	488/514	547	Fluka
JOE	488/525	554/557	PE Biosystems
Cy3-dCTP	512/552	565/615	Biological Detection Systems
ROX	–	565	–
HEX	488/529	566/560	PE Biosystems
TAMRA	488/542	582/568	PE Biosystems
Cy5	–	699	Biological Detection Systems

– = not defined

5.6 The Effect of Mismatches Between PCR Primer and Target

Mutations in the target sequence or incorrectly designed PCR primers may result in mismatches occurring during primer/target hybridisation resulting in incomplete base pairing between the two nucleic acid molecules. Such incomplete base pairing lowers the annealing temperature (T_m) of the two nucleic acid molecules, significantly reducing the yield of specific PCR amplification products. In general, the greater the number of base pair mismatches, the greater the effect on PCR yield, with base pair mismatches at the 3′-end of the primer having a larger impact than those near the 5′-end of the primer. This effect on PCR yield means that it is always wise to design a PCR protocol utilising primers that hybridise to known genomic conserved regions, a concept especially important during the development of clinically relevant diagnostic PCR tests. For example, regions of the HIV reverse transcriptase gene undergo mutation at a relatively high frequency, which makes these regions good targets for studying antiviral resistance but not good targets for developing reproducible diagnostic tests. Of course, conserved regions on either side of a region that displays a high frequency of mutation may also be considered as possible targets for PCR primer design.

Several strategies have been developed which actually utilise variable regions of DNA to look for clinically relevant mutations, as the use of a very high-stringency, carefully optimised PCR protocol allows mutations to be specifically detected within a disease causing gene, e.g. β-thalassemia or phenylketonuria [Sueoka et al., 2000]. If wild type sequence specific and mutant specific primers are used in separate very high-stringency PCRs, then the presence of amplification products using one or other of the primer pairs will be indicative of a homozygous wild type, homozygous mutant, or heterozygote genotype. Alternatively, wild type and mutant PCR primers may be "multiplexed" and incorporated into the same PCR mix so that both wild-type and mutant PCR amplification products will be formed in the same reaction mix. If the mutation gives rise to a particular restriction site sequence, post-PCR restriction of the amplification products will indicate the presence/absence of this restriction site and hence the presence or absence of the mutant, wild type or heterozygote genotypes.

The same sort of analyses may also be performed by first generating labelled PCR products using allele specific oligonucleotide hybridisation (ASO) techniques, and then probing the target DNA using for example direct hybridisation [Kapelari et al., 1999], or reversed hybridisation [Dahle et al., 2003]. Mismatches between oligonucleotides and target DNA may also be used in non-PCR-based single nucleotide polymorphism (SNP or SNiP) analyses, using for example allele specific oligonucleotide ligation, allele specific hybridisation or primer extension techniques.

5.7 Primer Concentration

Theoretical calculations indicate that a typical PCR containing 1–2 pmol of primer and an initial quantity of 10 pg template DNA should be sufficient to synthesise visible quantities (μg) of amplified DNA upon gel electrophoresis and ethidium bromide staining. In general, this means that a 0.1–0.5 μM concentration of primer in a PCR mix is sufficient for successful and accurate PCR amplification (in a 50 μl reaction volume). This implies that primers are present in a 10,000,000:1 ratio compared to the average quantity of target DNA sequence at the start of PCR amplification. This excess of primer is required because the number of target molecules increases exponentially during PCR amplification and after 25 cycles of PCR, many millions of copies of target DNA may be present. The presence of such a large excess of primers at the start of PCR amplification may occasionally lead to the preferential amplification of non-specific PCR products, in which case, increasing the annealing temperature or lowering the initial primer concentration may help increase the specificity of amplification. If non-specific products are still being amplified, the optimum primer concentration versus magnesium ion concentration for that particular PCR protocol may need to be determined. This is achieved by using a range of primer concentrations (e.g. 0.05–1 μM) versus a range of different magnesium ion concentrations (e.g. 0.5–5.0 mM) in a "checkerboard" titration format. If non-specific products are still being generated, then the PCR primers may need to be re-designed. In the vast majority of PCR protocols, the two PCR primers are included in the reaction mix at similar concentrations, though in at least one PCR protocol (namely "asymmetric PCR"), one primer is included in excess of the other resulting in an asymmetric amplification reaction and the preferential (non-exponential) amplification of single stranded DNA.

References

Afonina I, Zivarts M, Kutyavin I, Lukhtanov E, Gamper H, Meyer RB. 1997. Efficient priming of PCR with short oligonucleotides conjugated to a minor groove binder. Nucleic Acids Res 25:2657–2660.

Ailenberg M, Silverman M. 2000. Controlled hot-start and improved specificity in carrying out PCR utilising touch-up and loop incorporated primers (TULIPS). Biotechniques 29:1018–1020.

Altschul SF, Gish W, Miller W, Meyers EW, Lipman DJ. 1990. Basic local alignment search tool. J Mol Biol 215:403–410.

Atamas SP, Luzina IG, Handwerger BS, White B. 1998. 5′-degenerate 3′-dideoxy-terminated competitors of PCR primers increase the specificity of amplification. Biotechniques 24:445–450.

Aubele M, Smida J. 2003. Degenerate oligonucleotide primed PCR. Methods Mol Biol 226:315–318.

Ayyadevara S, Thaden JJ, Shmookler Reiss RJ. 2000. Discrimination of primer 3′-nucleotide mismatch by Taq DNA polymerase during PCR. Anal Biochem 284:11–18.

Baldino F, Chesselet MF, Lewis ME. 1989. High resolution in situ hybridisation histochemistry. Methods Enzymol 168:761–777.

Benavides GR, Hubby B, Grosse WM, McGraw RA, Tarleton RL. 1995. Construction and use of a multi competitor gene for quantitative RT-PCR using existing primer sets. J Immunol Methods 181:145–156.

Breslauer KJ, Frank R, Blocker H, Marky LA. 1986. Predicting DNA duplex stability from the base sequence. Proc Natl Acad Sci USA 83:3746–3750.

Buchanan AV, Risch GM, Robichaux M, Sherry ST, Batzer MA, Weiss KM. 2000. Long DOP PCR of rare archival anthropological samples. Hum Biol 72:911–925.

Cheng JY, Chen HH, Kao YS, Kao WC, Peck K. 2002. High throughput parallel synthesis of oligonucleotides with 1536 channel synthesiser. Nucleic Acids Res 30:e93.

Cheung VG, Nelson SF. 1996. Whole genome amplification using a degenerate oligonucleotide primer allows hundreds of genotypes to be performed on less than a nanogram of genomic DNA. Proc Natl Acad Sci USA 93:14676–14679.

Dahle UR, Sandven P, Heldal E, Caugant DA. 2003. Continued low rates of transmission of *Mycobacterium tuberculosis* in Norway. J Clin Microbiol 41:2968–2973.

Freier SM, Kierzek R, Jaeger JA, Sugimoto N, Caruthers MH, Neilson T, Turner DH. 1986. Improved free-energy parameters for predictions of RNA duplex stability. Proc Natl Acad Sci USA 83(24):9373–9377.

Gorelenkov V, Antipov A, Lejnine S, Daraselia N, Yuryev A. 2001. Set of novel tools for PCR primer design. Biotechniques 31:1326–1330.

Grant SF, Steinlicht S, Nentwich U, Kern R, Burwinkel B, Tolle R. 2002. SNP genotyping on a genome wide amplified DOP PCR template. Nucleic Acids Res 30:125.

Hahn M, Wilhelm J, Pingoud A. 2001. Influence of fluorophor dye labels on the migration behaviour of PCR amplified short tandem repeats during denaturing capillary electrophoresis. Electrophoresis 22:2691–2700.

Hilali F, Saulnier P, Chachaty E, Andremont A. 1997. Decontamination of PCR reagents for detection of low concentrations of 16S rRNA genes. Mol Biotechnol 7:207–216.

Huang Y, Kong D, Yang Y, Niu R, Shen H, Mi H. 2004. Real-time quantitative assay of telomerase activity using the duplex scorpion primer. Biotechnol Lett 26(11):891–895.

Kaboev OK, Luchkina LA, Tretiakov AN, Bahrmand AR. 2000. PCR hotstart using primers with the structure of molecular beacons (hairpin-like structures). Nucleic Acids Res 28:94.

Kämpke T, Kieninger M, Mecklenburg M. 2001. Efficient primer design algorithms. Bioinformatics 17:214–225.

Kapelari K, Ghanaati Z, Wollmann H, Ventz M, Ranke MB, Kofler R, Peters H. 1999. A rapid screening for steroid 21-hydroxylase mutations in patients with congenital adrenal hyperplasia. Mutations in brief no.247. Online. Hum Mutat 13:505.

Korganow AS, Martin T, Weber JC, Lioure B, Lutz P, Knapp AM, Pasquali JL. 1994. Molecular analysis of rearranged VH genes during B-cell chronic lymphocytic leukemia: intraclonal stability is frequent but not constant. Leuk Lymphom 14:55–69.

Kraszewski A, Norris KE. 1987. Simple and rapid procedure for synthesis of deoxynucleoside 3′-2-cyanoethyl-N,N-diisopropyl-amino phosphites. Nucleic Acids Symp Ser 18:177–180.

Mackay IM, Arden KE, Nitsche A. 2002. Real time PCR in virology. Nucleic Acids Res 30:1292–1305.

Martin FH, Castro MM, Aboul-ela, Tinoco I. 1985. Base pairing involving deoxyinosine: implications for probe design. Nucleic Acids Res 13:8927–8938.

Marota I, Basile C, Ubaldi M, Rollo F. 2002. DNA decay rate in papyri and human remains from Egyptian archaeological sites. Am J Phys Anthropol 121:109–111.

Mazars GR, Theillet C. 1996. Direct sequencing by thermal asymmetric PCR. Methods Mol Biol 65:35–40.

Mitsuhashi M. 1996. Technical report: Part 2: Basic requirements for designing optimal PCR primers. J Clin Lab Anal 10:285–293.

Morgan GJ, Pratt G. 1998. Modern molecular diagnostics and the management of haematological malignancies. Clin Lab Haematol 20:135–141.

Mueller U, Muller YA, Herbst-Irmer R, Sprinzl M, Heimann U. 1999. Disorder and twin refinement of RNA heptamer double helices. Acta Crystallogr D Biol Crystallogr 55:1405–1413.

Ohno H, Sakai H, Washio T, Tomita M. 2001. Preferential use of some minor codons in bacteria. Gene 276:107–115.

Peuschel KE. 2000. New insights into the annealing behaviour of DNA. Med Hypotheses 54:624–625.

Riemersma W, van der Schee C, van der Meijden W, Verbrugh H, van Belkum, A. 2003 Microbial population diversity in the urethras of healthy males and males suffering from nonchlamydial, nongonococcal urethritis. J Clin Microbiol 41(5):1977–1986.

Rose T. 2005. CODEHOP-mediated PCR - a powerful technique for the identification and characterization of viral genomes. Virol J 2:20

Rychlik W. 1990. Selection of primers for polymerase chain reaction. Mol Biotechnol 3:129–134.

Rychlik W, Rhoads RE. 1989. A computer program for choosing optimal oligonucleotides for filter hybridization, sequencing and in vitro amplification of DNA. Nucleic Acids Res 11:8543–8551.

Rychlik W, Spencer WJ, Rhoads RE. 1990. Optimization of the annealing temperature for DNA amplification in vitro. Nucleic Acids Res 11:6409–6412.

SantaLucia J. 1998. A unified view of polymer, dumbbell, and oligonucleotide DNA nearest-neighbor thermodynamics. Proc Natl Acad Sci USA 95:1460–1465

Sayada C, Picard B, Elion J, Krishnamoorthy R. 1994. Genomic fingerprinting of *Yersinia enterocolitica* species by degenerate oligonucleotide primed PCR. Electrophoresis 15:562–565.

Shen Z, Liu J, Wells RL, Elkind MM. 2003. Direct sequencing with highly degenerate and inosine-containing primers. Methods Mol Biol 226:367–372.

Sueoka H, Nagao M, Chiba S. 2000. Rapid mutation screening of phenylketonuria by PCR-linked restriction enzyme assay and direct sequence of the phenylalanine hydroxylase gene: clinical application in Northern Japan and Northern China. Genet Test 4:249–256.

Thein SL et al. 1986. Human genetic diseases, a practical approach. IRL, Virginia.
Tucker ON, Dannenberg AJ, Yang EK, Zhang F, Teng L, Daly JM, Soslow RA, Masferrer JL, Woerner BM, Koki AT, Fahey TJ. 1999. Cyclooxygenase-2 expression is up-regulated in human pancreatic cancer. Cancer Res 59:987–990.
Van Belkum A, Scherer S, Van Alphen L, Verbrugh HA. 1998. Short sequence DNA repeats in prokaryotic genomes. Microbiol Mol Biol Rev 62:275–293.
Weirich G, Hornauer MA, Bruning T, Hofler H, Brauch H. 1997. Fixed archival tissue: purify DNA and primers for good PCR yield. Mol Biotechnol 8:299–301.
Woodward SR, King MJ, Chiu NM, Kuchar MJ, Griggs CW. 1994. Amplification of ancient nuclear DNA from teeth and soft tissues. PCR Methods Appl 3:244–247.
Zhou MY, Gomez-Sanchez CE. 2000. Universal TA cloning. Curr Issues Mol Biol 2:1–7.

Chapter 6

Deoxynucleotide Triphosphates and Buffer Components

Deoxynucleotide triphosphates (dNTPs) are the essential building blocks of nucleic acid molecules, and as such are necessary components of PCR mixes as no new (amplified) DNA could be generated without them. The four individual deoxynucleotides which make up a DNA sequence (i.e. deoxyadenosine triphosphate, dATP; deoxythymidine triphosphate, dTTP; deoxycytosine triphosphate, dCTP; and deoxyguanosine triphosphate, dGTP) are usually added to PCR and RT-PCR mixes in equimolar amounts, though if the target DNA sequence to be PCR amplified comprises mainly dATP/dTTP (AT ratio) or mainly dGTP/dCTP (GC ratio), then the molar ratio of dNTPs added to the PCR mix may be altered in order to take into account this imbalance. dNTPs are usually purchased either individually or as an (equimolar) mix from commercial suppliers (e.g. Roche, Promega, etc.), and are chemically stable when stored in slightly alkaline aqueous solutions at –20°C. However, it should be noted that dNTPs are naturally acidic in solution; hence working and stock solutions may have to be neutralized with alkaline compounds prior to long-term storage. Neutralized dNTP solutions are normally adjusted to 10 mM stock solutions by spectrophotometry, or by adding the correct volume of sterile water to the lyophilized product directly after its chemical synthesis. Some PCRs may profit from the use of highly purified "PCR-grade" dNTPs (Roche) which contain less than 0.9% deoxynucleotide diphosphates (dNDP), are >99% pure, and free from contaminants such as modified nucleotides and tetrapyrophosphate. These (more expensive) high-purity dNTPs may be helpful in successfully amplifying PCR DNA from preparations containing very low copy numbers of target DNA.

In general, a concentration of 20–200 µM of each dNTP in a 100 µl PCR reaction volume allows approximately 2.6 µg of amplified DNA product to be synthesized at maximum rate. This corresponds to approximately 10 pmol of a 400 nucleotide long double stranded DNA fragment. The 20–200 µM range is not an absolute set of values, as dependant on PCR reaction requirements (e.g. long and accurate PCR), or the specifics of polymerase activity (e.g. proofreading ability), optimization of the dNTP concentration above or below this range may be required. However, this value takes into account the fact that dNTPs are "used up"during PCR thermocycling via incorporation into newly synthesized DNA target, and that they may also suffer non-specific thermal degradation. Indeed, research suggests that during a 25 cycle PCR program up to 10% of the total input dNTPs may be lost due to non-specific heat

E. Van Pelt-Verkuil et al., *Principles and Technical Aspects of PCR Amplification,*

inactivation, whilst during a 40 cycle PCR program the net loss of dNTPs may be in the order of 50% of the original amount added to the PCR mix.

dNTPs carry a net negative charge at physiological pH 7.2 and at this pH have the ability to bind both mono- and divalent cations, including magnesium ions which are added to the PCR mix as essential co-factors for the correct functioning of thermostable DNA polymerase. Due to the fact that the pH of a PCR mix is buffered to approximately pH 7.0, a large proportion of these divalent magnesium ions will therefore be bound by dNTPs, a phenomenon which may necessitate the optimization of dNTP concentration with respect to magnesium ion concentration in the design of new PCR amplification protocols or when different primer pair sequences are to be used.

In some PCR applications and protocols, one of the four dNTPs may be replaced by an analogous, dNTP, e.g. inosine, 7-deaza-2′-deoxyguanosine, or a modified dNTP, e.g. a biotin-, fluorophore- or radioactively-labeled dNTP. Such modifications allow further downstream post-PCR processing applications to be performed.

6.1 Factors Affecting the Choice of dNTP Concentration

The concentration of dNTPs during PCR affects both the yield and specificity of the reaction. It is generally recommended to use a concentration of dNTPs that results in only a fraction of the initial amount added remaining at the end of the PCR cycling regime. Moreover, the specificity of PCR amplification is in part determined by the rate at which the DNA polymerase incorporates dNTP into the growing DNA chain, and if the dNTP concentration is partially limiting, then the rate of dNTP incorporation will be reduced, leading to a reduced rate of dNTP misincorporation into the newly synthesized DNA chain and increasing amplification fidelity (i.e. increasing the accuracy of complementary nucleotide incorporation at each nucleotide position). This effect is most pronounced when limiting dNTP concentrations are used in conjunction with low PCR primer concentrations though in most PCR protocols a dNTP concentration of 80 μM is sufficient for adequate fidelity. It should be noted that some thermostable DNA dependent polymerases (e.g. Taq), lack a 5′- 3′ directional "proofreading" activity, leading to a decreased fidelity (greater rate of misincorporation of nucleotides) when used. In this case, the prior optimization of the dNTP concentration (at dNTP concentrations of between 10 and 100 μM), should result in an increase in PCR fidelity and is recommended. Other PCR protocols, requiring a lower than normal dNTP concentration, are those involving the incorporation of modified nucleotides into the amplified DNA product. In such cases, some researchers advocate a dNTP concentration of only 2 μM, since this renders the PCR amplification process more specific [McEvoy et al., 1998]. In contrast, elevated dNTP concentrations favour the blunt end addition of non-specific, overhanging nucleotides to the 3′-end of the PCR primers or the completed PCR product by the thermostable DNA polymerase. This leads to the addition of non-target encoded nucleotides to the PCR primers or the completed PCR amplification product, a process that may interfere with correct primer/target hybridization and ultimately PCR specificity and sensitivity. At dNTP concentrations over 6 mM, substrate inhibition occurs, a process that actually helps inhibit PCR amplification.

6.2 Modified dNTPs and Their Applications

In a typical PCR protocol, equimolar amounts of all four dNTPs are added to the PCR mix. However, specially modified dNTPs may also be added to the PCR reaction mix under certain circumstances. The helix destabilizing nucleotide 7-deaza-2′-deoxyguanosine, helps to overcome amplification problems which may arise due to the presence of stable secondary structures within the template DNA and/or intended target DNA molecule. 7-Deaza-2′-deoxyguanosine acts by destabilizing hairpin stem-loop structures, rendering the copying of any GC-rich regions within the target DNA less problematic. It also acts to reduce "stacking" interactions in the helical structure (Section 5.2). This modified nucleotide may be especially useful in amplifying CpG islands (GC rich sequence found in the promoter sequences of many higher eukaryotes) from small amounts of poor quality DNA [Jung et al., 2002]. Unfortunately, however, DNA in which all of the guanosine nucleotides have been replaced by the 7-deaza-2′-deoxyguanosine is very difficult to stain using ethidium bromide. For this reason, if 7-deaza-dGTP is to be used in the PCR protocol, then it is recommended that the nucleotide dGTP is also added to the PCR mix at a concentration 25% of its normal value (for example 5 μM instead of 20 μM). Another disadvantage of the use of this modified 7-deaza-dGTP is the fact that their presence modifies restriction digest recognition sites, with the result that these sites may no longer be recognized by restriction enzymes.

Inosine is a naturally occurring nucleotide that is an important structural building block in certain transfer RNA (tRNA) molecules, and is often found as the 5′ nucleotide of the anticodon loop. The most important characteristic of the deoxyinosine base (at least with respect to PCR applications), is that it is capable of base pairing with any of the four natural (dATP, dCTP, dGTP and dTTP) nucleotides, and hence is frequently used in primer design to solve problems related to sequence variation within the target DNA. PCR primers designed to anneal to a broad range of targets often contain inosine, however it should be noted that DNA/DNA hybrids containing deoxyinosine are less stable than hybrids containing deoxyadenosine, deoxythymidine, deoxycytosine or deoxyguanosine. Hence, it is not feasible to design PCR primers that comprise regions of multiple deoxyinosine bases, since the annealing temperature of the primers will be adversely affected, thereby seriously affecting PCR specificity. Alternatively, another universal nucleoside (1-(2′-deoxy-β-D-ribofuranosyl)-3-nitropyrrole) has been described in the literature [Nichols et al., 1994] (Fig. 6.1). This nucleoside (an unphosphorylated nucleotide), is known under the generic name of "M-nucleoside" and its incorporation in PCR

Fig. 6.1 Chemical structure of the universal "M nucleoside" (1-(2′-deoxy-beta-D-ribofuranosyl)-3-nitropyrrole). Both inosine and M nucleoside may be used to compensate for mismatches during primer/DNA hybridization (Reproduced from Nichols et al., 1994. With permission from Macmillan Publishers Ltd.)

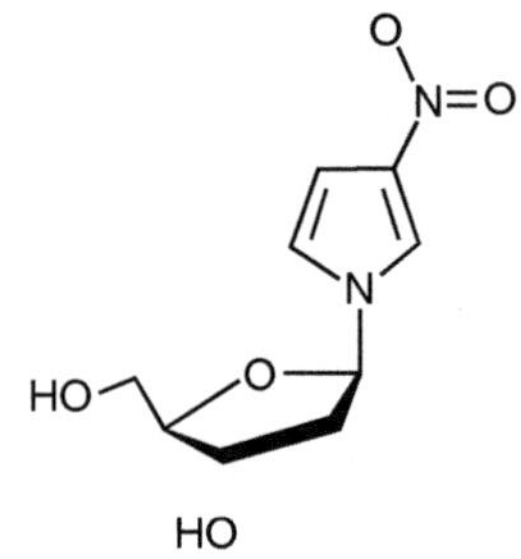

primers affects the T_m of primer annealing much less markedly than deoxyinosine yet still maintains the ability to hybridize to any of the four normal dNTPs which make up DNA molecules.

6.3 The PCR Buffer

PCR reaction buffer systems should facilitate optimum amplification activity of the thermostable DNA polymerase (optimum sensitivity), whilst simultaneously allowing optimum PCR primer/DNA hybridization (optimum specificity) to occur. Nowadays, much less attention needs to be paid to PCR reaction mix buffering, as it is usual for commercial suppliers of thermostable DNA polymerases to provide an aliquot of the necessary (10X concentrated) PCR reaction buffer "free" with each order of thermostable DNA polymerase purchased. To make the setting-up of PCRs even easier, PCR buffers may also be supplied with a separate vial of magnesium ions, or alternatively, the magnesium ions may be supplied already added to the to the buffer mix (dependent on the supplier). For large volumes of buffer, it may even be possible to order a PCR buffer mix containing a magnesium ion concentration tailored to suit your own PCR requirements. Some buffers are even designed to precipitate magnesium ions prior to PCR thermocycling in order to prevent the formation of primer dimers or non-specific products (e.g. RockStart buffer, DNA Polymerase Technology Inc.).

To optimize thermostable DNA polymerase activity, the pH conditions required for the enzyme should be maintained over a broad range of temperatures and should not be affected by the enormous changes that occur in PCR mix composition (e.g. the concentration of available dNTPs) as PCR thermocycling proceeds. Comparative analyses using several thermostable DNA polymerases have revealed that simple tris-HCl buffers perform best, with glycerin/NaOH systems as the second best option (e.g. Fig. 6.2). Phosphate buffers perform poorly, most probably because phosphate ions may disturb any physical interactions occurring between phosphorylated compounds present in the PCR mix (i.e. primers and dNTPs), and because it is possible for phosphate ions to chelate magnesium ions, thereby removing them

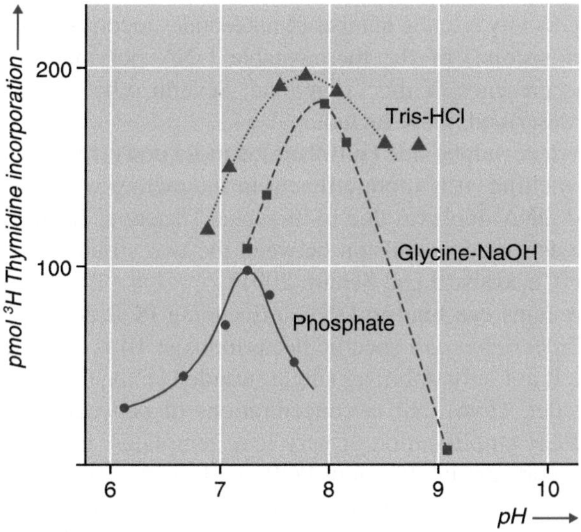

Fig. 6.2 Optimum curves for Taq polymerase activity in different buffer systems. Calf thymus DNA was used as template to monitor the incorporation of radioactive thymidine triphosphate at 74°C [Chien et al., 1976]. The concentration of all three buffers used was set at 25 mM. Optimum performance is achieved using Tris-HCl buffer, although the difference observed with glycin NaOH is marginal

from solution and inhibiting the correct functioning of the thermostable DNA polymerase. Though preferred, tris-HCl buffers tend to be rather sensitive to changes in temperature, with a reduction of 0.1 pH unit occurring for every 5°C rise in temperature. For this reason tris-HCl PCR buffers are formulated to exhibit a pH of 8.4 at room temperature, which means that at 74°C (the optimum temperature for DNA amplification for most thermostable DNA polymerases), the pH in the PCR mix will be approximately 7.4 (the optimum pH for most thermostable DNA polymerases).

Most PCR buffers contain a proteinaceous compound which helps stabilize thermostable DNA polymerases in the PCR reaction mix. In many cases, autoclaved gelatin or nuclease free bovine serum albumin (BSA) is used, with both compounds being included at a concentration of approximately 100 μg per ml. Non-ionic detergents are also usually included in PCR buffers in order to maintain the polymerase in the correct physical conformation, (shielding it from protease- or temperature-induced inactivation), whilst helping disassociate any secondary structure which may form within the primers or template. Popular non-ionic detergents added to PCR buffers include the "soap-like" compounds Tween-20, Triton X-100 or Laureth. Occasionally, dimethylsulphoxide (DMSO) is also included in the PCR buffer at a concentration of 2–10% (vol/vol), in order to help prevent primer/target mispriming. Although DMSO slightly inhibits thermostable DNA polymerase activity, the processivity (i.e. the number of nucleotides incorporated into the growing DNA chain per second) of the thermostable DNA polymerase is significantly enhanced in the presence of this compound. Several other PCR buffer additives have also been described, these include:

1. Tetramethylene sulphoxide (sulfolane) is more powerful than DMSO in facilitating dsDNA melting. It is more efficient in interacting with hydrogen bonds in double stranded DNA duplexes due to its planar structure. It may also help lower the steric and electrostatic repulsion between the two strands of the double helix DNA molecule [Chakrabarti and Schutt, 2001].

2. Several proteins can bind to DNA influencing PCR activity. The "gene 32" protein of the *Escherichia coli*-specific bacteriophage T4 is a single stranded DNA binding protein. It not only stabilizes single stranded DNA, thereby enhancing PCR efficiency [Kreader, 1996], but in concentrations of between 0.5–10 nM actually stimulates the PCR amplification of very long templates. Interestingly, this compound appears to prevent inhibition of PCR amplification by blood components or contaminants present in soil samples, an activity that is shared with several commercially secret preparations, e.g. AmpDirect (Shimadzu Corp). Other proteins such as lactoferrin, haem and IgG, specifically recognize and bind to double stranded DNA, inhibiting the activity of any contaminating DNase. α-2-Macroglobulin is a protease inhibiting enzyme, which can sometimes be used in combination with crude DNA extracts to prevent proteolysis of thermostable DNA polymerases [Al-Soud and Radstrom, 2000].

3. Pyrophosphatase, betain, ammonium sulphate or spermidine help to remove any inhibitory products, such as pyrophosphate, which may accumulate during PCR amplification. Betain (N,N,N-trimethylglycine carboxymethyl-trimethyl ammonium) is a Zwitterion (i.e. it contains both acidic and basic groups on the same molecule), that helps to equalize the thermodynamic stability contribution of AT- and GC-basepairs [Henke and Loening, 1998]. In addition, betain appears to enhance the hydration of GC-basepairs, leading to easier destabilization and disassociation of double stranded DNA helices [Hengen, 1997], as well as serving as an osmo-protectant, helping to prevent the denaturation of thermostable DNA polymerases in solutions containing high concentrations of salts.

4. Uracyl-N-glycosylase (UNG) is a hydrolase that is capable of digesting DNA that contains uracil, a ribonucleotide that is able to substitute for thymidine and pair with adenosine nucleotides. PCRs performed using dUTP instead of dTTP in the reaction mix, will generate amplimers that can be digested by the UNG hydrolase, thereby rendering the amplimers non-amplifiable in subsequent PCRs. This procedure can be used to reduce the chances of false positive PCR results occurring due to contamination of virgin batches of PCR mix by amplimers generated from previous PCR amplifications (a precaution which may be particularly necessary in the routine clinical diagnostic setting). Of course the contaminating PCR products themselves must contain dUTP instead of dTTP for this protocol to work.

5. Dithiothreitol (DTT) is a reducing agent that may be included in PCR buffers to help maintain the active conformational structure of the thermostable DNA polymerase enzyme. In the case of Taq polymerase, a working concentration of 5 mM DTT is used.

6. Formamide can be used in amounts between 2% and 5% (vol/vol) in PCR buffers. It destabilizes base pairing interactions and is reported to reduce the number of non-specific amplification products generated in PCRs. An alternative to formamide is tetramethyl ammonium chloride (TMAC) [Honore and Madsen, 1997].

7. Several specific compounds that bind to thermostable DNA polymerases have been developed. Elegant examples of these are the oligonucleotide inhibitors. These compounds bind firmly to DNA polymerases preventing their enzymatic activity. These compounds are not substrates for DNA synthesis, nor are they susceptible to 5′-3′ exonuclease activity, though they tend to completely disassociate at temperatures above 45°C. Two different classes of these "aptamer" oligonucleotide inhibitors exist [Lin and Patel, 1997], both being useful in fully inhibiting DNA polymerase activity during the heating step of the very first cycle of PCR thermocycling protocols (Fig. 6.3). The different classes of aptamers have differential affinity for RNA versus DNA molecules. By inhibiting DNA polymerase activity at low temperatures, these compounds help to prevent the extension of non-specific PCR products, which may occur due to non-specific primer/non-target DNA hybridization at low temperatures (e.g. LightCycler-RNA Master SYBR green I kit, Roche Diagnostics Corporation). Also, it has been shown that low amounts of tRNA can form specific complexes with oligonucleotides (short fragments of DNA), thereby shielding these oligonucleotides from non-specific priming interactions at low temperatures [Sturzenbaum, 1999].

8. Several companies market their own (secret) proprietary PCR mix additives, e.g. Q-solution (Qiagen), GC-melt (Clontech), Fail Safe PCR enhancer (Epicentre) and the Enhancer solution (Life Technologies), all of which are reported to significantly enhance (multiplex) PCR efficiency. The Enhancer solution (Life Technologies) is reported to be especially effective in increasing the reliability of PCRs that span repetitive regions of DNA, and renders the PCR reaction less susceptible to variation in the magnesium ion concentration.

Finally, it should be noted that there is no consensus about which of these additives should be included in PCR reaction buffers, as different thermostable DNA polymerase enzymes react differently to each of the additives. Further, some additives may actually increase the specificity of PCR amplification but concomitantly reduce PCR sensitivity or *vice versa*, and different PCR applications may require different sets of additives. For example, *in situ* PCR is frequently frustrated by the presence of single stranded nicks in the template DNA which can be the starting point for non-specific DNA repair by the Taq polymerase [Moretti et al., 1999; Lantz et al., 1998]. This phenomenon is not related to mispriming of PCR primers, as DNA synthesis proceeds in the presence of these single stranded nicks even when PCR primers are omitted from the reaction mix. Blocking of this type of enzymatic activity may be achieved using certain tRNA preparations [Sturzenbaum, 1999]. Ultimately, PCR optimization (especially magnesium ion concentration versus dNTP concentration optimization) should be regarded as a necessity for all PCR protocols whose buffers use these additives.

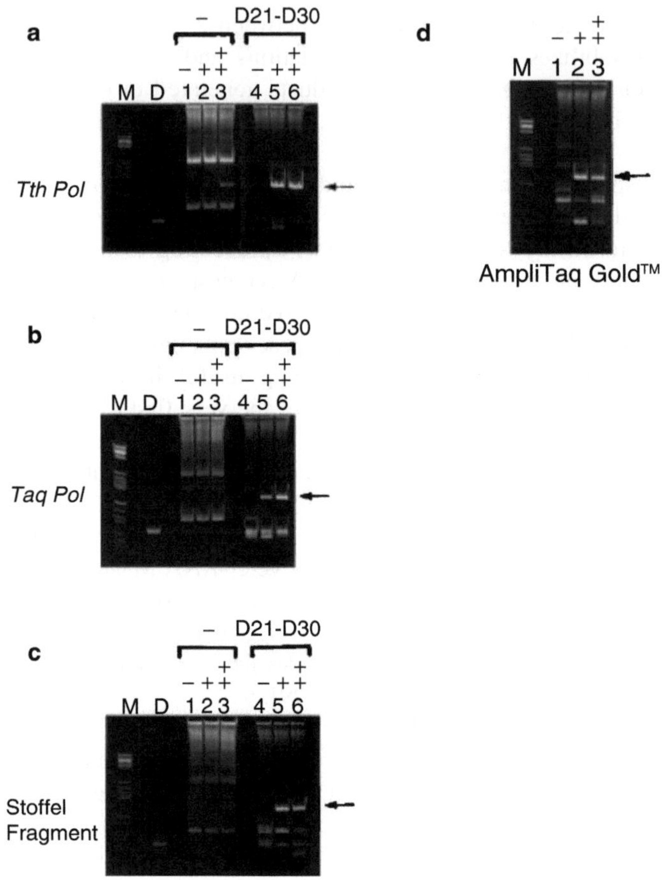

Fig. 6.3 The use of aptamers in inhibiting non-specific PCR amplification for a low copy number target using three different polymerases and the Stoffel fragment. M = molecular size marker, D = aptamers D21 and D30 alone, lanes 1–3 = amplification without aptamers, whilst lanes 4–6 = amplification performed in the presence of 50 nM D21-D30. Amplifications were performed using HIV-specific aptamer primers and 1.4 microgram human placental DNA (as a source of superfluous DNA). – = negative controls, human placental DNA only; + = 10 copies of template HIV DNA; ++ = 50 copies of template HIV DNA. The arrow on the right indicates the specific 142 bp PCR product. All amplifications were performed using buffers and thermocycling parameters recommended by the manufacturer (From Lin and Jayasana, 1997. With permission by Elsevier)

6.3.1 Monovalent Ions

Sodium (Na^+), potassium (K^+) and ammonium (NH^+_4) ions stimulate the activity of thermostable DNA polymerases and shield the negatively charged phosphate DNA backbone, thereby weakening the electro-repulsive forces between primer and target

DNA. Potassium ions have an optimum stimulatory effect on PCR DNA polymerases at a concentration of approximately 50 mM, though the stimulatory effect ceases at a concentration greater than 75 mM and inhibition of polymerase activity occurs at potassium ion concentrations greater than 100 mM. NH^{+}_{4} ions compete for the hydrogen bonds between DNA strands thereby facilitating specific primer/target DNA hybridization over a wider range of annealing temperatures. This reduces the number of incompletely synthesized amplimers and mispriming interactions. In some cases an optimized ratio of potassium and ammonium concentrations in the PCR mix renders optimization of the magnesium concentration unnecessary.

6.3.2 Magnesium Ions

The divalent magnesium ion is a very important ingredient of PCR reaction mixes as it acts as a co-factor for thermostable DNA polymerase activity, stimulating the enzymes in a concentration dependent manner [Park and Kohel, 1994; Ely et al., 1998]. Indeed, several "hot-start" PCR techniques rely on the sequestering of magnesium ions from the PCR reaction mix until an annealing temperature high enough to facilitate only specific primer/template annealing is achieved. It is therefore recommended that the optimum concentration of magnesium ions required for specific amplification in a newly developed PCR protocol is determined experimentally. Such optimization is frequently performed in combination with an experiment to determine the optimum dNTP concentration for the PCR protocol utilizing a "checkerboard titration" format, i.e. combining different concentrations of magnesium and dNTP in different experimental PCR reaction mixes. However, in general, a magnesium ion concentration range of between 0.5 and 5 mM is used in most PCR protocols, with a concentration of 2.5 mM being the most frequently used. Differences in optimal magnesium ion concentration between different PCR protocols may in part be attributable to different thermostable DNA polymerases being used and/or the presence of chemicals (e.g. EDTA and calcium ions) that may sequester or compete with the magnesium ions. These inhibitors tend to be accidentally added as impurities from template DNA or RNA extraction procedures, and their presence may in part be compensated for by the addition of extra magnesium ions in the PCR reaction mix. This means that all newly designed PCR protocols need to be optimized in order to define both the optimum magnesium and dNTP concentrations required for maximum specificity and sensitivity (Fig. 6.4).

It should also be noted that elevated magnesium ion concentrations inhibit PCR amplification, since at these concentrations the double stranded DNA structure is actually stabilized, which may result in the incomplete disassociation of double-stranded amplification products. This stabilization effect also acts to increase the stability of misprimed primer/target DNA interactions increasing the risk of non-specific product amplification. Furthermore, thermostable DNA polymerase activity *per se* is strongly inhibited by high magnesium concentrations, with enzyme

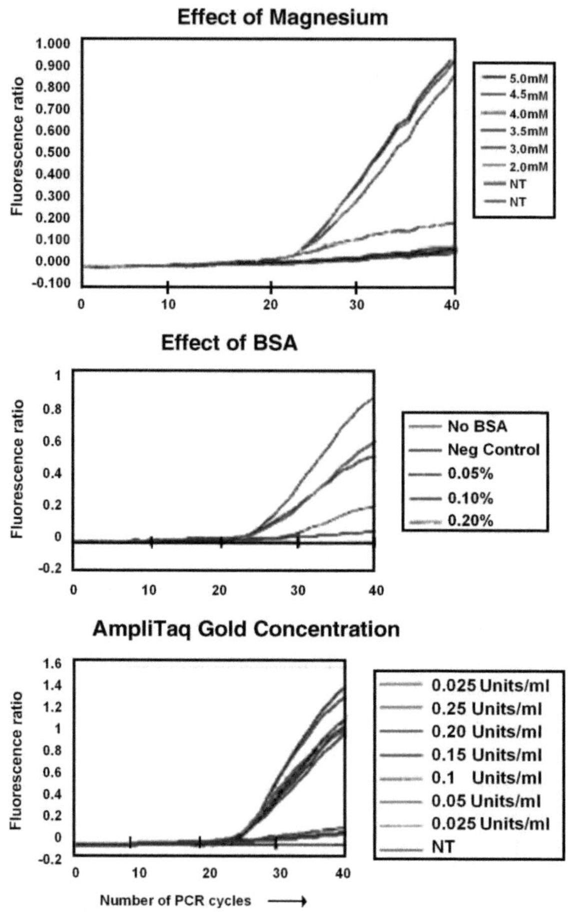

Fig. 6.4 Influence of divalent magnesium ions on real time PCR efficiency using thermostable Taq DNA polymerase. The efficiency (but not necessarily the specificity) of PCR increases with increasing Mg^{++} concentration. At approximately 10 mM Mg^{++} concentration, enzyme activity (PCR efficiency) is reduced. BSA and Taq polymerase concentration can also be two other important variables. This real time PCR test system involved the optimisation of a beta-actin specific PCR. In the figure, the X-axis indicates PCR cycle number and the Y-axis the fluorescence ratio (emission intensity of the reporter probe divided by the emission intensity of a passive reference) (Reproduced from Taylor et al., 1997. With permission from 'Oxford Journals') (*see Color Plates*)

activity being reduced by as much as 50% at magnesium ion concentration approaching approximately 10 mM. Finally, at high magnesium ion concentrations, the frequency of misincorporated nucleotides is increased, since the fidelity of the polymerase is generally reduced under these reaction conditions.

References

Al-Soud W, Radstrom P. 2000. Effects of amplification facilitators on diagnostic PCR in the presence of blood, feces and meat. J Clin Microbiol 38:4463–4470.

Chakrabarti R, Schutt CE. 2001. The enhancement of PCR amplification by low molecular-weight sulfones, Gene 274:293–298.

Chien A, Edgar DB, Trela JM. 1976. Deoxyribonucleic acid polymerase from the extreme thermophile *Thermus aquaticus*. J Bacteriol 127:1550–1557.

Ely JJ, Reeves-Daniel A, Campbell ML, Kohler S, Stone WH. 1998. Influence of magnesium ion concentration and PCR amplification conditions on cross-species PCR. Biotechniques 25(1):38–40, 42.

Hengen PN. 1997. Optimizing multiplex and LA PCR with betaine. Trends Biochem Sci 22:225–226.

Henke W, Loening SA. 1998. Recently, betaine has been introduced as an additive in different PCR strategies. Nucleic Acids Res 26:687.

Honore B, Madsen P. 1997. The tetramethylammonium chloride (TMAC) method for screening cDNA libraries with highly degenerate oligonucleotide probes obtained by reverse translation of amino acid sequences. Methods Mol Biol 69:139–146.

Jung A, Ruckert S, Frank P, Brabletz T, Kirchner T. 2002. 7-Deaza-2′-deoxyguanosine allows PCR and sequencing reactions from CpG islands. Molecular Patholog 55:55–57

Kreader CA. 1996. Relief of amplification inhibition in PCR with bovine serum albumin or T4 gene 32 protein Appl Environ Microbiol.62:1102–1106.

Lantz PG, Knutsson R, Blixt Y, Al Soud WA, Borch E, Radstrom P. 1998. Detection of pathogenic *Yersinis enterocoloitica* in enrichment media and pork by a multiplex PCR: a study of sample preparation and PCR inhibitory components. Int J Food Microbiol 45:93–105.

Lin CH, Patel DJ. 1997. Structural basis of DNA folding and recognition in an AMP-DNA aptamer complex: distinct architectures but common recognition motifs for DNA and RNA aptamers complexed to AMP. Chem Biol 4:817–832.

Lin Y, Jayasana SD. 1997. Inhibition of multiple thermostable DNA polymerases by a heterodimeric aptamer. J Mol Biol 271:100–111.

McEvoy CR, Seshadri R, Firgaira FA. 1998. Large DNA fragment sizing using native acrylamide gels on an automated DNA sequencer and GENESCAN software. Biotechniques 25:464–470.

Moretti S, Pinzi C, Spallanzani A, Berti E, Chiarugi A, Mazzoli S, Fabiani M, Vallecchi C, Herlyn M. 1999. Immunohistochemical evidence for cytokine networks during progression of human melanocytic lesions. Int J Cancer 84:160–168.

Nichols R, Andrews PC, Zhang P, Bergstrom DE. 1994. A universal nucleoside for use at ambiguous sites in DNA primers. Nature 369:492–493.

Park YH, Kohel RJ. 1994. Effect of concentration of MgCl2 on random-amplified DNA polymorphism. Biotechniques 16(4):652–656.

Sturzenbaum SR. 1999. Transfer RNA reduces the formation of primer artifacts during quantitative PCR. Biotechniques 27:50–52.

Taylor TB, Winn-Deen ES, Picozza E, Woudenberg TM, Albin M. 1997. Optimization of the performance of the polymerase chain reaction in silicon-based microstructures. Nucleic Acids Res 25:3164–3168.

Chapter 7

Taq and Other Thermostable DNA Polymerases

The commercial availability of thermostable DNA dependent DNA polymerase enzymes has been the major factor in facilitating the success of PCR amplification. Initially, the Klenow fragment of DNA-dependent DNA polymerase I (involved in replication and repair in the bacterium *Escherichia coli*), was employed in PCR amplification [Saiki et al., 1985; Mullis et al., 1992; Saiki et al., 1986]. The Klenow fragment is actually a hydrolytic product of the native *E. coli* DNA-dependent DNA polymerase enzyme which lacks a 5′-3′ exonuclease activity (Fig. 2.3). The Klenow fragment polymerase was used in the first PCR protocols developed, but has the huge disadvantage of exhibiting an optimum reaction temperature at 37°C and being heat labile at the temperatures used in PCR thermocycling reactions. This lability meant that originally, fresh enzyme had to be added after each and every PCR cycle, making the PCR procedure time consuming, labour intensive and highly prone to contamination. Further, it was impossible to generate PCR fragments longer than 400 bp and further processing (using for example Southern blotting or dot spot hybridization), was usually required to identify the presence/absence of specific amplification products from the mixture of amplimers produced.

A breakthrough was achieved however in 1976, when Chien et al. [1976] described a 94kD thermostable DNA-dependant DNA polymerase derived from a eubacterium called *Thermus aquaticus*, whose natural habitat is hot thermal springs (with ambient temperatures of 70–75°C). This thermostable DNA polymerase or "Taq" enzyme was found to possess similar properties to *E. coli* DNA dependant DNA polymerase I, with strong homology being found at the amino acid level so: the 3′-OH nucleotide addition site, the dNTP/DNA binding sites, and the 5′-3′ exonuclease sites of the two enzymes (Fig. 7.1). Unfortunately however, though this new Taq polymerase did exhibit some endogenous reverse transcriptase activity, it was found that the 3′-5′ proofreading domain (whereby misincorporated nucleotides are removed from double stranded DNA and replaced by the correct complementary nucleotide) was missing. Moreover, despite the obvious thermoresistant qualities of Taq polymerase, it took a further 2 years for the enzyme to be included in a published PCR protocol [Saiki et al., 1986], after which, there was an explosive increase in interest in PCR mediated amplification protocols.

To date, many different types of thermostable DNA dependent DNA polymerases have been discovered in many different thermophilic organisms (Section 7.5), with

E. Van Pelt-Verkuil et al., *Principles and Technical Aspects of PCR Amplification*,

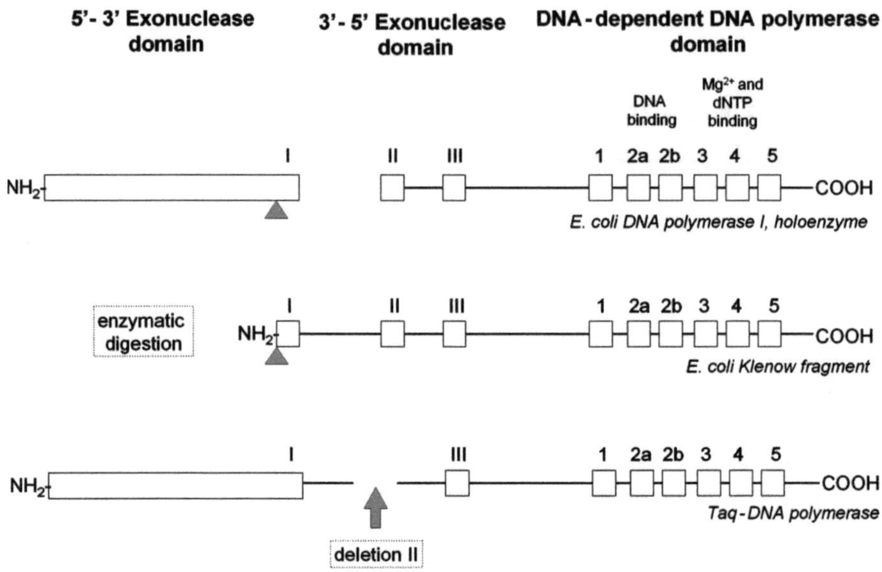

Fig. 7.1 Structural homologies between various (forms of) DNA polymerases. The Klenow fragment can be generated enzymatically, whereas the Stoffel fragment is produced as a recombinant protein

the majority of these enzymes having their own particular characteristics and being available from commercial suppliers. However, Taq thermostable DNA polymerase (along with its variants) still remains the most widely used polymerase in PCR reactions performed today.

7.1 The Advantages and Disadvantages of Taq over Klenow Fragment DNA Polymerase

Taq thermostable DNA dependent DNA polymerase has a large number of advantages over *E. coli* Klenow DNA polymerase, which was originally used in PCR protocols. These include

1. The enzyme is heat stable with a half-life of 130 minutes at a temperature of 92.5°C. Further, high-stringency PCR protocols where the temperature during an entire PCR protocol does not drop below 70°C, are also tolerated by the enzyme. However, at higher temperatures the half-life of Taq decreases rapidly. For example, at a constant 95°C the half-life of the enzyme is 40 minutes, but at 97.5°C the half-life is only 5 – 6 minutes. This means that in a 50-cycle PCR protocol with a denaturation temperature of 95°C for 20 seconds, approximately 50% of the Taq enzyme will theoretically be denatured at the end of PCR thermocycling.

2. The PCR reaction vial does not have to be opened in order to add fresh enzyme after each PCR cycle, thereby greatly reducing the risk of contamination during thermocycling.
3. The enzyme has an elevated optimum temperature compared to normal DNA polymerases, with an optimum reactivity (Vmax) occurring between 70°C and 80°C and significant residual activity between 55°C and 70°C [Chien et al., 1976] (Fig. 7.2). This elevated temperature for optimum polymerase activity helps to prevent any secondary or tertiary structure present within the template DNA or RNA interfering with DNA amplification.
4. The high temperature for optimum activity of the enzyme helps to prevent mispriming of the PCR primer pair to non-complementary or partially complementary regions present within the template DNA [Saiki et al., 1988] (Fig. 7.3).
5. The enzyme has increased processivity at elevated incubation temperature. Processivity measures the number of nucleotides that are incorporated into a growing polynucleotide chain by the enzyme in a certain period of time. For Taq polymerase this value is approximately 60 nucleotides per second at 72°C, but only 1.5 nucleotides per second at 37°C, though PCR processivity is strongly affected by secondary structure elements in the template nucleic acid molecule. Hence, Taq enzyme exhibits maximum processivity at temperatures that help to disassociate any secondary or tertiary structure that may be present in the target DNA to be amplified. The higher incorporation rates observed for Taq polymerase compared to Klenow polymerase also results from the fact that Taq polymerase does not exhibit 3′-5′ proofreading exonuclease activity, this proofreading activity actually slows the rate of nucleotide incorporation into the growing DNA chain.
6. The Taq polymerase has the capacity to synthesize relatively long PCR products. Though the Klenow fragment experiences problems with the amplification of fragments longer than 400 basepairs in length, Taq polymerase can routinely synthesize fragments up to 4,000 bp in length. Moreover, if the PCR conditions are chosen carefully, and if the quality of the template DNA is high, then amplimers up to 8,000–10,000 base pairs in length may be synthesized
7. Taq polymerase generates significantly higher yields of amplification products than the Klenow fragment. The Klenow fragment only being able to amplify a typical DNA target 200,000 times at maximum rate during a standard PCR protocol, whilst a value ten times higher (2,000,000 times) has been documented for Taq polymerase.
8. The Taq enzyme frequently adds an overhanging nucleotide, usually an adenosine, to the 3′-OH end of blunt ended double stranded DNA. This property has been used to develop fast and efficient PCR cloning strategies using vectors equipped with matching thymidine overhangs (known as "TA-cloning", e.g. TOPO TA cloning kits, Invitrogen).
9. Finally, Taq polymerase has a number of miscellaneous properties that have also significantly contributed to its successful application in PCR amplification protocols, including the fact that the enzyme is not inhibited by chemical contaminants remaining after nucleic acid extraction, e.g. chloroform or citrate.

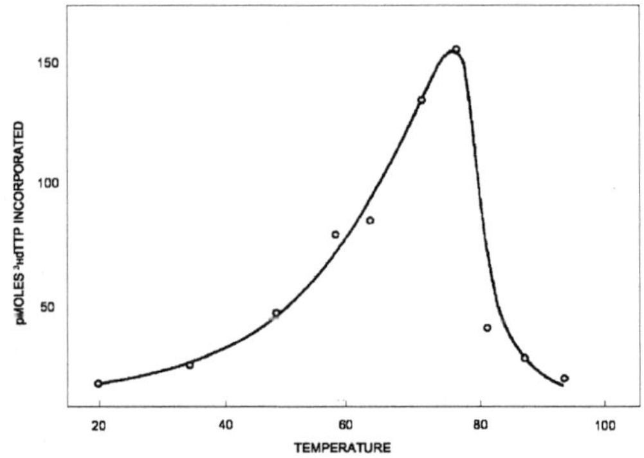

Fig. 7.2 Assessing the optimum reaction temperature for Taq polymerase. The incorporation of radioactive thymidine to calf thymus DNA at different temperatures generates an a-symmetric curve [Chien et al., 1976]. Note that the optimum incorporation rate of the Taq polymerase occurs at approximately 80°C, though significant activity can still be measured between 40°C and 70°C

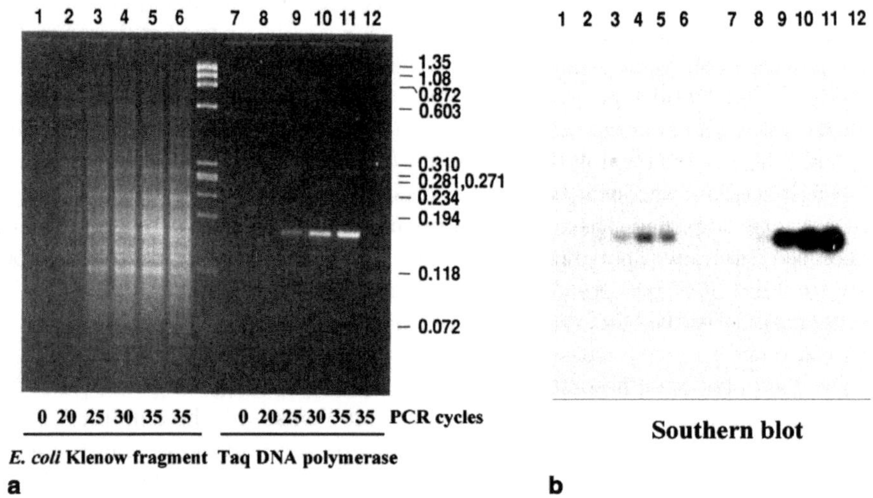

Fig. 7.3 Comparative analysis of PCR performed with *E. coli* DNA polymerase versus *T. aquatus* DNA polymerase. The PCR is targeted towards the human β-globin gene. Panel A, displays the results of post-PCR gel electrophoresis, whereas panel B shows the same analysis after Southern hybridization with a β-globin specific DNA probe. Lanes 1–6 (Klenow enzyme) and 7–12 (Taq polymerase) display the products obtained after 0, 20, 25, 30 and 35 cycles, respectively. Lanes 1 to 5 and 7 to 11 = human cell line MOLT4. Lanes 6 and 12 = human cell line GM2064 (possesses a homozygous deletion of the entire beta-globin gene complex). The length of the various molecular weight markers is indicated in kilobasepairs (Reprinted from Saiki et al., 1988. With permission from AAAS)

As well as advantageous properties, Taq polymerase also displays a number of disadvantageous properties. These include

1. Taq polymerase is relatively susceptible to proteolytic degradation. For this reason, some PCR protocols advise that the Taq polymerase be added only after

an initial heat treatment of the PCR reaction mix has occurred, as most proteases are completely denatured during the initial heating step of typical PCR protocols. In practice however, the use of commercial nucleic acid extraction kits has greatly reduced problems associated with protease contamination of extracted nucleic acid material.

2. Taq polymerase is inhibited to a greater or lesser extent by a wide variety of compounds used in biochemical or molecular biological experiments. These include compounds such as DMSO, EDTA, guanidinium HCl, urea, agar and agarose. Also inhibitory, are various compounds frequently used in DNA isolation protocols *per se*, these include (as vol/vol ratios), >2% phenol, 1% ethanol, 1% isopropanol, >5 mM sodium acetate (NaAc), 0.005% cetyl trimethyl ammonium bromide (CTAB), 0.005% sodium dodecyl sulphate (SDS), or >20% formamide, which are all completely inhibitory at the given concentrations. Other important inhibitory factors include the RNase inhibitor aurintricarboxylic acid (ATA) [Finke et al., 1993] and the blood anticoagulant polyanetholsulphate, which inhibits PCR amplification at concentrations as low as 0.01 μg/ml. Plant polysaccharides such as glycan, acid mucopolysaccharides, polyphenols, rice starch, pectin, dextransulphate and β-glucans can also exhibit inhibitory activities. Finally, UV irradiation of polystyrene-based plastics, sometimes used to sterilize microtitre plates for example, may liberate water-soluble Taq polymerase inhibitors [Burgess and Hall, 1999].

7.2 Misincorporation of Nucleotides and Fidelity of DNA Synthesis by Taq Polymerase

Both intrinsic and extrinsic factors contribute to the error rate (also known as the "fidelity"), of Taq mediated PCR amplification:

1. *Misincorporation of nucleotides due to intrinsic factors*: Under optimized conditions, Taq polymerase can specifically amplify a single copy of a gene from a background of several micrograms of miscellaneous template DNA. However, the enzymatic extension of a DNA fragment by thermostable DNA polymerases occasionally results in the (mis)incorporation of an incorrect nucleotide to the growing DNA chain (i.e. insertion of a nucleotide that is not complementary to the nucleotide on the template DNA strand). *In vivo*, such "mutations" are corrected by the 3′-5′ proofreading activity which most thermolabile DNA polymerases possess. However, due to the lack of 3′-5′ proofreading activity in Taq DNA polymerase, these errors are not corrected and may therefore be apparent in the final amplification products. Even enzymes with proofreading activity (Section 7.3), may occasionally incorporate the wrong nucleotide into the growing DNA chain. The misincorporation rate of a polymerase enzyme may be determined mathematically and is represented by the term "error rate" (Table 7.1). The error rate per nucleotide per cycle of a DNA polymerase

enzyme is denoted by the letter "E" [Flaman et al., 1994], with the E values between different thermostable polymerases differing by as much as a factor of 100 or more. Subtle test systems for the experimental enumeration of E have been described, though in most cases the laclOZ-α test is used [Eckert and Kunkel, 1993; Cline et al., 1996]. The exact value of E may be calculated using the simple formula:

$$\mathbf{E = \%mutations / d \times del \times 100}$$

where **d** is the overall number of amplimers present in the final reaction mix, and **del** is the length of the amplified product (amplicon length).The E value for Taq polymerase is in the order of 1.8×10^{-4} and for Klenow fragment somewhere in the region of 9×10^{-6}, indicating that approximately 1 in 6,000 bases is wrongly incorporated by Taq polymerase and approximately 1 in 12,500 wrongly incorporated by the Klenow fragment. Another frequently used term is "accuracy" which is the inverse of the error rate:

$$\mathbf{Accuracy = 1/E}$$

The error rate is defined as the mutation frequency per basepair synthesized, whereas the accuracy describes the number of correct nucleotide incorporations before a misincorporation event occurs. The error rates and accuracy for a number of commonly used PCR DNA polymerases are shown in Table 7.1. For Taq, it has also been calculated that the chance of a frameshift mutation occurring (i.e. a

Table 7.1 Fidelity and error rates for various thermostable DNA dependent DNA polymerases

Enzyme	Error rate ($\times 10^{6}$)	Accuracy ($\times 10^{-5}$)	3′-5′ Exonuclease activity
Klenow fragment	80	–	Yes
Pfu	1.3	7.7	Yes
Platinum *Pfx*	1.6	–	Yes
Pfu Turbo	2.4	–	–
DeepVent	2.7	3.6	Yes
Proofstart	3.6	–	Yes
Tli(Vent)	2.8	3.6	Yes
Tfu	90	–	Yes
Taq	180	1.2	No
UL*Tma*	55	0.2	Yes
Thermal Ace	60	–	Yes
FailSafe	8	–	Yes
MasterAmp	32	–	Yes

Some of the enzymes have been tested in the presence of stabilizing compounds. The error rate may deviate by approximately ±20%. The error rate is defined as the mutation frequency per basepair synthesized, whereas the accuracy describes the number of adequate incorporations before a misincorporation event occurs. Note that these rates are extremely sensitive to variation in the experimental conditions. (– = value not determined) change in the reading frame of a gene encoding DNA fragment), due to deletion or insertion of a base by Taq polymerase, is approximately 1 in 41,000.

PCR mixes that utilize thermostable DNA polymerases lacking a 3′-5′ proofreading activity, induce a cumulative mutation rate (i.e. the number of misincorporated nucleotides within the PCR amplification mix gradually increases in number after every cycle). These misincorporations may or may not seriously affect the overall sequence of the amplification product. Misincorporations present in only a few hundred amplimers out of a background of many millions of copies of correctly amplified amplimers at the end of PCR thermocycling, will most likely not affect the outcome of any downstream applications (e.g. cloning, sequencing, hybridization reactions, etc.). However, misincorporations that occur early on in the PCR, during the first few cycles of amplification, will be carried through to become millions of incorrectly amplified PCR amplimers in the final reaction mix, almost certainly affecting downstream applications. Though very difficult to detect upon gel electrophoresis, the presence of multiple misincorporated amplimers, derived from several different nucleotide misincorporation events, may often be recognized by the presence of "fuzzy" or "unfocussed" amplification products. Alternatively, sequence analysis will indicate the presence of multiple nucleotides (two or more chromatogram peaks of similar height), at the same sequence position. Though several thermostable DNA polymerases that exhibit 3′-5′ proofreading ability are commercially available (3′-5′ proofreading ability helps to limit the rate of nucleotide misincorporation into PCR amplification products), the processivity of these enzymes, i.e. the length of DNA sequence that may be PCR amplified, is somewhat lower than that of Taq and other non-proofreading enzymes (Section 7.5). In some cases however, the lack of proofreading ability of Taq polymerase may be put to good use. Under certain conditions, the E value for Taq polymerase may be manipulated to be as high as 2%, facilitating the discovery of DNA-structure-dependent mutational "hot" and "cold" spots, i.e. areas where point mutations are more or less likely to occur. This approach has helped to confirm the enhanced susceptibility of simple repeat sequences and homopolymeric tracts of DNA sequence to mutation [Saunders et al., 2000].

2. *Misincorporation of nucleotides due to extrinsic factors*: Extrinsic factors also influence Taq PCR fidelity, including the length of time and temperature of the elongation step within each PCR cycle (which should be kept short to prevent the extension of mis-hybridized primers and occur at a relatively high temperature), and the use of an optimized quantity of Taq polymerase and other reaction components in the PCR mix (e.g. the amount of input template, optimum concentration of magnesium ions/primer concentration, ramp rate, elongation time, etc.). If however, the fidelity of the PCR remains low despite PCR optimization (e.g. when the initial target DNA is <100 copies), then an increase in PCR fidelity may be facilitated using a specially adapted PCR protocol, where the non-thermostable DNA dependent DNA polymerase of bacteriophage T4 or T7 is manually added to the initial first few cycles of PCR thermocycling. After these first few cycles have been completed, and the target DNA concentration has

been increased, thermostable Taq polymerase may be added to the reaction mix to allow conventional PCR thermocycling to proceed. This approach may be particularly useful in PCR protocols requiring either the detection of rare point mutations, for the assessment of homo- or heterozygosity, or for amplification products that are to be cloned and sequenced. In such cases however, it may be necessary to work with an alternative thermostable polymerase other than Taq (Section 7.5), for example a thermostable polymerase with a 3′-5′ proofreading ability or even a cocktail of enzymes (Section 7.6). These different enzymes/cocktails will all exhibit different error rates and fidelities during PCR amplification.

As well as Taq polymerase mediated nucleotide misincorporation errors, template-dependent DNA sequence variation may also occur. These template-dependent mistakes occur when nucleotides in the original genomic DNA sequence are either lost or enzymatically, physically, or chemically, modified by external factors, e.g. by aggressive chemical reactants and/or UV irradiation (e.g. T dimers). Alternatively, extensive secondary structure of the template DNA may prevent access to regions of the target DNA by the polymerase enzyme. These effects may give rise to random nucleotide incorporation events, or even nucleotide deletions. Repetitive DNA is notorious in this respect, with "stutter bands" often appearing upon gel electrophoresis, caused by DNA polymerase mediated replication errors which leads to the addition or deletion of repeat unit sequences [Saunders et al., 2000; Pestoni et al., 1995].

7.3 Taq DNA Polymerase and Its Modifications

Initially, Taq polymerase enzyme was not generally available from commercial suppliers. However, with increasing demand, large-scale cultivation and purification technologies for the enzyme were developed. Moreover, advances in genetic engineering meant that commercial quantities of the Taq enzyme could be obtained by cloning and expressing the Taq polymerase gene in the more easily cultivable *Escherichia coli* bacterium, thereby reducing the cost of manufacture. Indeed, though "native" Taq enzyme is still commercially available, the largest market share of Taq (and other thermostable DNA polymerases), belongs to enzymes cloned and expressed in *E. coli*. Though many commercial companies offer different versions of Taq polymerase (as well as thermostable polymerases from other micro-organisms), all of these enzymes and variants require a patent license from Roche Applied Science – the current patent holder for the PCR amplification process. Some of the most popular Taq variants and non-Taq enzymes are mentioned below.

AmpliTaq is a 94 kD recombinant Taq polymerase variant expressed in *E. coli* which lacks 3′-5′ proofreading activity but does exhibit 5′-3′ exonuclease and DNA polymerase activity. The major advantage of AmpliTaq is that it can very easily be purified and that the preparations contain no traces of contaminating thermostable nucleases or hydrolases. This recombinant enzyme also shares the 3′ blunt-end nucleotide addition activity associated with native Taq polymerase. Unfortunately however, AmpliTaq preparations may be contaminated with traces of *E. coli* DNA, which could possibly interfere with the specific amplification of rare targets if the target sequence (or a close homologue thereof) is also present in *E. coli*. More extensively purified versions of the enzyme can also be supplied by the manufacturer (Perkin Elmer). In addition to the classical thermostable DNA polymerases such as Taq, heat activated enzymes have also become available. These enzymes require a pre-treatment step of approximately 5–10 minutes at 95°C before their polymerase activity is activated, a process which allows a "hot-start" step to be incorporated into PCR thermocycling protocols. This hot-start heat activation step (also known as "time release" PCR) helps prevent amplification of non-specific products during the warming-up period (the initial step whereby the PCR mix is heated from room temperature to 95°C), before the first cycle of PCR thermocycling *per se* (Fig. 7.4). This means that any mis-hybridization between primers and non-target DNA during the warming-up period will not lead to amplification of non-specific product (product which could act as template for subsequent rounds of PCR amplification if the Taq enzyme was already functional). The hot-start PCR protocol may be varied by reducing the recommended time for initial hot-start

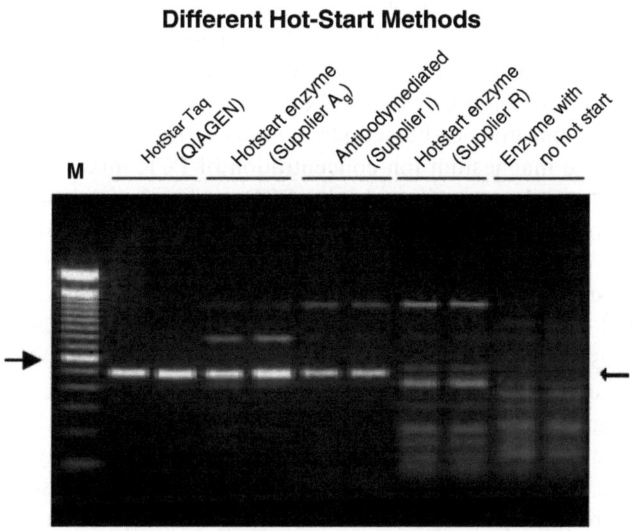

Fig. 7.4 Hot-start (time release) PCR and its effect on PCR quality. Lane 1 = molecular size marker; lanes 2 and 3 = the result of a time-release PCR of the HIV polymerase gene performed in duplicate; lane 4 = the same PCR without the time-release Taq formulation. Fifty copies of template were used in the presence of 1 μg of human DNA and the amplimers were analysed on a 2% agarose gel. The correct PCR product is 497 bp in length. The arrow on the left identifies a 600 basepair long marker in the lane marked M (From Critical Factors for Successful Real-Time PCR, www.qiagen.com)

activation, resulting in the enzyme being gradually activated during the first five or so cycles of PCR thermocycling. This may have advantages with regard to PCR specificity, as low initial concentrations of Taq can reduce the number of mis-hybridized primer/template combinations amplified [Zimmermann and Mannhalter, 1998]. However, it should be noted that even after incubation for 5–10 minutes at 95°C, not all of the hot-start enzyme will have been activated.

The mechanism of action of hot-start thermostable DNA polymerases tends to vary. Some polymerases have been genetically modified to fold into a non-active structural conformation, which after heat-treatment re-organizes into an enzymatically active conformation. Examples of this type of thermostable DNA polymerase include *Thermo-start* (ABGene) and *Proofstart* (Qiagen) enzymes. Another mechanism facilitating hot-start PCR is to add a polymerase inhibitor that binds to the active site of the polymerase. The *HS TaQuant-OFF* (Q-Biogene) is one such compound, which can be added to PCR mixes prior to PCR and starts to dissociate from the blocked active site of the DNA polymerase once the temperature of the PCR mix rises above 48°C. Other polymerases are inhibited at their active sites by specific monoclonal antibodies, e.g. *Taq Platinum* (Life Technologies); *TaqStartAntibody* (ClonTech); *Fast Start Taq* (Roche); *JumpStart* (Sigma); *RedHot* DNA Polymerase (ABIgene), where the initial heating step denatures the bound antibody exposing the active site of the polymerase. The *AmpliTaq Stoffel* fragment is a truncated 61 kD form of Taq polymerase (Perkin Elmer) which lacks 5′-3′ exonuclease activity. It is well suited for copying GC-rich regions, as it does not destroy templates that have stable secondary structure and is even more thermotolerant than Taq polymerase (with a half-life at 97.5°C of 20 minutes as compared to 5–6 minutes for Taq). The Taq *Stoffel* fragment can be used to amplify regions from circular molecules such as plasmids or ligated DNA mini-circles and is far less susceptible to variations in the magnesium ion concentration of PCR mixes. This latter property is an important advantage in multiplex PCR applications where several primer pairs are incorporated in a single PCR mix and where the magnesium concentration for one or more of these primer pairs may not be fully optimal. However, its somewhat limited processivity rate (5–10 nucleotides per second) makes the *Stoffel* fragment less well suited for the synthesis of long PCR products. Taq *Stoffel* fragment may offer advantages in real time PCR assays [Wilhelm et al., 2001].

Other companies have developed similar low molecular weight versions of Taq polymerase, e.g. *AdvanTaq* (Clontech) and *KlenTaq* (Sigma). Using molecular engineering, the 5′-3′ exonuclease activity of Taq polymerase may be neutralized by inserting the relevant nucleotide point mutation into the gene sequence for the enzyme. This prevents the enzyme from damaging the 5′-end termini of amplification products and renders the enzyme more suited to the synthesis of large amplimers than the Taq *Stoffel* fragment. Such enzymes are often used in PCR cycle sequencing, e.g. *TaqDNA polymerase Sequencing Grade* (Promega).

Enzymes such as the recombinant *T. thermus* rTth DNA polymerase XL (Applied Biosystems), or a combination of enzymes derived from *T. thermus* and *T. lithoralis* (*SuperTaq Plus* (Ambion)), have been reported to support the PCR mediated amplification of fragments up to 20,000 bp (20 kb) in length. These specially adapted enzymes allow the amplification of long PCR amplimers with reasonable fidelity and yield using specially developed "long and accurate" PCR protocols. To help in the amplification of long PCR products (10–30 kb), Applied BioSystems and Sigma have developed the *Extender PCR Additive* kit and the *Long PCR Core Kit* respectively, which should be used in combination with specific thermostable DNA polymerases (as recommended by the manufacturer of each kit). The exact nature of the Taq Extender additive present in these kits is a commercial secret.

Some manufacturers supply thermostable DNA polymerase or ready-made PCR mastermixes that contain an inert red dye which does not interfere with either polymerase activity, purification of the amplified PCR product, sequencing or further downstream processing applications. These dye-related products include *REDAccuTaq, REDTaq* (Sigma) and *Red Hot* Taq (Abgene), etc., allowing the actual addition and mixing of polymerase or mastermix to be visually controlled as well as facilitating visual tracking of the PCR amplification products upon gel electrophoresis. In effect, these dyes act as a molecular weight marker corresponding to a 125 base pair DNA fragment.

7.4 Taq Polymerase Unit Definition and Working Concentrations

The activity of native and recombinant Taq polymerases is expressed in units per microlitre, where one unit comprises the amount of enzyme that generates a fixed amount of acid precipitable DNA in an activated salmon sperm DNA assay. This assay normally comprises one femtomole salmon sperm DNA in a pool of 10 nanomoles of dNTPs and is performed in a 50 µl volume under standard reaction conditions of 74°C for 30 minutes (see individual Taq suppliers' technical data sheets). The concentration of polymerase to be used for each new set of PCR primers designed, and/or PCR application developed, should ideally be optimized, with values usually lying between 0.5 and 5 units per 100 µl PCR mix. If the concentration of the polymerase in the reaction mix is too low, then the yield of amplimers will be comparatively low, whereas excessive amounts of polymerase will facilitate the amplification of non-specific products.

7.5 Other Thermostable Polymerases and Their Applications

Homologous thermostable DNA polymerases have now been purified from several other *Thermophilus* species as well as *Thermophilus aquaticus* (Taq), including *T. thermophilus* (Tth), *T. flavus* (Tfl) and *T. brokianus* (Tbr). Moreover, homologous thermostable DNA polymerases have also been isolated from other bacterial genera; examples include *Thermogata maritima* (Tma), *Sulfolobus solfataricus*(Sso), *Bacillus stearothermophilus* (Bst), *Pyrococcus furiosus* (Pfu), *Pyrococcus woesei*

(Pwo), *Pyrococcus abysi*, *Pyrococcus* spp (Pfx), *Pyrolobus fumarus*, *Thermococcus lithoralis* (Tli) and *Thermococcus fumicolans* (Tfu). All of these enzymes share significant amino acid sequence homology with thermolabile *E. coli* DNA dependent polymerase I, and many are currently commercially available, marketed under such exotic names as Replinase, Replitherm, Thermalase, Dynazyme, Vent, DeepVent, Ultima, ThermalAce, etc. Within this mixture of homologous enzymes, DeepVent polymerase is the only polymerase not derived from a eubacterium or archaebacterium, as it was initially purified from a worm-like organism called *Carboxydothermus hydrogenoformans*, which lives in hot volcanic water. The enzymes Vent (Tli), DeepVent, DeepVent-exo, Pfu polymerase, ThermalAce and Thermalase are recombinant proteins with an extended thermostability, with DeepVent polymerase having a half life of 23 hours at 95°C and Pfu having a half life of 13 hours at 95°C. The improved thermostability exhibited by these enzymes allows for PCR protocols that incorporate: (i) extended denaturation periods, (ii) denaturation at higher temperatures and (iii) a higher number of thermocycles (up to 70 steps). Examples of applications requiring a DNA polymerase with an extended thermostability include PCR protocols for amplifying cDNA from reverse transcribed ribosomal RNA (where the enzyme DNase I is used to remove any contaminating DNA prior to reverse transcription, and which therefore incorporate a 94°C for 50 minute incubation period in order to fully denature the DNase I enzyme [Hilali et al., 1997]), as well as "long and accurate" PCR protocols for the amplification of DNA from 10–30 kb in length, which require very long elongation times (sometimes greater than 30–45 minutes per cycle!). A survey of some of the key characteristics of some of these PCR enzymes is shown in Table 7.2.

If PCR amplification protocols are to be used in mutation detection (e.g. in PCR sequencing protocols), site directed mutagenesis or cloning strategies, then it is important that the error rate (nucleotide misincorporation) of the enzyme used be as low as possible. This is also extremely relevant if the PCR product is to be used as a template in downstream applications involving *in vitro* protein synthesis [Rungpragayphan et al., 2003]. The fewer errors introduced during PCR amplification, the lower the chance of incorrect "stop" codons or incorrect amino acids being incorporated into the resultant transcribed and translated protein. The error rates for some frequently used thermostable DNA polymerases are listed in Table 7.1. However, some enzymes, such as Taq, Tfl and Tth polymerases, completely lack the 3′-5′ proofreading enzyme domain and are therefore not suited for the "high-fidelity" PCR applications mentioned above. In contrast, several commercially available polymerases do exhibit significant 3′-5′ proofreading activity, these include Vent and DeepVent (Tli; New England Biolabs), Pwo (Boehringer; Hybaid), Tli (Promega), Tfu Direct (Qbiogene), Pfu Turbo (Stratagene), Ultima (Perkin Elmer), *Pyrococcus* spp (Qiagen Proofstart kit; Invitrogen), Failsafe (Epicentre) and ThermalAce (Invitrogen), though the fidelity of these enzymes may vary dependent on variations in pH, magnesium ion concentration and the amount of dNTP present in the PCR mix [Cline et al., 1996]. In general, amplimers synthesized by polymerases containing 3′-5′ proofreading activity contain far fewer errors than those generated using non-proofreading polymerases, though the overall processivity rate of such

Table 7.2 General characteristics of commercially available thermostable DNA dependent DNA polymerases

Enzyme	Thermostability (min at 95°C)	5′-3′ Proofreading	3′-5′ Proofreading	Extension speed (nucleotides/second)	pH optimum	Mg^{++} optimum ($\times 10^{-3}$)	Mol weight
AmpliTaq	40	Yes	No	75	7–7.5	1–4	94
Stoffel	80	No	No	>50	–	–	61
Ampliterm	–	No	No	–	–	1.5	–
Pyra	>300	No	No	–	7–7.5	3	90
TTH 94	20	Yes	No	>33	7–7.5	1.5–2.5	
Tfl	40	Yes	No	16	7–7.5	1–4	94
Tfu	300	No	Yes	64	–	–	–
DeepVent	400	No	Yes	>80	–	–	–
Vent	1380	No	Yes	–	–	–	–
Tli	400	No	Yes	33	7–7.5	2–6	90
Proofstart	>240	No	Yes	–	–	Broad	–
PFU 92	>120	No	Yes	60	8–9	2–6	
Pfx	–	–	Yes	–	–	–	–
Pwo	–	No	Yes	–	–	–	–
UL*Tma*	>50	No	–	–	–	–	70
ThermalAce	>4	–	Yes	–	–	–	–

–: not known; opt: optimum

3′-5′ proofreading enzymes is somewhat lower than that of non proofreading polymerases. Further, many proofreading polymerases can modify or even degrade single stranded primer DNA molecules or primer/template mismatches, possibly leading to mis-hybridization and subsequent amplification of non-specific amplimers due to hybridization of partially degraded primers or amplimer products to unintended DNA sequences. Hot-start protocols may be utilized to help reduce this effect by using especially available chemically modified versions of the proofreading polymerases. As a final note, an important feature of proofreading polymerases is that they generate blunt-ended PCR amplification products without 3′-end overhangs. This means that amplimers generated using these enzymes are not suitable for "TA-cloning" applications without further amplimer processing, e.g. incubating the amplimers with one unit of Taq at 72°C for 8–10 minutes to add nucleotide overhangs. Alternatively, a mixture of Taq and proofreading polymerases may be used (e.g. the commercially available Expand or eLONGase mixtures), as long as the Taq polymerase/proofreading polymerase concentration is at least 10:1 to make sure that the overhang is correctly added.

Some thermostable DNA polymerases also exhibit a degree of reverse transcriptase (RT) activity (the ability to generate a copy DNA molecule from an RNA molecule). DNA polymerases from *T. thermus* (Tth), *T. flavus* (Tfl) and *C. hydrogenoformans* (Chy), have the most prolific RT activity of the currently commercially available enzymes, with the recombinant Tth DNA polymerase having the advantage of being especially resistant to the effect of RT inhibitors. However, though all of these enzymes are reasonably efficient at reverse transcribing RNA into cDNA at elevated incubation temperatures and in the presence of manganese ions (N.B. not magnesium), it should be remembered that the "native" RT enzymes MMLV (sometimes also written as M-MLV, M-MuLV or even Mo-MuLV) and AMV, are approximately one hundred times more efficient at reverse transcribing RNA into cDNA than these thermostable DNA polymerases. This difference in efficiency is related to the fact that the native enzymes MMLV and AMV are actually derived from the Moloney Murine Leukaemia virus and Avian Myeloblastosis virus retroviruses respectively, and reverse transcription activity is their main function in nature. Also, thermostable DNA polymerases that exhibit RT activity lack proofreading activity and are consequently less accurate during reverse transcription than MMLV and AMV, which exhibit error rates in the range of 1 in 30,000 bases and 1 in 17,000 bases respectively as compared to an error rate of approximately 1 in 5,000–10,000 for Tth and Tfl. Indeed, there exist differences in the reverse transcription error rates observed between the different thermostable DNA polymerases themselves. The major advantage of thermostable DNA polymerases in reverse transcription reactions is that both reverse transcription and cDNA PCR amplification may be performed in the same reaction tube and at relatively high temperatures of approximately 74°C, thereby helping to remove any secondary structure which may be present (particularly within RNA molecules), and increasing the specificity of primer/template hybridization in both the reverse transcription and PCR amplification reaction. In contrast, MMLV RT is typically used at 37°C whilst AMV is usually employed at 42°C (but can remain active up to 58°C).

Finally, it should be noted that, each of the different brands of commercially available thermostable DNA polymerases do exhibit different characteristics with regard to specific PCR inhibitors or PCR facilitators, e.g. betain, BSA, gp32, etc. [Al Soud and Radstrom, 1998; Al Soud et al., 2000; Al Soud and Radstrom, 2001].

7.6 Mixtures of Thermostable Polymerases

Mixtures of thermostable polymerases have been designed to help reduce the disadvantages that individual polymerases may exhibit in particular PCR protocols. Some examples of these blends are listed below, though it should be noted that the number of commercially available mixtures is still growing. Clontech and Sigma for example, sell a mixture of the Klentaq-1 enzyme in combination with a separate DNA polymerase that has 3′-5′ exonuclease proofreading activity. This preparation combines thermostable DNA polymerase processivity with proofreading activity. ThermalAce by Invitrogen is an enzyme mix with similar characteristics.

In order to efficiently generate PCR amplimers using RNA as template material, viral RT enzymes (MMLV or AMV) may be combined with various thermostable DNA polymerases or even with heat-labile DNA polymerases. Several companies simply sell "combination packages" comprising their own brand of RT and polymerase enzymes, e.g. MULV/Taq by Promega, AMV/Tlf by Promega, AMV/AccuTaq by Sigma.

The use of proofreading and exceptionally thermostable DNA polymerases are extremely important in the PCR amplification of DNA of between 5–40 kb ("long and accurate" PCR protocols), where the extension step of the PCR cycle may be as long as 30–45 minutes. However, proofreading Vent polymerase and rTth polymerase XL, specifically require magnesium acetate instead of magnesium chloride ions in the PCR mix, and the processivity of proofreading polymerases are generally low (a maximal dNTP incorporation rate of approximately 1 kb per minute), as they tend to "fall" off the target DNA relatively easily. This low processivity may be compensated for by the inclusion of a polymerase with a high processivity (e.g. non proofreading Taq polymerase) in the PCR mix, and commercial combinations of such enzymes are currently available, e.g. eLONGase (Gibco BRL); SuperTaq Plus (Ambion); ProofExpander (Hybaid); AccuTaq (Sigma) and Expand (Boehringer). These mixes are reported to be able to generate amplimers as long as 45 kb in length. Some single enzymes which reportedly combine all the necessary qualities required for long and accurate PCR include Pwo (Roche) and Isis DNA polymerase (Qbiogene), an enzyme from *Pyrococcus abyssi* GE5. Various enzyme mixes are also available to help generate long PCR amplimers from mRNA molecules. In these mixtures, thermostable DNA polymerases with a high processivity rate are combined with thermostable DNA polymerases that exhibit 3′-5′ proofreading exonuclease activity, some examples include AccuTaq LA DNA polymerase (Sigma), eLONGase (Gibco), Advantage 2 (Clontech), FailSafe and MasterAmp (Epicentre and ArrowTaq (Qbiogene).

Mixes that are specifically designed to amplify DNA fragments of 5 kb or less with a low frequency of errors may also be purchased. Examples of such mixes include Thermozyme (Invitrogen) and Masteramp Tth polymerase (Epicentre).

References

Al Soud W, Johnson LJ, Radstrom P. 2000. Identification and characterization of immunoglobulin G in blood as a major inhibitor of diagnostic PCR. J Clin Microbiol 38:345–350.

Al Soud W, Radstrom P. 1998. Capacity of nine thermostable DNA polymerases to mediate DNA amplification in the presence of PCR inhibiting samples. Appl Environ Microbiol 64:3748–3753.

Al Soud W, Radstrom P. 2001. Purification and characterization of PCR-inhibitory components in blood cells. J Clin Microbiol 39:485–493.

Burgess LC, Hall JO. 1999. UV light irradiation of plastic reaction tubes inhibits PCR. Biotechniques 27(2):252–256.

Chien A, Edgar DB, Trela JM. 1976. Deoxyribonucleic acid polymerase from the extreme thermophile *Thermus aquaticus*. J Bacteriol 127:1550–1557.

Cline J, Braman JC, Hogrefe HH. 1996. PCR fidelity of Pfu DNA polymerase and other thermostable DNA polymerases. Nucleic Acids Res 24:3456–3451.

Eckert KA, Kunkel TA. 1993. Fidelity of DNA synthesis catalyzed by human DNA polymerase alpha and HIV-1 reverse transcriptase: effect of reaction pH. Nucleic Acids Res 21:5212–5220.

Finke J, Fritzen R, Ternes P, Lange W, Dolken G. 1993. An improved strategy and a useful housekeeping gene for RNA analysis from formalin fixed paraffin embedded tissues by PCR. Biotechniques 14:448–453.

Flaman JM, Frebourg T, Moreau V, Charbonnier F, Martin C, Ishioka C, Friend SH, Iggo R. 1994. A rapid PCR fidelity assay. Nucleic Acids Res 22:3259–3260.

Hilali F, Saulnier P, Chachaty E, Andremont A. 1997. Decontamination of PCR reagents for detection of low concentrations of 16S rRNA genes. Mol Biotechnol 7:207–216.

Mullis K, Faloona F, Scharf S, Saiki R, Horn G, Erlich H. 1992. Specific enzymatic amplification of DNA in vitro: the polymerase chain reaction. Biotechnol 24:17–27.

Pestoni C. Lareu MV, Rodriguez MS, Muniz I, Barros F, Carracedo A. 1995. The use of the STRs HUMTH01, HUMVWA31/A, HUMF13A1, HUMFES/FPS, HUMLPL in forensic application: validation studies and population data for Galicia (Spain). Int J Legal Med 107:283–290.

Rungpragayphan S, Nakano H, Yamane T. 2003. PCR-linked in vitro expression: a novel system for high-throughput construction and screening of protein libraries. FEBS Lett 540:147–150.

Saiki RK, Bugawan TL, Horn GT, Mullis KB, Erlich HA. 1986. Analysis of enzymatically amplified beta-globin and HLA-DQ alpha DNA with allele specific oligonucleotide probes. Nature 324:163–166.

Saiki RK, Gelfand DH, Stoffel S, Scharf SJ, Higuchi R, Horn GT, Mullis KB, Erlich HA. 1988. Primer directed enzymatic amplification of DNA with a thermostable DNA polymerase. Science 239:487–494.

Saiki RK, Scharf S, Faloona F, Mullis KB, Horn GT, Erlich HA, Arnheim N. 1985. Enzymatic amplification of beta-globin genomic sequences and restriction site analysis for diagnosis of sickle cell anemia. Science 230:1350–1354.

Saunders NJ, Jeffries AC, Peden JF, Hood DW, Tettelin H, Pappuoli R, Moxon ER. 2000. Repeat-associated phase variable genes in the complete genome sequence of *Neisseria meningitidis* strain MC58. Mol Microbiol 37:207–215.

Wilhelm J, Pingoud A, Hahn M. 2001. Comparison between Taq DNA polymerase and its Stoffel fragment for quantitative real time PCR with hybridization probes. Biotechniques 30:1052–1056.

Zimmermann K, Mannhalter JW. 1998. Comparable sensitivity and specificity of nested PCR and single-stage PCR using a thermally activated DNA polymerase. Biotechniques 24:222–224.

Chapter 8

Important Considerations for Typical, Quantitative and Real-Time PCR Protocols

8.1 The Typical PCR Amplification Protocol

The typical PCR amplification protocol comprises four main phases involving: (1) an initial nucleic acid template melting/thermostable DNA polymerase activating (hot-start) phase of 90–95°C for 2–10 minutes, (2) a user defined amplification phase involving pre-programmed thermocycling, (3) an extension step of 70–75°C for 5–10 minutes (in order to ensure the complete amplification of all the many millions of target DNA copies present in the PCR mix at the end of PCR thermocycling), and finally (4) a cooling phase to either ambient room temperature or 4°C for temporary storage of reaction mixes within the PCR thermocycler prior to downstream processing, e.g. gel electrophoresis (Fig. 3.3).

The most important phase of the PCR protocol is the amplification phase, comprising a pre-determined number of amplification cycles (or thermocycles). In the vast majority of cases, a typical thermocycle comprises three distinct but sequential steps. In sequential order, these steps include: (1) a template nucleic acid/amplified PCR DNA melting step, required to dissociate double stranded DNA ready for primer hybridisation (usually 90–95°C for 30–60 seconds), (2) a primer annealing step performed at a temperature that has been calculated to be optimal for the PCR primer pair being used (usually 40–70°C for 30–60 seconds), and (3) a final DNA chain extension step at an optimum temperature for the enzyme of choice to promote the full extension and amplification of the DNA sequence (usually around 70–75°C for 60–120 seconds). It should be noted that both annealing and elongation steps of each thermocycle are performed at "stringent" temperatures, facilitating a reduction in the likelihood of non-specific hybridization between primers and regions of DNA which contain similar but not identical DNA sequences. Exceptions to this rule do exist for certain PCR protocols, e.g. PCR using "degenerate" primers. Finally, some PCR thermocycle protocols utilise only two (instead of 3) steps, by combining the annealing and DNA chain extension steps into a single step using a single annealing and chain extension temperature.

The yield of target DNA amplification product generated during a typical PCR follows a sigmoidal S-shaped curve (Fig. 8.1), with the amplimers remaining essentially undetectable during the first few cycles of PCR, and becoming detectable

E. Van Pelt-Verkuil et al., *Principles and Technical Aspects of PCR Amplification,*

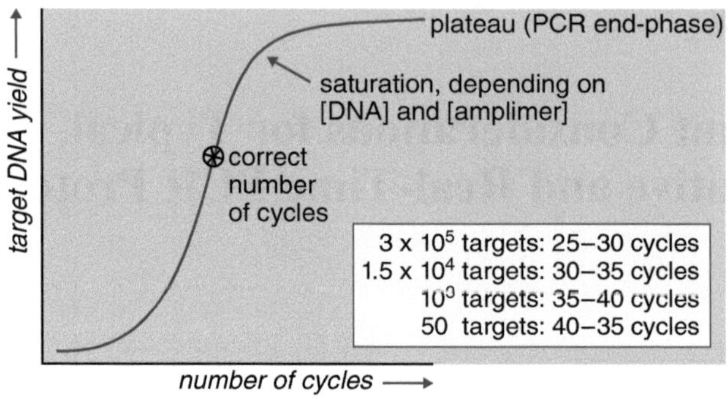

Fig. 8.1 The amount of PCR product generated during a typical PCR protocol accumulates as an S-shaped curve. The "amplification plateau" is reached once one of the PCR components becomes limiting. Ideally, the number of PCR amplification cycles should be limited so as to stop PCR amplification in the early plateau phase, as this will help prevent the accumulation of incomplete and non-specific PCR products

(using traditional agarose gel electrophoresis and ethidium bromide staining), only after approximately 10–20 cycles of thermocycling. With reference to Fig. 8.1, it can be seen that the greatest rate of amplification occurs during the exponential phase of the curve. Mathematically, the rate of amplification during the exponential phase may be calculated using the equation:

Rate of DNA amplification = 1.85^n (where n = number of cycles) [Saiki et al., 1988].

This equation also shows that a very large proportion of all of the amplimers present in the final PCR reaction mix are actually synthesised in the final few cycles of the PCR protocol (i.e. when "n" is large and many millions of amplimer copies are being generated). Also, even in a fully optimised PCR, amplification inefficiencies during PCR thermocycling act to reduce the theoretical exponential amplification factor per cycle, reducing the amplification efficiency from 2 to 1.85. Moreover, once one of the PCR mix reaction components becomes limiting, or when the number of amplified DNA fragments are so large that they interfere with exponential amplification, then a "plateau phase" (the flattened part of the S-shaped amplification curve) is reached, meaning that amplification no longer proceeds in an exponential manner. The cycle at which the plateau phase is reached depends to a large extent on the number of template molecules present at the start of PCR thermocycling. Moreover, the presence of large quantities of amplified DNA, coupled to an imbalance in one or more of the essential ingredients of the PCR, greatly increases the chance of amplifying non-specific (often incomplete) amplimers during the plateau phase of the PCR protocol. This explains why PCR protocols, which include an excessive number of thermocycles frequently generate non-specific products (often characterised by the appearance of a "smear of DNA" upon gel electrophoresis).

8.1.1 Denaturation (Melting) of the Template DNA

Denaturation or melting of template nucleic acid basically means the separation of the double stranded DNA into its single stranded components, thereby allowing efficient hybridisation of the shorter single stranded PCR primers to their target sequences (if present). The first phase of a PCR protocol, as well as the first step of every PCR thermocycle, involves temperature dependent DNA template denaturation. Ideally, the vast majority of double stranded DNA molecules should be melted into single strands prior to initiation of PCR thermocycling *per se*, with 10 minutes of heating at 95°C during the first phase of a PCR protocol usually being sufficient to achieve complete denaturation of template DNA (this process is also useful in activating many hot-start enzymes (Section 7.3). This 95°C melting temperature is then used for every subsequent PCR thermocycle, though some researchers prefer a denaturation temperature of 98°C during the first five or so thermocycles of the PCR followed by 95°C for subsequent cycles. However, the average melting temperature of DNA is in the order of 82°C, which means that heating to 95°C (and not 98°C) should be sufficient to achieve complete denaturation within 10 minutes, especially if thin-walled reaction tubes and small PCR volumes (up to 25 µl) are used. Circular DNA template molecules may require extra heating over and above the usual 10 minutes at 95°C in order to achieve full denaturation. This arises due to the fact that circular DNA renatures more easily than linear DNA, and because short fragments of bound DNA (including hybridised PCR primers) may be displaced by re-hybridisation of the complementary circular strands. If necessary, this effect may ultimately be circumvented by enzymatic linearization of the double stranded circular molecule using restriction enzymes prior to PCR amplification. Other problems possibly affecting the rate and efficiency of DNA denaturation include: (1) the use of nucleic acid template material from formalin fixed tissue samples, (2) the presence of "sticky" biological materials such as mucus, and (3) GC-rich template DNA. Complete protein hydrolysis prior to PCR amplification may help to alleviate problems associated with contaminating biological materials, and the addition of DNA denaturing agents which do not greatly interfere with PCR amplification (e.g. formamide, DMSO, and several common detergents such as Triton X-100) to the PCR mix may help in facilitating DNA denaturation.

Complete denaturation of the template DNA is especially important in quantitative PCR protocols such as in "LightCycler" (Roche) or "TaqMan" (Applied Biosystems) assays, due to the fact that inefficient amplification of target molecules during the first few cycles of the quantitative PCR protocol, will result in a final underestimation of the initial number of target molecules present in the sample. Indeed, a decrease in DNA amplification efficiency by only a factor of two or three during the first few cycles of a quantitative PCR protocol may seriously compromise the cycle threshold (Ct) value, i.e. the number of PCR thermocycles required for a positive fluorescent amplification signal to be achieved which exceeds the background "noise" for that particular PCR assay [Wilhelm et al., 2000] (Fig. 8.2).

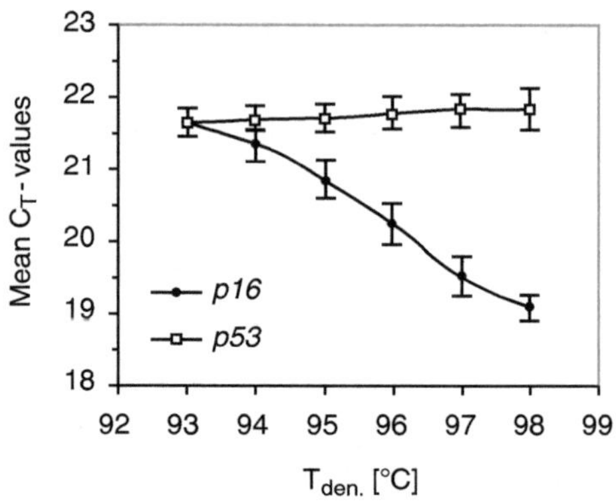

Fig. 8.2 The effect of denaturation temperature on the quantity of PCR products amplified using a LightCycler PCR machine. Two targets (the tumour repressor genes p16 and p53) were amplified using the appropriate primer pairs and denaturation temperatures of 93°C–98°C during the first five thermocycles. All subsequent PCR cycles were performed at a denaturation temperature of 95°C. Increasing the pre-PCR melting temperature ($T_{den.}$ [°C]) during the first five PCR cycles up to 98°C results in a significant increase in the amount of PCR product generated (i.e. a decrease in Ct value) for the p16 PCR but not the p53 PCR. This effect is attributable to differences in the melting temperature of genomic sequences in the immediate vicinity of the p16 and p53 genes. All experiments were performed in tenfold and bars identify the standard deviation (Reproduced from Wilhelm et al., 2000. With permission from American Association for Clinical Chemistry)

8.1.2 Annealing (Hybridisation) of PCR Primers

Under optimised conditions, the annealing (hybridisation) step of each PCR cycle should ensure that the PCR primers bind specifically to their complementary target site (if present). During this second step of PCR thermocycling, the temperature of the reaction mix is gradually lowered from the denaturing temperature of approximately 95°C to the calculated annealing temperature (T_m) required for specific primer/target hybridisation. Usually, the PCR annealing temperature used approximates to the mathematically calculated primer annealing temperature minus 5°C (for a more detailed explanation of T_m calculation, see Section 5.2), which means that the primer should specifically bind to its complementary target DNA and will not readily disassociate once hybridised. Moreover, primer dimer formation (which occurs when the PCR primer pair (partially) hybridise to each other) is suppressed so that the primers only hybridise to 100% homologous target sequences. In a typical PCR protocol, the annealing step of the PCR cycle is programmed to continue for 15–60 seconds, which (due to the vast excess of primers present and their inherently designed sequence specificity), is usually more than enough time to allow hybridisation of the majority of primers to their complementary target sequence. As a single primer usually hybridises to its target sequence within a few seconds, under optimal conditions and with excellent heat conduction and low sample volumes, it is actually possible to reduce the annealing incubation time to only a few seconds, though it should be noted that this short annealing time coupled to the changing balance between primer and amplification product concentrations, may seriously affect the quantity of PCR amplification products obtained during the later cycles of PCR thermocycling. Short and GC-rich primers exhibit the best hybridisation kinetics, though it is advisable to use longer primers in typical PCR protocols, as the melting temperature of a 10- to 15-mer primer may be up to 15°C lower than that of a 20-mer primer targeting the same template region. Sometimes, it may be possible to design and adjust the GC-content and primer length so that the primer annealing step and extension step of the PCR cycle can be performed at the same temperature, e.g. 70–75°C (two-step PCR).

The choice of annealing temperature is especially critical during the first few cycles of PCR amplification, as any non-specific annealing during this period will result in the amplification and accumulation of large quantities of non-specific products at the end of thermocycling. Especially important in this respect is the first phase of the PCR protocol, whereby template DNA is heated from room temperature to the melting temperature of DNA prior to PCR thermocycling *per se*. During this initial heating step, there occurs a period of time when the temperature of the PCR mix is below that of the temperature required for specific primer annealing and where there is an increased risk of non-specific primer hybridisation to non-homologous regions of DNA. This problem may be especially prevalent with

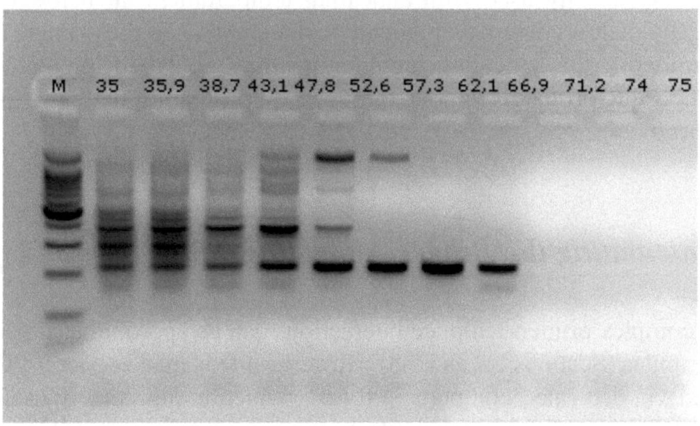

Fig. 8.3 Optimising the annealing temperature of PCR thermocycling greatly enhances the specificity of primer hybridisation and PCR products. In this "Factor V Leiden" gene PCR (the Factor V Leiden gene is involved in a common blood coagulation disorder), the effect of increased annealing temperature (T_m) on PCR specificity can be clearly seen. Between 35°C and 57°C a mixture of specific and non-specific products are amplified, whilst between 57.3°C and 62.1°C a single specific PCR product only is amplified. At annealing temperatures of 66.1°C or greater, PCR amplification ceases due to the inability of the Factor V Leiden primers to hybridise to the template DNA (i.e. the kinetic energy of the mix is too great to allow primer hybridisation)

AT-rich DNA sequences, as partial melting of these regions can occur at temperatures as low as 40°C. In order to greatly reduce the chances of such mis-annealing occurring during the initial heating step, "hot-start" procedures or "touchdown" PCR protocols may be used [Gupta et al., 2000; Ault et al., 1994]. The importance of the correct annealing temperature can be seen from Fig. 8.3, where PCR has been performed using a range of annealing temperatures.

In PCR protocols using "degenerate" primers, it is recommended that the annealing temperature be adjusted to lie between 10°C and 20°C below the lowest mathematically calculated melting temperature for the set of degenerate primers to be used. As a second recommendation, the PCR thermocycler ramp rate after the annealing step (i.e. the increase in temperature per unit time between the annealing step and the extension step), should be slower during the first few PCR cycles in order to prevent premature melting of degenerate primer/target hybridised DNA, which could lead to a possible reduction in PCR yield. After these first few cycles have been completed, the annealing – extension ramp rate may be increased, since enough specific target sequences should have been amplified to act as specific targets for further PCR amplification. If short primers are to be used (e.g. 12–15 nucleotides in length), it is recommended that the annealing temperature (as well as the annealing – extension ramp rate) should also be lowered during the first few cycles of the PCR protocol. Most modern PCR thermocyclers allow such changes to be programmed into the PCR cycle. Primers that are specifically intended to hybridise at multiple genomic locations, e.g. "DOP" primers, primers aimed at degenerate repeat regions, or primers designed to bind to variable DNA target regions, may be designed to be very short in length (e.g. ten nucleotides or even shorter) and are usually used with annealing temperatures far below (sometimes more than 20°C below) the mathematically calculated T_m of the primers to be used. In these protocols, the annealing temperature may be set to remain low for all cycles of the PCR protocol without the need of programming a two-phase PCR annealing protocol.

8.1.3 Calculating the Primer Annealing Temperature (T_m)

The T_m for complex primer–template interactions can be approximated using several different formulas (Sections 5.2 and 5.3). However, life is made easier if PCR primers are designed via dedicated computer software, which are extremely useful tools in primer design for (i) recognising possible primer complementarity, (ii) controlling for secondary structure features (e.g. hairpin loops), (iii) de-selecting PCR primer pairs which possess calculated T_m values that differ by more than 1–2°C, etc. Moreover, these criteria may be quickly and easily identified for both individual primers *per se* as well as for mixtures of different primer sets (e.g. as used in multiplex PCR applications). Such primer design software programs are commercially available to buy or evaluate for a certain period of time (e.g. Vector NTI, Novel software, and Lasergene, DNAstar software, etc), or may be freely available for use on the Internet, e.g. Primer3 (*http://frodo.wi.mit.edu/cgi-bin/primer3/primer3_www.cgi*). The use of PCR primer design software is recommended for PCR primer design.

If desired, once a primer pair has been designed and synthesised, the mathematically calculated T_m for that particular primer pair (or set of primer pairs) may be experimentally verified using an optimised PCR (optimised for primer concentration and magnesium concentration) at a variety of annealing temperatures, using for example a 50–64°C range of annealing temperatures with ±1–2°C difference between the range. The synthesis (or lack thereof) of the intended amplification target may then be determined using gel electrophoresis. Using experimental observations, it is not uncommon to find that the observed experimental T_m differs somewhat from the theoretically calculated T_m, with the experimentally determined optimal primer annealing temperature usually being slightly higher than the theoretically calculated optimal primer annealing temperature [Albretsen et al., 1988; Tolstrup et al., 2003]. Finally, there is some evidence to suggest that modifying the annealing temperature during PCR may actually increase the yield of PCR product obtained for products less than 1 kb in length, though the same thermocycling parameters may also lead to a reduction in yield for PCR fragments greater than 1kb in length [Rychlik et al., 1990].

8.1.4 DNA Chain Extension/Elongation

The elongation step of the PCR cycle is necessary in order to extend (amplify) the 5′-end hybridised PCR primer towards the 3′-end of the target DNA sequence. As with the melting and annealing steps of a PCR cycle, the elongation step is composed of a temperature and a time element. The optimal elongation temperature to be used is defined by the actual type of thermostable DNA polymerase used in the PCR protocol and the stability of the primer/template hybrids. For most enzymes (including Taq polymerase), this temperature lies around 72°C. The stability of primer/template hybridisation should be high if higher elongation temperatures are to be used, as increased elongation temperatures may facilitate partial or complete dissociation of any primer/template hybrids before complete extension of the target sequence is achieved. In general, the elongation temperature should be at least 5°C below the melting temperature of the entire amplimer, though it is possible to use higher elongation temperatures alongside primers that are specifically designed to contain a high GC-content, as guanosine – cytosine hydrogen bonding interactions are more stable than adenosine-thymidine hydrogen bonding interactions.

The second element to consider in the elongation step is the time of incubation, which should be long enough for the DNA polymerase to finish amplifying a complete copy of the target sequence, but short enough to prevent the complete amplification of any non-specific products that are longer than the target sequence. For targets up to 3 kbp, a 1-minute elongation time should be sufficient. For multiplex PCR protocols, the elongation time should correspond to the longest target to be amplified. If the amplification products are extremely long, e.g. between approximately 8–50 kbp (as amplified in "long and accurate" PCR protocols), then the elongation time needs to be suitably increased, especially if slower, proofreading thermostable DNA polymerases are included in the protocol. Indeed, dependant on the length of the product to be amplified, 2 minutes (3 kb fragment), 8 minutes (10 kb fragment), or even 20–45 minutes (30 kb fragment) of elongation time may be required for successful amplification. Moreover, during the later cycles of "long and accurate" PCR protocols (i.e. beyond cycle 15), the elongation time may need to be increased during every successive cycle, e.g. 5–20 seconds extra added per successive cycle, in order to guarantee that all targets being amplified are copied to completion. This gradual lengthening of the elongation time can be conveniently programmed into most modern PCR machines.

8.1.5 PCR Cycle Number

A typical PCR program usually employs 30–50 cycles of amplification, with the exact number of cycles being dependent on the initial quantity of target DNA added to the PCR mix. It is generally advisable to stop PCR amplification before the end of the exponential amplification phase is reached (the linear part of the "S-shaped" amplification curve, Fig. 8.1) though it should be remembered that sufficient PCR cycles should be programmed to allow the detection of low copy numbers of target DNA by the end of PCR thermocycling. If an initial copy number of 50 target molecules or less is present, then approximately 40 cycles of PCR amplification will usually be required in order to reach the plateau phase of amplification. If however, the initial template concentration is relatively high, then the end of the exponential amplification phase may be reached after only 25 cycles. In between, say for example when 3×10^5 DNA sequence targets are present at the start of PCR amplification, then somewhere in the order of 35 cycles are necessary to bring the PCR to completion. It may sometimes be necessary to increase or decrease the number of PCR amplification cycles utilised for a particular PCR protocol, for example when the amount of target cDNA present varies considerably. This may occur for example in reverse transcription-PCR (RT-PCR) protocols where the quantities of RNA to be reverse transcribed may vary for different sample types, for different treatments, or during different stages of the cellular growth cycle.

The "accessibility" of the target DNA is also an important issue, as tissues that have been fixed and embedded in paraffin, usually require a minimum of 40 cycles of PCR amplification in order to obtain visible product on gel electrophoresis. When developing new PCR protocols, the optimum number of PCR cycles required for target detection in a particular specimen or range of specimens should be first determined experimentally, preferably using the current "gold-standard" testing methodology as a control against which PCR sensitivity and specificity may be compared.

8.1.6 *The "Plateau Phase" and Final Stages of PCR Thermocycling*

PCR cannot proceed indefinitely in an exponential fashion as, after approximately 30–50 cycles of amplification, a "plateau phase" of amplification is reached where additional numbers of thermocycles do not significantly contribute to a further increase in the quantity of amplimer generated. Several different factors play an important role in facilitating the change from exponential amplification to the plateau phase of amplification. These factors include the fact that: (1) DNA synthesis becomes less efficient as the activity of the thermostable DNA polymerase is reduced (as a result of the polymerase gradually becoming thermally inactivated, or as a result of the number of amplified target molecules available exceeding the number of DNA polymerase molecules available), (2) the amount of dNTPs available for DNA chain amplification may become limiting in later cycles, (3) complete amplimers may start to compete with the PCR primers for target annealing sites, (4) dNTPs and primers may undergo gradual physical destruction, (5) accumulation of amplimers and pyrophosphate liberated during dNTP hydrolysis has an inhibitory effect on DNA synthesis, and (6) "shielding" of the thermostable DNA polymerase substrate binding pocket by non-specific amplification products may occur so that any DNA polymerase interaction with primer/template hybrids is severely reduced [Kainz, 2000]. For most PCR protocols, it is possible to amplify enough specific target molecules to be visible by gel electrophoresis and ethidium bromide staining without reaching the plateau phase, simply by optimising the primer/magnesium/dNTP concentration and the number of PCR cycles used, or optimising a combination thereof. The number of cycles required to reach the plateau phase of PCR amplification may be experimentally determined, by simply performing PCR for 50 cycles whilst removing aliquots of PCR product after every 2–5 cycles. The yield of target DNA in each aliquot may then be determined using for example gel electrophoresis with/without densitometry measurements.

After completion of PCR thermocycling, and whether the plateau phase is reached or not, most typical PCR protocols include a final extension phase of approximately 5–10 minutes at approximately 72°C. This helps to ensure that the majority of the millions of target DNA molecules generated will be completely amplified at the end of PCR thermocycling. At this point, some PCR protocols recommend the addition of 10 mM EDTA to the final amplification reaction mix in order to chelate the magnesium ions and block any residual replicase or exonuclease activity of the polymerase. However, this is seldom necessary and could actually interfere with downstream PCR applications (e.g. PCR product cloning). After PCR amplification, it is usual to examine the PCR products by subjecting a sample of the PCR amplification mix to agarose gel electrophoresis (unless "real-time" PCR protocols, e.g. LightCycler or Taqman assays are being employed), in order to obtain a qualitative and quantitative impression of the success or failure of the PCR. In some cases, the amplification products may then be purified for downstream processing, e.g. purification using commercially available silica spin columns for eventual cloning or sequencing reactions. Sometimes the 3′-end overhangs of amplification products (introduced by non-proofreading DNA polymerases, e.g. Taq) require "filling-in", using for example T4 DNA polymerase in the presence of dTTP, to yield blunt ended PCR products.

8.1.7 PCR Sensitivity

Estimating the sensitivity of PCR amplification is not a straightforward procedure. Theoretically speaking, PCR should be able to amplify a single target molecule via exponential amplification into many millions, if not hundreds of millions, of target copies. However, optimization of both the PCR mix components as well as the thermocycling reaction conditions are essential in facilitating efficient exponential amplification. Moreover, sufficient copies of the target sequence should ideally be present at the start of PCR amplification, as low target copy numbers at the start of PCR amplification greatly increases the probability of amplifying non-specific products (NB. In this case, "Booster" PCR, "Touchdown" PCR, "Nested" PCR or other more "exotic" methodologies [Grace et al., 1998], may help to reduce non-specific amplification).

As a final note, PCR *per se* is an exquisitely sensitive means of amplifying target DNA molecules, and this exquisite sensitivity coupled to the very large quantities of amplimers that are generated, means that contamination of work spaces by amplimers from previous PCRs is a very real problem. Therefore, if PCR is to be used in the diagnostic/clinical setting, then thorough and specific anti-contamination protocols and procedures should be implemented, in order to prevent such "carry-over" contamination generating false positive results.

8.2 Quantitative PCR Protocols

Quantitative PCR is a methodology used for assessing the initial number of DNA, RNA or mRNA target molecules in a nucleic acid extract. The methodology is useful for investigating for example the effect of mutations on gene expression, the effects of various drug treatments on cellular death, or the effect of antiviral treatment on virus copy number [Clementi et al., 1996]. The method can be used to define "absolute" or "relative" levels of gene expression on the basis of reverse transcribed cDNA (using RT-PCR) quantification. However, it is important to choose conditions that allow for the measurement of a broad analytical range of RNA expression, in order to generate reliable data and to be able to detect both over-expression and repression of gene expression. However, minor differences in amplification efficiencies between different PCR reaction vessels may seriously influence the final concentration of PCR product generated, and hence the estimated level of gene expression/viral copy number, etc. Indeed, the presence of PCR inhibitors, clinical samples that contain a broad range of target molecules, or slight differences in PCR machine performance between different PCR runs, may all result in changes in PCR amplification efficiency and lead to inaccurate quantification results. Further, the use of external standard reference curves (where the initial concentration of target material is estimated by comparing the final concentration of PCR products to a standard curve of initial target concentration versus final PCR concentration), or limiting dilutions of samples (where serial dilutions of the sample are PCR amplified and the first dilution yielding a positive signal provides an estimate of the original target concentration), are not sufficiently accurate for most clinical purposes [Mehmet et al., 2001]. Because of these inherent problems, several different control methods have been developed to monitor PCR amplification efficiencies in RT-PCR-based quantification protocols.

8.2.1 Quantitative PCR Controls

Quantitative PCR control methodologies may be divided into those using “external” calibration or those using “internal” calibration. External calibration makes use of standard curves comparing PCR cycle number of the unknown sample to a graph relating PCR thermocycle number (y-axis) to the logarithm of the initial number of target DNA/RNA molecules present in the PCR mix prior to thermocycling (y-axis) [Higuchi et al., 1992; Pfaffl and Hageleit, 2001]. Internal calibration methodologies may make use of “exogenous” (artificially added DNA or RNA) or “endogenous” (housekeeping genes) controls. In essence, internal calibration methodologies are more complex to set up than external calibration methodologies, with external methodologies not taking into account variations in amplification efficiency that may occur between the individual cycles of an amplification reaction and between individual samples. Therefore, external calibration methodologies are generally regarded as being less accurate than quantitative PCRs that utilise internal controls [Joyce, 2002; Haberhausen et al., 1998; Piatak et al., 1993].

8.2.1.1 Internal Exogenous Standards (Competitor/Mimic)

This approach utilises a single primer pair and an exogenously added nucleic acid fragment which is of approximately equivalent size to the target sequence to be quantified (i.e. restricted to between 10–20% size difference) in order to keep the amplification efficiencies between competing nucleic acid and target nucleic acid within a reasonable range [Monteiro et al., 1997]. The length of the target and competing amplimers should ideally also be designed to be between 100 and 300 basepairs in length. To avoid discrepancies and inaccurate quantification, the final quantity of PCR product obtained for both competing and target DNA must lie within the exponential part of the amplification curve for the dilution series used. In the non-linear phases of the PCR (e.g. the plateau phase), the accumulation of products may differ between target and reference leading to over- or underestimation of the initial amount of target. Further, the linearity of exponential amplification is usually lost when the ratio between target and competing DNA exceeds 1:10 or is less than 10:1. Outside of this range, competition between the target and competing DNA will result in significant inhibition of amplification of the DNA present in the lower concentration. This inhibition is least obvious when the ratio between target and competing DNA templates is 1:1. Whether equivalent amplification efficiencies exist for competing and target sequence may be experimentally investigated by performing two "singleplex" PCRs in parallel, with one group of identical PCR mixes containing primers specific for the competing sequence and another group containing primers for the target sequence. During every fifth cycle of exponential amplification (during PCR cycles 15–40), one PCR vessel is removed from each group and set aside. At the end of thermocycling, all products are electrophoresed on an agarose gels and the products quantified by densitometry. Both competing and target PCR products should accumulate at comparable rates if equivalent amplification efficiencies exist in the reaction mixes.

In such assays, it is essential that a dilution series of the competitor/mimic is always co-amplified. The competing nucleic acid fragment may consist of a shortened version of the intended target sequence (known as a competitor, Fig. 8.4), or may have little sequence identity with the intended target sequence apart from the termini, which are identical to the primer sequences used to amplify the target DNA (known as a mimic). Unfortunately, methodologies using competitors/mimics of equal size may ensure equivalent amplification efficiencies between competitor/mimic and target, but also usually necessitate a post-PCR processing step in order to discriminate between the two different amplimers. There are several methods available for generating quantitative competitors, including the addition of sequence insertions or deletions into plasmids containing the target sequence (e.g. QuickChange Site-Directed Mutagenesis Kit, Stratagene) or ligation of extra sequences into restriction digested plasmids containing the target sequence followed by PCR amplification of the mutated sequence for use as a competitor. After amplification, hybridisation tests, (denaturing) HPLC analysis, denaturing gel electrophoresis or temperature gradient gel electrophoresis may be used to quantify the target. Alternatively, if a restriction site is engineered into the competitor, then simple restriction digestion analysis may be sufficient to distinguish between control and competitor amplimers. Care should be taken when designing competitor sequences homologous to the target sequence, as co-hybridisation may result in heteroduplex formation, which could interfere with subsequent amplification. Heteroduplex formation is generally not a problem when heterologous competitors are used, though the competitor DNA used in such assays must usually be constructed *in house* via recombinant DNA technology. Further, every time a new protocol is developed, a new competitor control has to be designed and tested. Heterologous competitors may be prepared in bulk amounts *in house* using straightforward PCR protocols, or alternatively commercial kits may be purchased for generating competitors, e.g. Competitive DNA Construction kit (Takara). However, the use of RNA competitors are preferred over cDNA competitors in quantitative

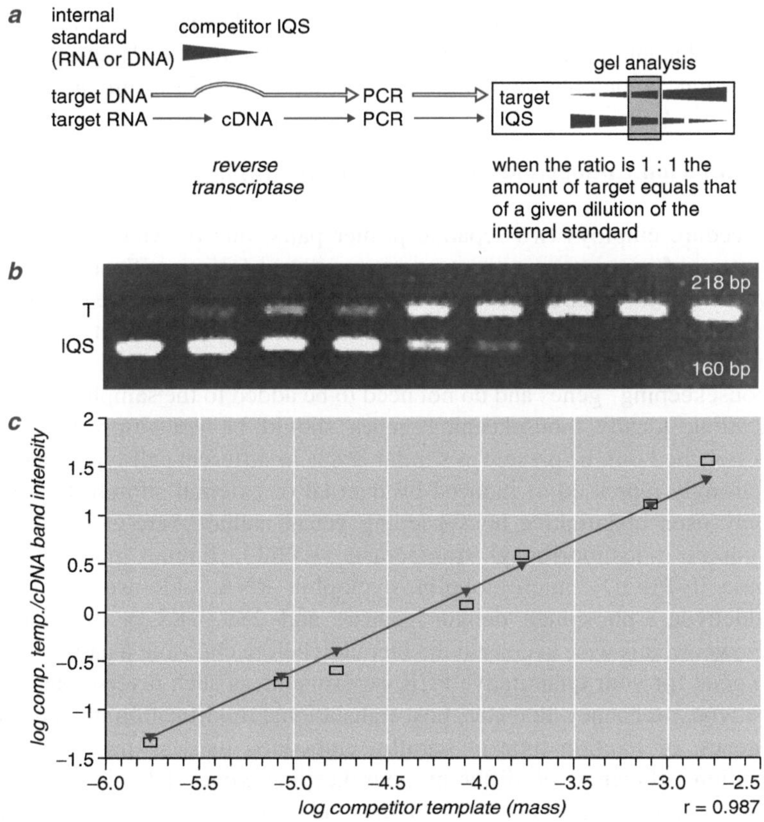

Fig. 8.4 Panel a: Scheme of a competitive PCR test using an internal standard. Panel b: Gel electrophoresis of products from a competitive quantitative PCR. IQS = amplification products from competitive standard. T = amplification products from target. In this example, sequential dilutions of the standard have been added to nine aliquots containing the non-quantified target and then PCR amplified. The target and standard "compete" for reagents during this PCR amplification. The dilution that generates PCR product of equivalent concentration to the non-quantified PCR product, is the dilution at which the starting concentrations of both standard and non-quantified cDNA were approximately equal. Panel c: This illustration relates the amount of competitor template to the cDNA band intensity

RT-PCR assays, as RNA controls yield more accurate results than DNA controls (only a small percentage of full-length mRNA transcripts are converted to cDNA during reverse transcription, hence the use of DNA controls may underestimate the initial quantity of mRNA by up to 100-fold). RNA competitors may be produced by transcribing RNA from plasmids or PCR products containing the competitor DNA sequence with a 5′-end T7 or T4 RNA polymerase transcription start sequence and a transcription stop sequence. Commercial kits may be purchased for generating such RNA competitors, including the RT-PCR Competitor Construction Kit,

(Ambion). Many companies now also supply "ready-made" competitors, e.g. Competitive Quantitative RT-PCR Kits (Ambion), for frequently quantified gene transcripts, e.g. human interleukin genes, TNF-alpha and IFN-gamma.

8.2.1.2 Internal Endogenous Standard (Housekeeping Gene)

This procedure employs two separate primer pairs, one of which targets a non-homologous reference gene as a comparison control for quantification and yields "relative" rather than "absolute" quantitative values (i.e. an increase or decrease in expression relative to a stably expressed gene, rather than a distinct value of gene copy numbers). Internal endogenous controls or reference genes are often referred to as "housekeeping" genes and do not need to be added to the sample prior to PCR amplification. Ideally, housekeeping genes should be constitutively expressed (always switched on), be present at similar levels in different cell types, and should not be strongly repressed or induced by internal or external stimuli. Examples of commonly used quantitative housekeeping genes include beta-globin, hypoxanthine-guanine phosphoribosyl transferase (HPRT), human porphobilinogen deaminase (h-PBGD), microglobulin, cyclophin RNA, aldolase A, beta-actin, glyceraldehyde-3-phosphate dehydrogenase, and 28S, 18S or 16S ribosomal RNA. However, it is wise to consult the literature before choosing a particular house-keeping gene for your quantitative PCR experiments as such diverse phenomenon as tissue type, pregnancy, diabetes, post-translational modification, the presence of pseudogenes, or fixation time of paraffin embedded tissues, may influence the amplification efficiency of different housekeeping genes [Bustin et al., 2000; Suzuki et al., 2001; Foss et al., 1994; Steele et al., 2002]. Nowadays, commercially

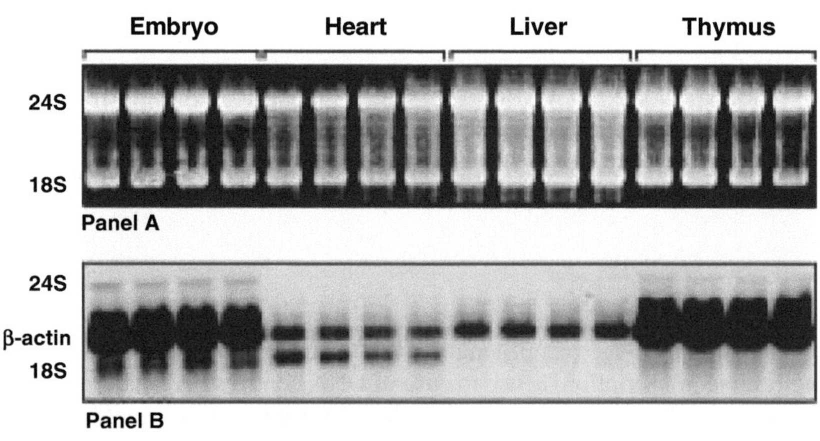

Fig. 8.5 Comparison of the expression levels of the 28 and 16S ribosomal genes (panel A) and β-actin (panel B) in various tissue samples. The differential expression of these RNAs within different tissue types may influence their suitability as endogenous "housekeeping" gene controls in certain quantitative RT-PCR protocols (From Ambion Notes Technical Bulletin 151.)

available kits for the quantification of frequently studied mRNA species of human, mouse and rat are also available (Ambion). Relative quantitative PCR assays essentially establish a relative increase or decrease in the target concentration as compared to the internal standard and as such, strict quality requirements for this type of assay are necessary. These include: that (1) the two primer pairs utilised do not non-specifically hybridise to each other, (2) amplification of control and target sequences should proceed with equal efficiency, and (3) amplification should be restricted to the exponential phase (Fig. 8.5). Also, the relative levels of housekeeping and target RNA are very important, especially in quantifying rare RNA transcripts, as an excess of housekeeping RNA will tend to inhibit target amplification. One way of avoiding this problem is to use "competimer" molecules in the RT-PCR in order to attenuate internal (housekeeping) RNA amplification (e.g. QuantumRNA Competimer Technology (Ambion)). PCR experiments using "relative" quantification methods and internal endogenous standards may be performed as separate PCR reactions, or may be multiplexed to the same PCR reaction tube (e.g. using real-time PCR).

8.3 Real-Time PCR Protocols

Quantitative PCR is most frequently performed using real-time PCR machines and real-time PCR protocols. Real-time quantitative PCR comprises a technology, which facilitates the visualisation of amplimers during PCR by means of continuous, on-line fluorescence measurements, and is regarded as the most sensitive and reproducible form of PCR-based quantification. Both the Roche LightCycler and the Applied Biosystems' ABI PRISM 7000/7700 are machines that use capillaries or tubes with translucent lids to detect an increase in the quantity of amplimers during PCR amplification on the basis of a concomitant increase in fluorescence. Using these machines, any change in amplimer concentration is reflected in a change in the fluorescence intensity measured, a process that is largely independent of the amplimer size. This means that post-PCR endpoint detection is not necessary, thereby reducing the chance of carry-over contamination whilst increasing throughput numbers. Real-time reverse transcription (RT-PCR) protocols offer a much wider "dynamic range" of possible target concentrations that may be accurately quantified (i.e. a range of 10^7 instead of 10^3 offered by conventional RT-PCR protocols). This usually results in more accurate quantification. Real-time PCR relies on the detection and quantification of a fluorescent "reporter" molecule, which can be in the form of a simple intercalating fluorescent dye (e.g. SYBR-green or ethidium bromide, Chapter 9), or DNA probes specifically designed to hybridise to an internal section of the PCR amplimer being amplified, and which may or may not be hydrolysed after each hybridisation. However, it should be noted at this point that specifically designed hybridisation probes have distinct advantages over fluorescent dyes alone, as fluorescent dyes are able to bind to any non-specific amplification products that are co-amplified during PCR thermocycling,

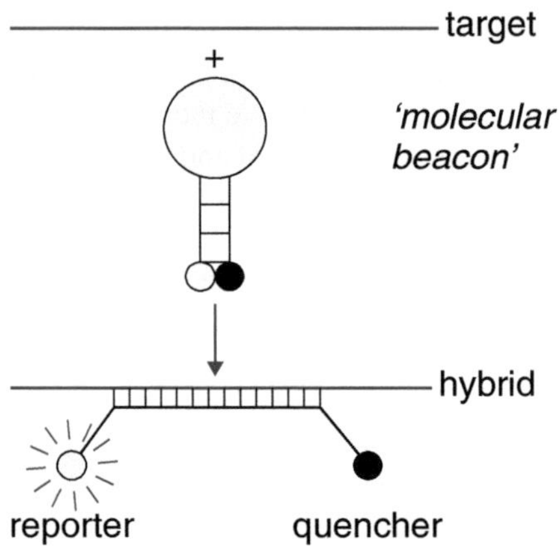

Fig. 8.6 General structure of a molecular beacon. The molecular beacon oligonucleotide folds into a hairpin loop structure under native conditions. Upon PCR thermocycling, the beacon loses its hairpin loop structure and becomes available for (specific) hybridisation to another DNA molecule (i.e. the PCR product). Upon hybridisation, the reporter and quencher molecules are spatially separated. This limits the ability of the quencher molecule to absorb reporter molecule fluorescence, rendering the reporter detectable via fluorescence measurement systems

resulting in false positive signal and an incorrect estimation of target copy number. Further, intercalating fluorescent dye protocols require extensive optimisation, post-PCR processing (e.g. melt point curve analysis), and generate a stronger signal for longer target amplimers (i.e. the degree of fluorescence is in part dependent on the amplimer length). If simple intercalating fluorescent dyes are to be used, non-specific DNA synthesis may be (partially) prevented by using specially designed and labelled primers harbouring a stem-loop structure (molecular beacons, Fig. 8.6 and Section 5.1.1). Alternatively, protein nucleic acid (PNA) probes may also be evaluated [Wolffs et al., 2001]. The most important part of real-time PCR thermocycling occurs during the exponential phase of PCR amplification, where an increase in fluorescence intensity directly correlates to the quantity of amplimer being synthesised. The advantage of real-time PCR methods is that PCR amplification may be readily followed (in "real time" of course!).

Currently, the most popular method for performing real-time PCR utilizes the 5′-nuclease property of Taq polymerase and specifically designed (Taqman) probes, which hybridize to an internal region of the PCR amplimer during the annealing step of each PCR cycle. A fluorochrome is attached to the 5′-end of each Taqman probe (e.g. FAM, VIC, NED) and a quencher molecule to the 3′-end of the same probe, usually 6-carboxytetramethylrhodamine (TAMRA). The quencher extinguishes the fluorescence of the fluorochrome so long as the Taqman probe remains intact [Lee et al., 1993]. However, upon specific hybridisation to a PCR amplimer, the probe becomes vulnerable to 5′-3′ exonuclease activity of Taq polymerase during target strand replication. The subsequent hydrolysis of the Taqman probe leads to the fluorochrome and quencher becoming physically separated so that the fluorescent signal from the reporter is no longer absorbed. This means that during PCR cycling the fluorescent intensity measured will gradually increase as the amount of specific amplimer also increases. The specificity of Taqman probe hybridisation contributes to the specificity of the Taqman reaction, as the probe will not hybridise to non-specific amplification products. Taqman technology is suited for both DNA and RNA. It has to be emphasised that Taqman-based protocols are not fundamentally different from any typical PCR protocol, and as such should be performed using all of the normal PCR controls (positive, negative, inhibition controls, etc). A very good example of the approach required in setting up a real-time PCR assay is presented by Tucker et al., 2001. If real-time Taqman multiplex PCR methodologies are required, then different reporter dyes (exhibiting different emission spectra peaks) may be attached to the different Taqman hybridisation probes to be used. Combinations of FAM λ_{max} ~518 nm; VIC λ_{max} ~550 nm; and NED λ_{max} ~580 nm are recommended.

Genes in eukaryotes often contain internal sequences that are not present in the final translated mRNA (intron sequences). Eukaryotic quantitative RT-PCR assays may be designed to utilise primers that span one or more of these intron sequences. By designing a real-time reporter hybridisation probe that spans the two exon boundaries (i.e. a probe whose sequence comprises the end of one exon and the beginning of the other exon without the intervening intron sequence), then only intron-less reverse transcribed cDNA will be detected, though it should be noted that intron-containing genomic DNA as well as intron-less reverse transcribed cDNA will still be amplified by the real-time PCR primers, which will result in a decrease in amplification efficiency and hence assay sensitivity. One way of alleviating this problem is to design a real-time PCR primer that hybridises to the two-exon regions simultaneously (instead of the real-time reporter hybridisation probe). In this protocol, only reverse transcribed cDNA will be amplified, as one of PCR primers will lack sufficient complementarity to successfully bind to the intron sequence.

For real-time measurement purposes, two important cut-offs are defined, namely the baseline and the threshold cycle (C_t-value). The baseline represents a line fitted to a defined range of PCR thermocycles (usually 3–15 cycles), where any amplification signal tends to lie within the "background noise" of the system. This baseline is important with respect to calculating the threshold cycle (C_t) as C_t measurements are not calculated until the real-time fluorescent signal rises above the baseline value for a given PCR run. The C_t is a value set within the exponential part of the PCR amplification curve and above the baseline value, and is given as a number (the number of PCR cycles at which the threshold has been set). Real-time PCR software can now calculate baseline and C_t values automatically for every PCR amplification reaction within a batch of PCR amplifications, setting the C_t value halfway between the calculated exponential part of the amplification curve (above the baseline value) and the beginning of the plateau phase of PCR amplification. The minimum C_t value is actually dependant on the amount of target present at the beginning of PCR amplification (fewer starting copies means a greater number of PCR cycles are required to generate an exponentially increasing fluorescent signal significantly above the background "noise", which in turn means that the C_t-value will be correspondingly higher, Fig. 8.7). In the majority of real-time PCR protocols, the Ct-value varies between 17 cycles (approximately 10,000,000 initial target copies) and 37 cycles (approximately 1 initial target copy).

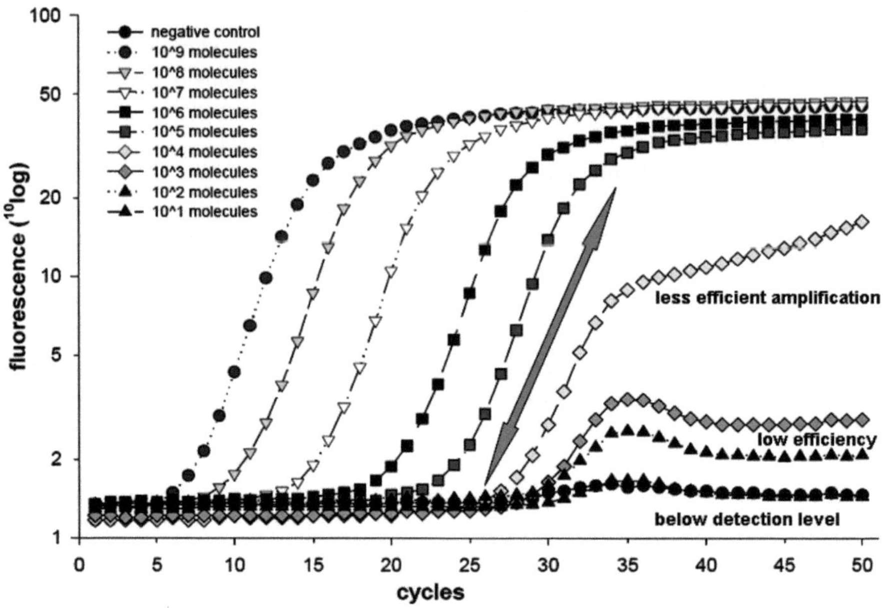

Fig. 8.7 In real-time PCR, the number of PCR cycles required to reach threshold level (C_t-value) of fluorescence is dependent on the concentration of target DNA present at the start of PCR thermocycling. The higher the initial target DNA concentration, the fewer number of PCR cycles required for the amplification signal to cross the C_t threshold (i.e. the flat part of the curve observed during the first few cycles of PCR amplification that is indistinguishable from background "noise" present in the system). In this figure, the S-shaped curves from left to right represent real-time amplification results obtained using a decreasing amount of starting template DNA. The arrow indicates the part of the curve where exponential (approximately twofold) amplification has occurred, and is the region used to calculate the threshold (C_t) value in real-time PCR assays (Pfaffl M.W. 2001)

During real-time PCR, the amplification efficiency should ideally lie between 90% and 100% ($-3.6 > \text{slope} > -3.1$), which may be calculated using the equation:

$$\text{Efficiency} = 10^{(-1/\text{slope})}\text{-}1$$

where: slope = the slope of the graph of C_t value plotted against initial copy number of target DNA present.

The amplification efficiency itself may be affected by several factors, including length of the target amplimer, any secondary structure in the target and the quality of the primers used.

For "absolute" quantification (when target gene copy numbers are required), absolute standard curves are most often used, i.e. a dilution series containing known concentrations of target molecules are amplified within each PCR run and the C_t value of the unknown sample is calculated based upon the number of target molecules necessary to reach this same C_t value in the control dilution series [Giulietti et al., 2001; Mengelle et al., 2003]. However, other methods for absolute quantification are available, including the "fit points" method, and the "second derivative maximum" method [Swillens et al., 2004; Luu-The et al., 2005]. For "relative" quantification (when target gene expression is measured as a multiple value of expression (e.g. 2×, 5×, 0.33×) to a "calibrator" sample (e.g. liver cell expression versus kidney cell expression)) then by far the most common method used is the comparative threshold (or $\Delta\Delta C_t$) method [Livak and Schmittgen, 2001], not least because it eliminates the use of standard curves. However, this method of calibration requires that the PCR efficiency of both target and endogenous control genes be similar over a dynamic rage of approximately 10^5 copies. If this is not the case then a relative standard curve method may be utilised, where target and endogenous controls are first normalised to each other via standard curves, and then the normalised target values from sample and calibrator are compared (see *http://docs.appliedbiosystems.com/pebiodocs/04371095.pdf* and *http://www.dorak.info/genetics/realtime.html*). If PCR efficiencies between target and endogenous (housekeeping) control are not comparable, then it is generally recommended to try a different endogenous control rather than adopting the relative standard curve methodology.

Finally, advances and optimisation of real-time 5′-nuclease PCR assay protocols are ongoing, including the increased use of minor groove binding hybridisation (MGB) probes in preference to currently favoured probes (Taqman probes included). MGB probes possess increased specificity and require shorter sequences when used in hybridisation-based protocols (Kutyavin et al., 2000).

8.4 RNA Extraction and Treatment

Typical RT-PCR protocols, quantitative and real-time RT-PCR protocols all require an initial input of target (and control) RNA for reverse transcription and amplification. This means that for most protocols the removal of contaminating genomic DNA is essential in obtaining accurate quantitative results. However, no RNA extraction methodology generates RNA free of DNA contamination; for example, silica-based RNA filter extraction methodologies typically yield RNA contaminated with 1–10% DNA. The most common method for removing contaminating DNA is to use DNase I digestion of the DNA followed by either subsequent inactivation of the enzyme (typically by heat-treatment at 75°C for 5 minutes), or alternatively subsequent proteolysis and removal of the enzyme using proteinase K treatment followed by organic extraction. Further, some DNase preparations may be contaminated with difficult to remove RNases. Therefore, it is recommended that commercially available reagents for the removal of DNA contamination are used, preferably reagents not requiring a heating or proteolytic digestion phase, e.g. the DNA-*free* DNase Treatment and Removal Reagent (Ambion) or the Absolutely RNA kits (Stratagene). DNase I digestion also helps eliminate the possible influence of genomic pseudogenes on RNA quantification.

References

Albretsen C, Haujanes BI, Aasland R, Kleppe K. 1988. Optimal conditions for hybridisation with oligonucleotides: a study with myc-oncogene DNA probes. Anal Biochem 170:193–202.

Ault GS, Ryschkewitsch CF, Stoner GL. 1994. Type specific amplification of viral DNA using touchdown and hot-start PCR. J Virol Methods 46:145–156.

Bustin SA, Gyselman VG, Siddiqi S, Dorudi S. 2000. Cytokeratin 20 is not a tissue specific marker for the detection of malignant epithelial cells in the blood of colorectal cancer patients. Int J Surg Investig 2:49–57.

Clementi M, Menzo S, Bagnarelli P, Valenza A, Paolucci A, Sampuolesi R, Manzin A, Varaldo PE. 1996. Clinical use of quantitative molecular methods in studying human immunodeficiency virus type 1 infection. Clin Microbiol Rev 9:135–147.

Foss RD, Guha-Thakurta N, Conran RM, Gutman P. 1994. Effects of fixative and fixation time on the extraction and PCR amplification of RNA from paraffin-embedded tissue: comparison of two housekeeping gene from mRNA controls. Diagn Mol Pathol 3:148–155.

Giulietti A, Overbergh L, Valckx D, Decallonne B, Bouillon R, Mathieu C. 2001. An overview of real-time quantitative PCR: applications to quantify cytokine gene expression. Methods 25:386–401.

Grace MB, Buzard GS, Hughes MR, Gore-Langton RE. 1998. Degradable UMP outer primers in merged tandem (M/T) nested PCR: low and single copy DNA target amplification. Anal Biochem 263:85–92.

Gupta AK, Kohli Y, Summerbell RC. 2000. Molecular differentiation of seven *Malassezia* species. J Clin Microbiol 38:1869–1875.

Haberhausen G, Pinsl J, Kuhn CC, Markert-Hahn C. 1998 Comparative study of different standardization concepts in quantitative competitive reverse transcription-PCR assays J Clin Microbiol 36(3):628–633.

Higuchi R. Dollinger G, Walsh PS, Griffith R. 1992. Simultaneous amplification and detection of specific DNA sequences. Biotechnol 10:413–417.

Joyce C. 2002. Quantitative RT-PCR: a review of current methodologies. Humana, Totowa, NJ.

Kainz P. 2000. The PCR plateau phase – towards an understanding of its limitations. Biochim Biophys Acta 1494:23–27.

Kutyavin IV, Afonina AI, Mills A, Gorn VV, Lukhtanov EA, Belousov ES, Singer MJ, Walburger DK, Lokhov SG, Gall AA, Dempcy R, Reed MW, Meyer RB, Hedgpeth J. 2000. 3′-Minor groove binder-DNA probes increase sequence specificity at PCR extension temperatures, Nucleic Acids Res 28:655–661.

Lee LG, Connell CR, Bloch W. 1993. Allelic discrimination by nick-translation PCR with fluorigenic probes. Nucleic Acids Res 21:3761–3766.

Livak KJ, Schmittgen TD. 2001. Analysis of relative gene expression data using real-time quantitative PCR and the 2(-Delta Delta C(T)) Method. Methods 25:402–408.

Luu-The V, Paquet N, Calvo E, Cumps J. 2005. Improved real-time PCR method for high-throughput measurements using second derivative calculation and double correction. Biotechniques 38:287–293.

Mehmet D, Ahmed F, Cummins JM, Martin R, Whelan J. 2001. Quantification of the common deletion in human testicular mitochondrial DNA by competitive PCR assay using a chimaeric competitor. Mol Human Rep 7:301–306.

Mengelle C, Sandres Kaune K, Pasquier C, Rostaing L, Mansuy JM, Morty M, Da Silva I, Attal M, Massip P, Izopet J. 2003. Automated extraction and quantification of human cytomegalovirus DNA in whole blood by real-time PCR assay. J Clin Microbiol 41:3840–3845.

Monteiro L, Hua J, Birac C, Lamouliatte H, Megraud F. 1997. Quantitative polymerase chain reaction for the detection of *Helicobacter pylori* in gastric biopsy specimens. Eur J Clin Microbiol Infect Dis 16(2):143–149.

Pfaffl MW. 2001. Development and validation of an externally standardised quantitative Insulin like growth factor-1 (IGF-1) RT-PCR using LightCycler SYBR® Green I technology In: Rapid Cycle

Real-time PCR, Methods and Applications (Editors: S. Meuer; C. Wittwer & K. Nakagawara) Springer Press, Heidelberg, ISBN 3-540-66736-9, p281–291.

Pfaffl MW, Hageleit M. 2001. Validities of mRNA quantification using recombinant RNA and recombinant DNA external calibration curves in real-time RT-PCR. Biotechnol Lett 23:275–282.

Piatak M, Saag MS, Yang LC, Clark SJ, Kappes JC, Luk KC, Hahn BH, Shaw GM, Lifson JD. 1993. Determination of plasma viral load in HIV-1 infection by quantitative competitive PCR. AIDS 7:S65– S71.

Remacle JE, Kraft H, Lerchner W, Wuytens G, Collart C, Verschueren K, Smith JC, Huylebroeck D. 1999. New mode of DNA binding of multi-zinc finger transcription factors: delta EF1 family members bind with two hands to two target sites. EMBO J 18:5073–5084.

Rychlik W, Spencer WJ, Rhoads RE. 1990. Optimization of the annealing temperature for DNA amplification in vitro. Nucleic Acids Res 18:6409–6412.

Saiki RK, Chang CA, Levenson CH, Warren TC, Boehm CD, Kazazian HH, Erlich HA. 1988. Diagnosis of sickle cell anemia and beta-thalassemia with enzymatically amplified DNA and non-radioactive allele-specific oligonucleotide probes. N Engl J Med 319:537–541.

Steele BK, Meyers C, Ozbun MA. 2002. Variable expression of some "housekeeping" genes during human keratinocyte differentiation. Anal Biochem 15:341–347.

Suzuki S, Chuang LF, Doi RH, Bidlack JM, Chuang RY. 2001. Kappa-opioid receptors on lymphocytes of a human lymphocytic cell line: morphine induced up-regulation as evidenced by competitive RT-PCR and indirect immunofluorescence. Int Immunopharmacol 1:1733–1742.

Swillens S, Goffard JC, Marechal Y, de Kerchove d'Exaerde A, El Housni H. 2004. Instant evaluation of the absolute initial number of cDNA copies from a single real-time PCR curve. Nucleic Acids Res 32(6):e53.

Tolstrup N, Nielsen PS, Kolberg IG, Frankel AM, Vissing H, Kauppinen S. 2003. OlogiDesign: optimal design of LNA (locked nucleic acid) oligonucleotide probes for gene expression profiling. Nucleic Acids Res 31:3758–3762.

Tucker RA, Unger ER, Holloway BP, Swan DC. 2001. Real-time PCR-based fluorescent assay for quantification of HPV types 6, 11, 16 and 18. Mol Diagn 6:39–47.

Wilhelm J, Hahn M, Pingoud A. 2000. Influence of DNA target melting behaviour on real-time PCR quantification. Clin Chem 46:1738–1743.

Wolffs P, Knutson R, Skoback R, Radstrom P. 2001. PNA based light-up probes for real-time detection of sequence specific PCR products. Biotech 31:769–771.

Chapter 9

Analysis of PCR Amplification Products

The analysis of PCR amplification products (amplimers) is an essential step in determining the quality (i.e. the presence or absence of non-specific amplimers) and quantity (relative values or exact values) of the DNA target that has been amplified. In general, the techniques used to analyze PCR products may be divided into two distinct groups: (1) the *ex-vitro* techniques where analysis is performed outside of the PCR reaction vessel (for example analysis using gel electrophoresis, DNA hybridisation, etc.), and (2) the more recently developed *in-vitro* techniques, where analysis is performed inside the PCR reaction vessel and during PCR thermocycling *per se*. Both the *ex-vitro* and *in-vitro* techniques require that the PCR products be first visualized in some way prior to analysis, with some of the methodologies used for PCR product visualization being shared by both groups. Ultimately however, the "gold-standard" in PCR amplimer analysis is direct sequencing of the amplimer to determine its actual nucleotide composition.

9.1 Visualizing PCR Amplification Products

Two main methods are available for visualizing PCR amplification products, namely (1) staining of the amplimer double stranded DNA using chemical dyes or silver ions, which insert (intercalate) between the two strands of the duplex, or (2) labelling of the PCR primer or dNTP nucleotides with fluorescent dyes (fluorophores) prior to PCR amplification, which results in the labels becoming directly incorporated into the PCR product during thermocycling.

9.1.1 Intercalating Chemical Dyes and Silver Ions

Intercalating chemical dyes insert between the organic bases within duplex strands of DNA facilitating a large increase (up to 1,000 times) in their capacity to fluoresce, and as such are useful for determining DNA concentrations ranging between 0.01 and 20μg per ml [Srinivasan et al., 1993]. Many intercalating dyes useful in PCR assays are currently commercially available, e.g. SYBR Gold (Molecular Probes) and VistaGreen (Molecular Dynamics). SYBR Gold is especially useful for staining ssDNA (e.g. in single stranded conformational polymorphism (SSCP) PCR protocols). OliGreen (Molecular dynamics) is especially useful in staining ssDNA of shorter lengths (e.g. PCR primers and oligonucleotide probes), whilst RiboGreen, PicoGreen (Molecular Probes) and silver staining can be used to measure very small quantities (down to approximately 25 pg/ml) of RNA and DNA respectively. However, probably the best known and most widely used intercalating dye is 2,7-diamino-9-phenyl-10-ethyl phenanthridium bromide, better known as ethidium bromide (E_{max} at 605 nm).

For quantitative DNA calculations, ethidium bromide can be used to detect and quantify DNA within the concentration range of 20 ng/ml–20 µg/ml DNA, with the relationship between staining intensity and the quantity of DNA being linear across this range of DNA concentrations. Another commercially available dye, Hoechst 33258, may be used to quantify even lower concentrations of DNA, down to 10 ng/ml of DNA. This compound preferentially intercalates within AT-rich regions, with minimal binding to either RNA or individual nucleotides or single stranded DNA (e.g. PCR primers). Also, Hoechst 33258 does not bind significantly to proteins, salts, chelating agents or glycerol. These attributes help to explain in part the higher sensitivity associated with this particular fluorescent dye. A further advantage of Hoechst 33258, is that the excitation-emission wavelengths of the intercalated dye differ from that of the freely dissolved non-intercalated dye, (excitation/emission wavelengths values being 365/458 nm and 356/492 nm respectively). This allows for precise DNA concentration measurements at 458 nm. Other intercalating dyes such as SYBR Gold and VistaGreen also stain DNA more intensely than ethidium bromide, diffuse very quickly into agarose or acrylamide matrices and do not show significant background staining when in the non-DNA intercalated form. Further, all three of these dyes can be removed by ethanol precipitation, thereby circumventing many of the laboratory-based environmental contamination problems raised by ethidium bromide.

For routine applications however, fluorescent intercalating dyes are used mainly in a qualitative manner as a means to define the position of amplimers within agarose or acrylamide gels. This provides a rapid and easy method for determining whether the actual size of the amplified PCR product corresponds to the expected size of the PCR product generated, and provides a simple method for assessing PCR specificity. For qualitative purposes, all of the fluorescent intercalating dyes mentioned above may be either added directly to the gel prior to electrophoresis, or the unstained gel may be immersed in a solution of the dye immediately after electrophoresis.

Chemical agents that intercalate within the DNA double helix disturb its regular structure and are usually potent carcinogens (Fig. 9.1). These compounds need to be treated with care, and the necessary precautions taken when handling such hazardous chemicals (see manufacturers' safety sheets for further information). Several protocols for the decontamination of these mutagens and teratogens have been described, with decontamination methodologies using activated charcoal being particularly effective and perhaps most frequently used. As well as visualizing amplimers using

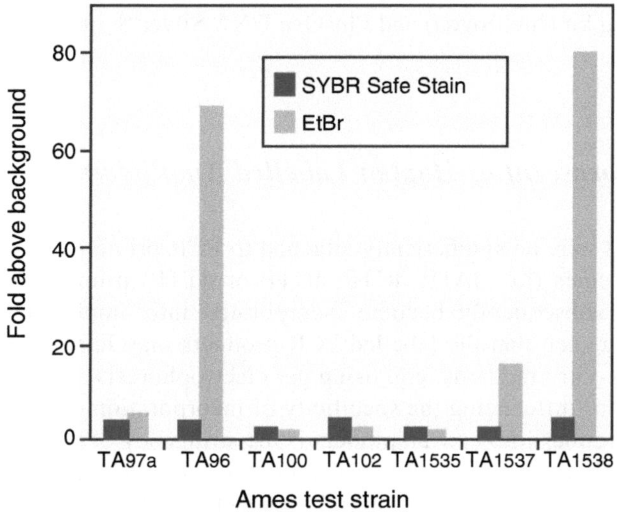

Fig. 9.1 Comparison of the mutagenicity of SYBR Green I and and ethidium bromide (EtBr) intercalating nucleic acid stains as assessed by the Ames mammalian microsome reverse mutation assay. The mean increase in the number of revertants per plate is shown for seven *Salmonella typhimurium* test strains (TA97A–TA1538), grown in the presence of mammalian microsomal enzymes from PCB-induced rat liver with either SYBR Green I or EtBr. Note the greater mutagenic potential of ethidium bromide compared to SYBR Green I (From Singer et al., 1999)

intercalating dyes, agarose and polyacrylamide gels may be subjected to the more permanent process of silver staining. This procedure may be used in both high-sensitivity protein staining protocols and also for detecting nanogram and even picogram quantities of DNA (where it is approximately five times more sensitive than ethidium bromide for short DNA fragments). The principle of the staining procedure is as follows; DNA molecules have an affinity for silver ions and impregnation of a gel via immersion in a silver nitrate solution leads to binding of metallic silver to any DNA molecules present in the gel. Bound silver atoms then serve as "precipitation nuclei" facilitating the deposition of more silver. After staining and washing for background reduction, the process generates recognizable banding pattern profiles, with the DNA staining a brownish black colour due to the retention of the silver ions. Stained DNA fragments may be cut from the gel and used for re-amplification since limited contaminating amounts of the metallic silver does not seem to significantly inhibit optimized PCR protocols. The silver staining procedure requires a higher degree of care and attention during agarose/ polyacrylamide gel preparation than intercalating dye-staining protocols, due to the fact that silver ions have a significant affinity for fatty acids and other contaminants that may be inadvertently introduced into the gel during preparation. Ultimately, such contamination leads to an increase in background staining of the gel. Silver stained gels can be dehydrated and stored *in situ* for prolonged periods of time, at least several years. Commercial silver staining kits available include SilverXpress

Silver Staining kit (Invitrogen) and PlusOne DNA Silver Staining Kit (Amersham Biosciences).

9.1.2 Fluorescent or Hapten Labelled Amplimers

Various labels may be synthetically attached to PCR primers or even individual nucleotide species (i.e. dATP, dCTP, dGTP or dTTP) prior to thermocycling. These labels subsequently become incorporated into amplimers during PCR thermocycling such that the labelled PCR products may later be detected post-PCR using ex-vitro methods, e.g. using gel electrophoresis and laser excitation. The main factor influencing the specificity of incorporation of PCR primers and labelled nucleotides into new amplimers is the stringency of the PCR itself.

The use of fluorochrome dye labels that fluoresce at different wavelengths during laser excitation allows the use of several separately labelled nucleotide species (as used in certain PCR sequencing protocols for example), or multiply labelled PCR primers where each primer pair is separately labelled (as used in certain multiplex PCR protocols). Laser excitation and analysis of fluorescently labelled amplimers may be performed with the help of a series of dedicated machines, the most popular of which are marketed by Perkin Elmer (ABI-Prism electrophoresis systems). These machines utilize laser-induced fluorescence and multicolour CCD detection and are designed to be able to simultaneously detect four fluorochromes in a single gel lane, with the data being displayed as a nucleotide sequence, an electropherogram (peak pattern) or as a numerical table dependant on the application. Other examples of nucleotide labels include infrared labels (IRDyes, LICOR Biosciences, Nebraska, USA), which require the use of their own particular dedicated detection system.

In the simplest machine, the ABI Prism 377, DNA fragments are optically detected after having migrated through a 50 cm long polyacrylamide gel under denaturing conditions. A laser is situated near the bottom of this gel, which is able to excite the fluorescent labels present within the amplimers, with fluorescence detection being achieved using a digital camera detection system. As soon as any DNA fragment migrates into the path of the laser beam, a fluorescent dye-specific colour signal is generated and recorded. In this system, 96 samples may be analysed in a single gel run, with each run taking approximately 3 hours to complete, using a maximum fragment size of approximately 500–700 bp depending on the quality and quantity of the amplimer used. Dedicated software for data analysis and presentation is supplied upon purchase of the machine.

The ABI Prism 310 Genetic Analyzer and the 377 DNA sequencer have been developed for the high-resolution analysis of complex DNA banding patterns and are used in DNA sequencing, forensic PCR, and multiplex PCR applications, as well as the quality assessment of PCR products in general (e.g. [Wickham et al., 2000]) (Fig. 9.2). These particular systems are increasingly being used as an alternative to the more classical agarose gel electrophoresis/ethidium bromide staining.

The ABI Prism 3700 utilizes automated electrophoresis in small glass reaction vessels called capillaries to separate DNA fragments. It can be used with non-specific

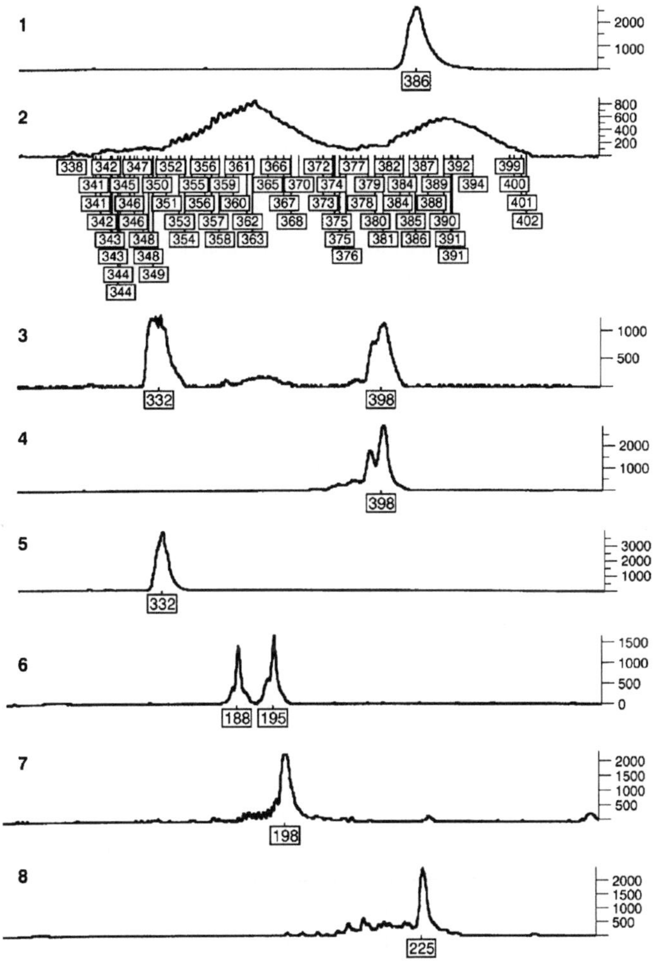

Fig. 9.2 Separation of PCR products using the ABI 377 DNA sequencer and data interpretation using GeneScan/GenoType Software. Presented are the results of a PCR protocol aimed at detecting gene rearrangements in human immunoglobulin genes using fluorescent R110-labeled dCTP as well as normal dCTP. This fluorescent labeling protocol facilitates discrimination between polyclonal disease states such as chronic lymphocytic leukemia and B-cell non-Hodgkin lymphoma. The resolution of the detection protocol for the expected range of amplimer molecular weights is in the order of three nucleotides (Reproduced from Wickham et al., 2000. With permission from 'BMJ Publishing Group Ltd.')

intercalating dyes, but most of the current applications involve multi-fluorescent dye labelled DNA applications. Capillary electrophoresis on the ABI 3700 machine also allows the analysis of PCR products that overlap in size, with estimation of the size of PCR amplification products being achieved by including differently coloured molecular weight markers (Fig. 9.3). A full run using the capillary approach takes approximately 30 minutes and has a fragment size coverage that equals that of other (non-capillary-based) ABI machines.

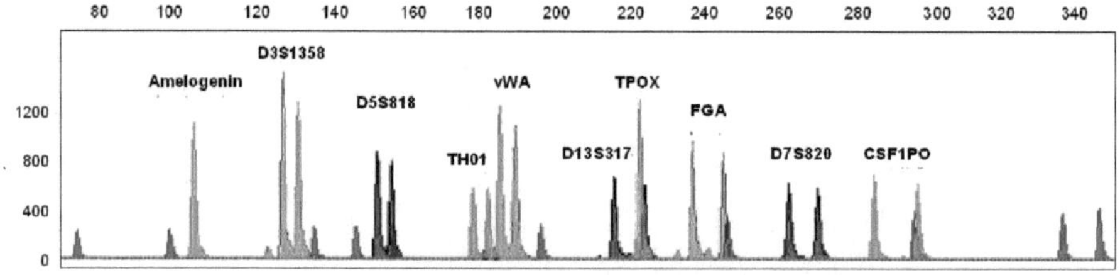

Fig. 9.3 Electropherogram showing the results from an "AmpFl STR Profiler PCR Amplification Kit (Applied Biosystems)" for nine STR loci and the amelogenin locus (Courtesy of Applied Biosystems)

Other haptens (other than fluorescent labels) may also be attached to PCR primers. One particular popular non-fluorescent label is biotin that binds very specifically and unusually strongly with streptavidin molecules. If the biotin coated PCR products are immobilised on a streptavidin coated support matrix (e.g. latex beads, ELISA plate wells), non-biotin labelled amplification products and other PCR mix ingredients may be washed away, effectively purifying the amplimer ready for further downstream PCR applications.

9.2 Post-PCR Electrophoretic Analysis of Amplimers

PCR amplimers are composed of DNA and as such exhibit a characteristic absorption peak at ultraviolet wavelengths of 260 and 280 nm (A260/A280). This means that the concentration of amplimers present within a solution may be measured using UV spectroscopy. However, direct measurement of amplimer concentration is not possible due to the fact that PCR amplification mixes contain reaction components (dNTPs and primers) that also absorb ultraviolet light at wavelengths comparable to that of DNA. Therefore, the amplimers should be either purified away from the remaining dNTPs and primers prior to spectrophotometric analysis or should be stained with an intercalating chemical dye. In fact, the sensitivity of methodologies involving DNA staining by intercalating dyes is low, though the large concentration of amplimers synthesised during PCR amplification easily compensates for this low sensitivity.

The most common methodologies used to assess the quality and quantity of PCR amplification products post-PCR utilize the relatively simple *ex-vitro* procedure of gel electrophoresis followed by staining of the gel using an intercalating dye. In effect, the electrophoretic procedure is first used to separate the amplification products according to size before staining and amplimer detection by UV transillumination. This allows an easy assessment to be made of the size of the amplification product compared to a standard molecular weight marker (which is added to an empty well of the gel). Any positive or negative PCR control samples (PCR amplified with and without specific target DNA respectively) should also be included in the same gel in order to further quality control the success or failure of the PCR. Alternative but time-consuming methodologies involve the hybridisation of amplimer and target DNA under varying stringency conditions (Southern blotting).

9.2.1 Gel Electrophoresis Methodologies

A wide variety of electrophoretic methods adapted for the differential analysis of PCR amplified DNA have been developed. The most common of these methods are described below.

(i) Simple agarose gel electrophoresis: The polysaccharide agarose (poly D-galactose 3,6-anhydro-L galactose; Fig. 9.4), is constructed from multiple disaccharide building blocks and may be purified from marine algae. It is a chemically stable solid compound at room temperature, which is commercially available in powder or granular form, and in a range of specifications (e.g. for the resolution of low molecular weight amplimers). For electrophoretic purposes, an agarose suspension is first made in water and heated up to at least 65°C (usually the suspension is heated to boiling point), until the agarose is dissolved, after which time, the agarose solution is allowed to cool to a temperature whereby the flask may be held in the hand without the risk of being burned "hand-hot" and then poured into a gel casting frame. After cooling to room temperature in the casting frame, the solution solidifies to an opalescent solid mass that remains solid as long as the temperature does not exceed 90°C. This solid agarose mass consists of a dense network of polymeric chains, the density of which is determined by the concentration and type of agarose used. During electrophoresis, the agarose matrix functions as a molecular sieve, with the driving force being an electric current applied to opposite ends of the gel to facilitate migration of negatively charged DNA molecules to the positively charged anode in the electrophoresis tank. If necessary, fluid agarose may be kept in solution for a few hours by incubating at 37°C before cooling and pouring of gels.

The average agarose gel used for molecular biology purposes contains between 0.8 and 4% agarose (1% = 10 g/l), with an agarose concentration of between 1%

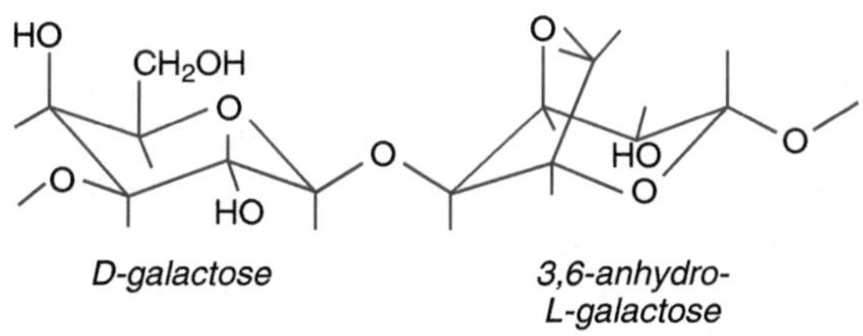

Fig. 9.4 Structural representation of the disaccharide D-galactose/3,6-anhydro-L-galactose, the building block of agarose (From Sambrook et al., 1990)

and 2% being most often used. This 1–2% matrix allows the separation of PCR amplimers of sizes between 100 base pairs and 10 kilobasepairs in length. However, the nature of agarose matrices does not allow the separation of molecules less than approximately 15 nucleotides in length. The higher the agarose concentration, the more dense becomes the polymer network such that even low molecular weight compounds of 100 bp or less find it difficult to penetrate and migrate through gel matrices of 4% or higher. In any case, agarose gel concentrations of 4% and above are very difficult to dissolve and pour due to their high viscosity. The type of electrophoresis buffer used may also affect the resolution of DNA separation, with TAE (Tris-Acetate-EDTA) buffer providing a better resolution of fragments larger than 4 kb, and TBE (Tris-Borate-EDTA) buffer providing better resolution of 0.1–3 kb fragments. TBE buffers also tend to have a greater buffering capacity and may be reused a few times before being replaced. For double stranded DNA molecules, the speed of migration through the agarose sieve is largely defined by the length of the DNA molecule in basepairs, whilst for single stranded DNA molecules both the length and the three-dimensional spatial conformation are important. As a general rule, the denser the agarose structure, the lower the mobility of the double stranded DNA through the agarose gel.

Agarose gels may be cast in various formats dependant upon the wishes of the individual researcher, though major differences in handling are required for horizontal versus vertical gel systems. PCR amplified DNA may be isolated after gel electrophoresis using commercially available kits (or enzymes e.g. beta-agarase), though if this is to be undertaken, it is recommended that the "purer" forms of agarose be used. These expensive agarose preparations produce gels with low endo-osmotic values, resulting in sharper bands upon electrophoresis.

Agarose gel electrophoresis can be performed under various physical conditions, ranging from "native" to "completely denaturing". If the electrophoresis is performed under denaturing conditions, e.g. in the presence of urea or formamide or in alkaline gel systems, then the physical conditions will destabilise the double stranded structure of the amplimer, allowing information to be obtained on the electrophoretic mobility of the individual DNA strands. The melting behaviour of double stranded PCR amplimers may also be studied in gradient agarose gel systems, where temperature gradients or gradients of chemical denaturants are applied to the gel. Denaturing Gradient Gel Electrophoresis (DGGE) and Single Strand Conformation polymorphism (SSCP) protocols utilise such gradients to analyse the melting behaviour of double stranded DNA and are discussed later in this chapter.

In order to assist in amplimer size determination, molecular weight markers are usually added to an empty well in the gel. A wide range of these molecular weight markers are available for purchase commercially (Fig. 9.5) and some gels may be purchased with the molecular weight markers already included.

In extreme cases, agarose gel electrophoresis yields a smear of amplification products, indicative of the amplification of many thousands of different amplification products.

Agarose gel smears may be caused by degraded PCR primers, carry-over contamination from previous PCRs, the use of an excessive quantity of DNA

S. cerevisiae chromosomal DNA	220,000 –2,000,000
Lambda Ladders	50,000 –1.000,000
Supercoiled DNA Ladder	2,000 – 10,000
Lambda *Hind*III Markers	125 – 23,130
Lambda *Hind*III / *Eco*RI Markers	125 – 21,226
Lambda *Eco*RI Markers	3,530 – 21,226
RNA Markers	281 – 6,583
BMV RNA Markers	816 – 3,234
pGEM DNA Markers	36 – 2,645
ΦX174 *Hae*III Markers	72 – 1,353
PCR Markers	50 – 1,000
ΦX174 *Hinf*I Markers	24 – 726
1 kb DNA Ladder	250 – 10,000
1 kb DNA Step Ladder	1,000 – 10,000
200 bp DNA Step Ladder	200 – 6,600
100 bp DNA Ladder	100 – 1,500
100 bp DNA Step Ladder	100 – 4.000
50 bp DNA Step Ladder	50 – 800
25 bp DNA Step Ladder	25 – 300
10 bp DNA Step Ladder	10 – 100

Fig. 9.5 Size distribution (in base pairs) of commercially available molecular weight markers which are commonly used as reference size indicators during agarose electrophoresis

thermostable polymerase, or a high magnesium ion concentration. Usually, the use of a newly designed PCR primer pair or further optimization of the PCR mix ingredients solves the problem. However, if the problem persists then either; (1) the number of cycles in the PCR assay may be too high, (2) the amount of template DNA added to the PCR mix may be too high, or (3) the DNA added to the PCR mix may be impure or degraded. Alternatively, the PCR amplification of very long target DNA fragments (e.g. long and accurate PCR protocols) in combination with too short a denaturation time and/or too short an elongation time within each thermocycle, may result in the amplification of incomplete target DNA fragments, possibly resulting in the presence of a smear upon gel electrophoresis. Sometimes, PCR amplification products may not migrate as sharp bands upon gel electrophoresis. This is usually caused by problems associated with gel preparation, with gels having been prepared too thin, too thick or electrophoresed before being sufficiently solid. "Band-tailing" or the "Venetian blind effect", showing a shaded region in front of and beyond the band of interest, may be caused by electrophoresis at excessive voltages, especially when the gel well has been overloaded with amplimer. Excessive voltages also generate relatively large quantities of heat, which may also affect the performance and appearance of PCR products after gel electrophoresis. If the electrophoresis power-packs and buffer tanks are not regularly inspected and/or technical maintenance not performed at regular intervals, then variations in the electrical field strength at different points within the gel may occur. This could possibly result in PCR products migrating in a non-uniform manner through the gel, leading to misinterpretation of amplimer size (PCR specificity) as well as a reduction in gel resolving capacity.

Recently, there has been a drive towards the development of automated gel systems and gel miniaturisation, in order to speed-up PCR amplimer identification. Indeed, "ready-made" gels of various formats can now be purchased, some of which allow for the rapid quality assessment of large numbers (up to 100 per gel) of PCRs. These advances in gel technology may soon be commonplace in many diagnostic laboratories where large numbers of specimens have to be processed.

(ii) Simple polyacrylamide gel electrophoresis (PAGE): An alternative to agarose gel electrophoresis is polyacrylamide gel electrophoresis (PAGE), a technique which has a greater "resolving power" for separating amplimers of similar sizes. Polyacrylamide gels are clear and not opaque like agarose gels and therefore do not scatter fluorescent light so readily, rendering the optical properties of polyacrylamide gels better than that of even the best agarose gels.

PAGE gels are composed of the polymer polyacrylamide, which is formed via an interaction between the monomeric neurotoxin acrylamide and the cross-linker bisacrylamide at a ratio of 19/1, respectively. Polymerisation is initiated by the addition of tetra-ethyl-methyl-ethylene-diamide (TEMED) in combination with the radical-supplying ammonium persulphate (APS). The TEMED molecule stabilises the APS-generated oxygen radicals required for the initiation of polymerisation and the formation of a polyacrylamide gel mesh. The density of the network is determined by the amount of acrylamide/bisacrylamide added, with the gel taking approximately 60 minutes to achieve complete polymerisation. The whole polymerisation reaction takes place between glass plates that define the dimensions of the gel and help to prevent the gel mix coming into contact with oxygen in the air, which acts as a polymerisation inhibitor. Polyacrylamide gels with a very dense cross-linked structure (achieved using elevated acrylamide concentrations), may be required to separate short DNA and RNA fragments that differ in size by only a single nucleotide. As both "native" and "denaturing" polyacrylamide gels may be prepared and used, the investigation of both single-stranded DNA, and double-stranded DNA and RNA may be performed. Denaturing polyacrylamide gels are prepared as for native gels but with the addition of 5–8 M urea or formamide and by running the gel under high voltage/current conditions, which heats the gel matrix up to 80°C facilitating complete double stranded DNA separation. Polyacrylamide gel electrophoresis can also be performed in two dimensions, where essentially, a complete migration track or lane is sliced out of a native gel after electrophoresis and placed across a gel containing 8 M urea. The 8 M urea denaturing gel is then electrophoresed so that the DNA present in the native gel is run at an angle of 90° compared to the first (native gel) dimension. This protocol has the advantage that the complete molecular range of RNA and DNA molecules can be analysed with a high the degree of resolution. This two-dimensional (2D) protocol has been utilised in the search for new bacterial, parasitic and viral pathogens, with several viruses and viroids having been provisionally identified using this method [Riesner et al., 1989]. The 2D protocol can also be adapted to include certain physical (temperature) or chemical (urea or formamide) gradients in one of the dimensions.

On the downside, appropriate safety measures should be taken when making PAGE gels and handling acrylamide in particular, as acrylamide is a potent neurotoxin. Laboratory coats, safety glasses, gloves and mouth protection should be worn during all manipulation with this chemical. Correct disposal procedures should be enforced. Working with solutions of acrylamide poses less of a risk than working with acrylamide powder. Polymerised acrylamide carries less neurotoxic risk than unpolymerised acrylamide.

Several complete polyacrylamide gel electrophoresis systems are commercially available. For example, the Pharmacia "PhastSystem" provides a complete package for polyacrylamide gel electrophoresis, allowing various DNA fragments to be analysed in detail and visualised by the silver-staining assay that is included. The PAGE gels in this system (either 12.5% or 20% polyacrylamide) are sold in a "ready-to-use" format and electrophoresis is completely standardised by using dedicated equipment. All physical parameters including temperature, power, current, etc. are automatically controlled. An advantage of systems like this is that the gels are supplied already polymerised (hence no manipulation of neurotoxic acrylamide monomer is required), the complete process is fully automated and quite reproducible, and radioactive labelling materials are not required. Other commercially available systems perform high-speed analyses using polyacrylamide, polyethylene or hydroxycellulose carriers within glass capillaries. This "capillary" electrophoresis methodology facilitates the analysis of samples containing low concentrations (in the nanomolar range) of DNA at high electric field strengths of 1,000 volts per cm [Holmila and Husgafvel-Pursiainen, 2006].

(iii) Restriction Fragment Length Polymorphism (RFLP) analysis and oligomer restriction: RFLP analysis is a methodology frequently used to identify genetic polymorphisms, including genetic polymorphisms detected using PCR amplification. The methodology in its simplest form involves the cutting of double stranded DNA molecules into smaller fragments using restriction enzymes. Restriction enzymes are enzymes that have been isolated from many different species of bacteria and are able to cut double stranded DNA by recognising specific sequences (usually between six and eight nucleotides in length) within DNA molecules. If the sequence of the DNA molecule is known, then the size and number of fragments generated using a particular restriction enzyme may be predicted. By running the digested DNA products on an agarose gel, the theoretically predicted size and number of fragments can be checked against the actual size and number of fragments obtained. If the observed sizes or number of fragments do not match that predicted or expected, then fragment length polymorphisms are seen, and variation (mutation) in the DNA sequence of that particular gene or amplimer exists. Using the RFLP technique, it is possible to screen regions of DNA, which are known to undergo (frequent) mutation. RFLP analysis combined with denaturing gradient gel electrophoresis in two-dimensional polyacrylamide gel systems may be used for high resolution mapping of human genetic variability [Uitterlinden and Vijg, 1991]. However in this case, the RFLP patterns present as highly complex clouds of spots, which may need to be interpreted using computer software.

(iv) Single Stranded Conformation Polymorphism (SSCP) analysis: SSCP analysis is a technique during which PCR-amplified double stranded DNA fragments are first denatured into their individual constituent (+) and (−) strands and then electrophoresed through a native, low polyacrylamide concentration, gel alongside the original non-denatured double stranded amplimer. The precise electrophoresis conditions of ionic strength, temperature, type of gel matrix, electrophoresis time, etc., should be determined empirically for each experimental protocol. However, the size of the PCR product to be analysed should optimally be around 300 bp, because the subtle mutation-dependent changes in electrophoretic mobility cannot be detected for larger fragments. The individual single DNA strands are obtained by (i) heating double stranded DNA molecules at 90°C for 5 minutes and then immediately quenching the products on ice or (ii) by using an alkaline NaOH treatment protocol. In both cases the separated strands will adopt a compact coil-like conformation. This coil-like conformation is unique for each DNA strand and is determined by the stability of intra-molecular interactions such as hairpins and repetitive stretches of nucleotides. Any single stranded DNA that contains mutations will adopt a slightly different structural conformation than the "wild type" DNA. These differences in conformation lead to differences in electrophoretic mobility during electrophoresis resulting in differences in distance travelled between any mutant and wild type strands.

The SSCP technique allows the detection of mutations in PCR amplimers without having to determine their complete nucleotide sequence (Fig. 9.6). Heterozygous mutations within paired genes results in two different PCR amplification products within the final PCR mix, and are observed as electrophoretic mobility differences due to the formation of homogeneous and heterogeneous DNA duplexes [Ikeda et al., 1995]. A very important quality requirement for SSCP is that the final PCR amplification product is homogeneous, i.e. free from non-specific amplification products.

Interestingly, SSCP can also be performed after multiplex PCR [Fregel et al., 2005], as long as the amplimers do not exhibit the same electrophoretic mobility. One possibility for enhancing the resolution of the SSCP protocol is to digest the PCR fragment prior to electrophoresis using an appropriate restriction enzyme. A larger number of small double stranded molecules are thereby created. These may then be separated into an even larger number of single stranded molecules so that SSCP analysis becomes easier than using a single large molecule alone. The detection of SSCP products and pattern determination within polyacrylamide gels was initially achieved using radioactive labels and autoradiography to reveal the position of the DNA fragments within the gel. Alternatively, silver-staining protocols may be used to detect the single stranded DNA molecules within the gel [De Sousa Menezes et al., 2003]. If sufficient quantities of single stranded DNA are present within the SSCP gel, then classical fluorescent dye staining (e.g. using ethidium bromide), may also be performed. However, this latter fluorescent protocol, exhibits a significantly lower sensitivity than either autoradiography or silver.

(v) Denaturing Gradient Gel Electrophoresis (DGGE): DGGE is a protocol designed to determine the presence of mutations between particular amplimers. As

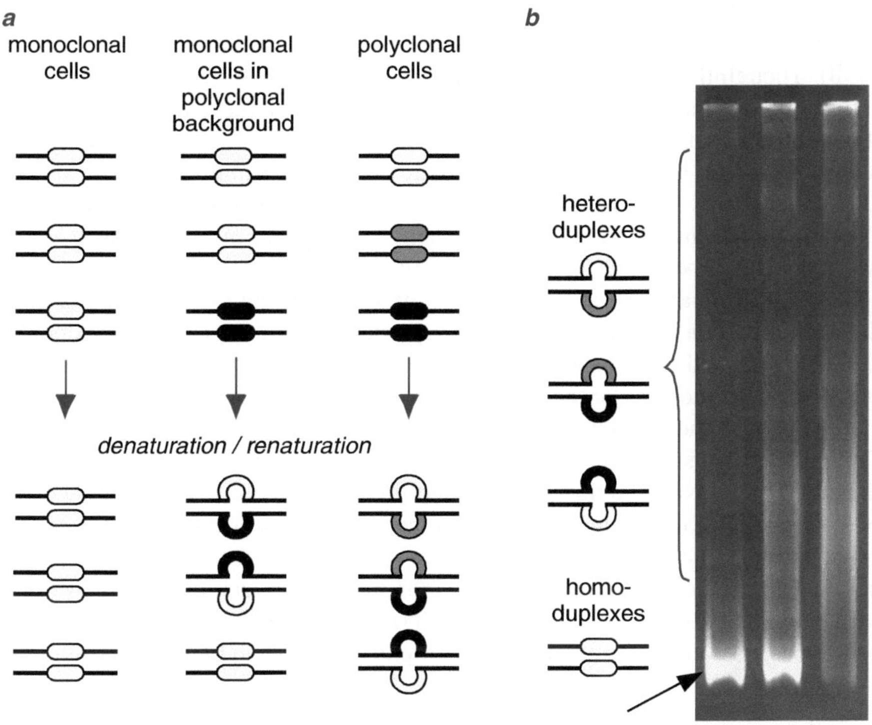

Fig. 9.6 The principle of PCR SSCP. Depending on the nature of the alleles of a given gene, various combinations of mutant and wild type alleles is possible. Such homo- and hetero-duplexes display varying electrophoretic mobility leading to banding pattern differences as displayed in panel b. The two lanes on the left show homoduplex presence (320 basepair long product highlighted by an arrow) (*see Color Plates*)

such, the protocol depends on the melting properties of a double stranded amplimer upon electrophoresis in an agarose or polyacrylamide gel, which is subject to a linear and increasing denaturing gradient. Once the electrophoresed amplimer reaches the portion of the gel where denaturation starts to occur, then changes in the electrophoretic mobility of the amplimer begin to become apparent and, dependent on the melting properties of that particular amplimer, migration through the gel slows or may even come to a complete stop. The behaviour of the amplimer after DGGE may be determined by simple ethidium bromide staining of the gel.

DGGE can be performed via two fundamentally different methodologies, with denaturation being induced by (i) the application of a temperature gradient across the gel (Temperature GGE or TGGE), or (ii) by the use of chemical gradients (urea or formamide or a combination of both). If the TGGE method is used, then the temperature gradient can be set parallel to the direction of electrophoretic migration or else positioned at 90°C to this direction. This requires a single large well on the gel and the final result will be a melting curve shaped electrophoresis pattern for the DNA sample under investigation. Other variants include two-dimensional DGGE (2D-DGGE), which enhances sensitivity and specificity [Dabora et al., 1998]. Though the 2D-DGGE technique is sub-optimally suited for GC-rich DNA fragments, the methodology may be used to process large numbers of samples [Fodde and Losekoot, 1994]. For the study of larger DNA molecules, restriction digestion prior to DGGE analysis is an option that may help to improve resolution. DGGE sensitivity can also be improved by modification of the PCR product, for example by the introduction of a "GC-clamp" at one end of the amplimer via the use of suitably modified PCR primers [Gejman et al., 1998]. This GC-clamp usually consists of 40 guanosine and cytosine bases whose prime characteristic is that it does not fully denature during the DGGE procedure. Consequently, it functions as a double stranded "pan handle" which due to its decelerating characteristics will prevent further mobility of the denatured PCR amplimer. However, the other (non-clamp) part of the DNA double helix will dissociate at increased temperatures. This method is usually sensitive down to single nucleotide polymorphisms, with the additional advantage that in PCR amplifications where both mutant and wild type sequences are present, four different heteroduplex molecules may be formed contributing to the sensitivity of the system (Fig. 9.7). Several computer programs helpful in predicting suitable DGGE conditions are commercially available. MELT87, MELT94, SQHTX and TGGE-STAR can generate estimates of melting conditions, as well as the estimated electrophoresis time and the best position and

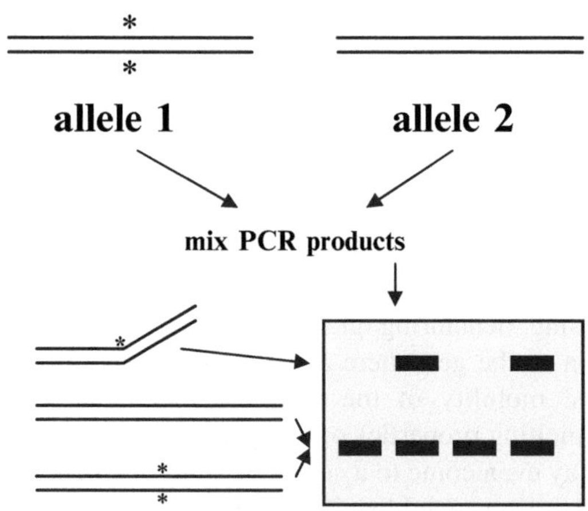

Fig. 9.7 Principle of the DGGE technique. A double stranded amplimer containing an artificially introduced GC-clamp at one of its termini migrates through a polyacrylamide gel which contains a chemical gradient of a denaturing agent (e.g. urea). Once the non-GC-clamp DNA domain melts, migration ceases. If normal and mutant alleles of a gene have been co-amplified, then four different homo- and heteroduplexes will be present in the final PCR incubation mix. These heteroduplexes will exhibit different melting behaviours with the heteroduplex complexes ceasing to migrate at different positions within the gel dependant on the nature of the mismatches present within the separate heteroduplexes

composition for a GC-clamp. These programs have been shown to be able to shorten the set-up time required in developing novel DGGE strategies. It should be noted that DGGE is not 100% sensitive to all mutations, as certain mutations may have little affect on the melting behaviour of the denatured single strands. However, DGGE is still a powerful method for mutation screening [Hovig et al., 1992].

(vi) Image analysis of PCR amplification products in gel electrophoresis systems: As previously mentioned, the most common method for examining the quality and quantity of PCR amplification products is to electrophorese the amplimers on an agarose or polyacrylamide gel and then visualise the amplimers by means of intercalating chemical dyes and ultraviolet transillumination. Intercalated dyes fluoresce in visible wavelengths when subjected to ultraviolet transillumination and this visible light may be recorded using an ordinary (Polaroid) camera and special photographic film. However, this approach is becoming a little outdated as elegant electronic systems for the analysis and documentation of PCR amplification products in agarose and polyacrylamide gels are becoming increasingly common in many laboratories. Using these systems the fluorescent signal is translated into an electrical signal which is collected on an electronic chip and then turned into a video signal. The image is then reconstructed on a monitor or video-screen from which the gel results can be analysed. The FluorImager (Molecular Dynamics), Eagle Eye (Stratagene), ChemImager (Biozym) and the BioProfil (Vilber Lourmat) systems for example, combine sensitive charged couple device (CCD) cameras with sophisticated software programs. Most of these programs can translate pictures into longitudinal scans, spectrographs or even numerical tables, with the CCD camera allowing the detection of very faint signals and providing unparalleled sensitivity.

Even using such sophisticated electronic analysis systems, various problems may arise in interpreting the electrophoresed amplification products. For example, for both agarose and polyacrylamide gels, the thicker the gel the higher the voltage required to separate the DNA fragments. Due to the fact that thicker gels generally require a higher voltage setting during electrophoresis and that the greater the voltage the larger the amount of heat generated (under constant resistance), the surface and/or inner layers of thicker gels may experience a more extensive temperature gradient than the surface and/or inner layers of thinner gels. Such uneven temperature gradients may lead to electrophoretic band deviations and result in a characteristic "smiling" band pattern being obtained, where amplimers loaded in the inner wells of a gel pass relatively more quickly through the gel matrix than amplimers loaded in the cooler inner wells. This effect can greatly hinder the interpretation of amplimer size and hence amplimer quality using gel electrophoresis systems.

(vii) Excising and cleaning PCR amplimers from electrophoresis gels: Electrophoretically separated amplimers can be excised from agarose or polyacrylamide gels and retrieved from the matrix using several described purification methods, the simplest of which involves cutting out the section of agarose containing the required DNA, immersing the gel "plug" in water or buffer, and incubating at 65°C until the agarose melts. The DNA will diffuse into the water phase and can be purified by phenol extraction and ethanol precipitation. If the solution is heated further, the agarose will dissolve more completely thereby increasing the quantity of DNA freed from the gel plug, though it should be noted that the quantity of inhibitory substances released from the plug will also be increased. Other methods include the use of "low melt point" agarose and beta-agarase I enzyme. Most laboratories now use commercially available and specific kits for the separation of DNA fragments from agarose or polyacrylamide gels, electrophoresed in both tris-acetate (TAE) and tris-borate (TBE) buffers. Agarose can be dissolved in high molarity sodium iodide (NaI) solutions by heating, and the liberated DNA be bound by silica particles. Once bound, the beads can be washed and the DNA finally eluted in a simple buffer. This principle is used in the GeneClean DNA extraction kit and is suitable for DNA fragments from 200 to 2,500 bp in length. Other kit assays have been developed, including assays which use chaotropic salts such as guanidine isiothiocyanate and silica filters, e.g. Zymoclean Gel DNA Recovery Kit (Zymo Research, California, USA). Other kits can be used to clean-up DNA from both gels and PCR reactions, e.g. Wizard SV Gel and PCR Clean-Up System, (Promega). For all extractions, it should be noted that for DNA fragments greater than 2 kbp in size, the larger the fragment the greater the risk of breaking the DNA into smaller fragments by "shear" forces during processing, e.g. by vortexing.

9.2.2 Probe Hybridisation Methodologies

Labelled DNA probes designed to be complementary to a particular nucleotide sequence, are able to specifically bind to that particular DNA sequence under the correct reaction conditions. Such probes have been widely applied to determine the presence or absence of specific DNA molecules (including PCR amplimers) for assessing DNA quantity and assessing the presence of mutations for nearly half a century. In fact, such probe hybridisation protocols are still of key importance today, especially for the determination of sensitivity and specificity of PCR amplification reactions [Marsh et al., 2000].

The methodology behind probe hybridisation protocols relies upon the specific annealing of a single stranded DNA or RNA probe to a single stranded target DNA or RNA sequence. For DNA identification, the double stranded DNA is first denatured into single strands (usually achieved by heating or by the addition of certain chemical denaturants such as urea or formamide), and an excess number of labelled single-stranded complementary probe molecules are added to the solution. The reaction conditions are then altered via the use of a series of "stringency" washing steps at pre-determined temperatures, so that the probe molecules are only able to bind to their specifically intended target molecule. Any specifically bound probe remaining is then detected via the attached label. The stringency of hybridisation is essentially determined by the temperature and the salt concentration (both magnesium and sodium ions play an important role), such that the higher the incubation temperature and the lower the salt concentration, the more stringent the hybridisation conditions. The annealing temperature (T_m) of probes ranging in size between 100 and 200 base pairs can be calculated using the formula developed by Baldino et al. [1989], and if the probe and target sequence are 100% homologous then very stringent hybridisation conditions may be applied. The smallest size for a probe is 13–15 nucleotides, with smaller probes (especially AT-rich probes) requiring the additional use of certain specific proteins, e.g. the "minor groove binding protein" [Kutyavin et al., 2000]. This particular protein can increase the hybridisation temperature of small probes by 15–20°C. In contrast, the largest DNA probes may be thousands of base-pairs in length, though these are rarely used in probe hybridisation protocols for the detection and identification of PCR amplimers.

The detection of hybridised probe molecules is of course an extremely important part of the identification of PCR products by probe hybridisation. Previously, DNA probes were labelled with radioactive materials, though this is nowadays an uncommon practice, not least because of the problems associated with the use and disposal of such materials. Several labelling strategies are currently available, including biotin, digoxygenin, fluorescein and enhanced chemical luminescence (ECL). Biotin is a compound that interacts very strongly with streptavidin, a protein that can be equipped with additional reporter molecules such as enzymes, colloidal gold particles or fluorochromes. Digoxygenin and other molecules may be the target for specific labelled antibodies [Sharma et al., 2006; DiCesare et al., 1993]. Classical protocols involving peroxidase and diazobenzidine (DAB)/hydrogenperoxide/imidazole or alkaline phosphatase and NBT/BCIP are still popular and frequently used. Peroxidase labels, (detected using tyramine signal amplification procedures), are currently one of the most sensitive non-radioactive procedures available to molecular biologists [Van Gijlswijk et al., 1997]. Of increasing importance are chemoluminescent labelling protocols using lumiphos, AMPPD, luminol/TMB or luciferin, which after activation, emit light which may be detected and recorded using light-sensitive film (Fig. 9.8). High-throughput PCR-based probe hybridisation protocols have also been described [Martin et al., 2000].

Though hybridisation reactions may be performed in solution, it is usual to immobilize either the probe or the target molecules to a solid phase such as charged nylon membranes or special microtitre plate wells (as used in PCR ELISA, Section 9.7 below). This helps simplify the stringency washing procedures.

Modern advances in chemistry and an increased knowledge of the structure of nucleotides have now allowed the development of artificial nucleotide-like analogues that mimic the naturally occurring nucleotides. Peptide or protein nucleic acids (PNA) are artificially created nucleotides where the peptide-linked backbone of the molecule has been modified to serve as an attachment site for nitrogen bases [De Koning et al., 2003]. PNA molecules hybridize with complementary nucleic acid molecules according to the Watson and Crick scheme [Tackett et al., 2002], but possess an affinity for hybridisation so strong that they can displace any complementary DNA strand. Further, PNA molecules can hybridize to double stranded DNA forming a very unusual triplex PNA/DNA structure. PNA probes have several advantages over DNA probes. Firstly, the PNA molecule is electrically neutral, with the result that the hybridisation of PNA molecules is less susceptible than DNA probes to electrostatic repulsion or changes in salt concentration. Secondly, hybrids which contain a PNA

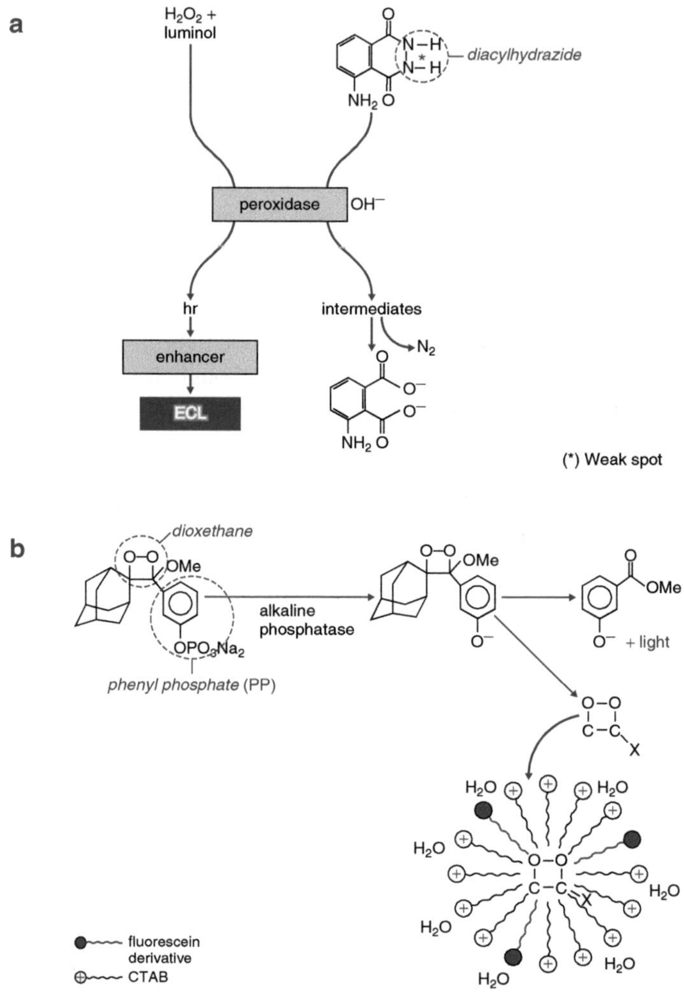

Fig. 9.8 Top panel A: the use of peroxidase in enhanced chemoluminescence (ECL) detection protocols. The oxidation of luminol under alkaline conditions results in the generation of blue light emitted at a wavelength of 428 nm (luminescence). Amplification of this signal with chemical enhancers may increase this signal by approximately 1,000-fold. The emitted chemoluminescent light may be detected by light sensitive film. Bottom panel B: the electrochemical detection of dioxetanes as a means for signal detection after nucleic acid hybridisation. Energy-rich peroxide compounds (such as 1,2 dioxetanes like "Lumigen" and "AMPPD") are water soluble and may be stabilized by the addition of chemical groups such as phenylphosphate. After removal of the phosphate, the compound degenerates through a set of reaction intermediates into a hydrophobic product generating photons in the process. By encapsulating the intermediates in cetyltriethylammoniumbromide/fluorescein micelles the energy content of the photons may be transferred to fluorescein molecules resulting in an approximate 400 times increase in the yield of light

molecule are much more stable than DNA/DNA, DNA/RNA or RNA/RNA hybrids, allowing the use of higher stringency conditions in hybridisation reactions, including PCR protocols [Wolffs et al., 2001; Jiang-Baucom et al., 1997].

(i) Southern blotting: Southern blotting is the name given to a technique first developed by Southern et al. [1975], in which either single-stranded or double-stranded DNA molecules are electrophoresed in an agarose or polyacrylamide gel and then transferred to a nylon membrane. For double-stranded DNA molecules, efficient blotting is usually achieved via diffusion of the double-stranded DNA in a denaturing solution, in order to simultaneously generate and blot individual single strands of the DNA duplex. After application, the DNA is fixed to the membrane by heating (baking) or by illumination under ultraviolet light. Solutions containing single-stranded, labelled, "DNA probes" with specifically designed sequences may then be applied to the surface of the membrane and hybridisation allowed to occur under conditions of varying stringency (achieved by varying the salt concentration, formamide concentration and incubation temperature). DNA probes that are complementary to a DNA sequence attached to the membrane will specifically hybridize to this complementary sequence (under the correct stringency conditions), such that the probe remains attached to the membrane during subsequent washing steps. After several washing steps, hybridisation of the probe may be established using the label attached to the probe. Southern blotting, probe hybridisation and detection protocols may be up to 100 times more sensitive than simple gel electrophoresis, allowing the detection of smaller quantities of DNA than agarose gel electrophoresis. Also, though Southern blotting/hybridisation protocols are more time consuming and technically demanding than agarose gel electrophoresis, the specificity of DNA probe hybridisation in Southern blotting allows specific and non-specific PCR products of similar sizes to be differentiated. Further, the use of commercially available and electrically charged Southern blot membranes allows the DNA probe to be stripped off of the membrane without denaturing or removing the attached target DNA molecules. This allows a single membrane to be used with several specific DNA probes, which may be sequentially hybridised and washed off of the membrane.

(ii) Dot spot analysis: A quicker method than Southern blotting involves the immediate "spotting" of DNA (including PCR amplimers) onto a solid phase membrane without first separating the products by molecular weight using gel electrophoresis. For PCRs, a few microliters of the post-PCR mix are denatured and spotted onto a hybridisation membrane, which after baking or ultraviolet illumination is used for probe hybridisation experiments. In this way, membranes containing the amplimers from many different PCRs can be quickly prepared facilitating the screening of large numbers of PCR amplimers all at the same time. Moreover, though the use of "dot spot" analysis results in loss of information regarding the size of the PCR amplimer, the methodology is much quicker than Southern blotting, with the information gained being sufficient for many screening protocols, e.g. the high-throughput identification of human papillomavirus (HPV) variants using broad spectrum PCR primers and species-specific probes [Kleter et al., 1999].

(iii) Oligomer restriction: This is a specific variant of RFLP (restriction fragment length polymorphism) analysis that is suited for the detection of point mutations that alter specific restriction sites. A prerequisite for this method is that mutations occur which have the capacity to generate differential RFLP patterns within the genome or gene of interest. Two distinct variants of the *oligomer restriction* methodology may be performed. In the first variant, the amplified PCR product is denatured and hybridised with a 5′-end labelled DNA probe containing the expected mutations or non-mutated wild type DNA under conditions so strict that the probe will only hybridize if there is 100% complementarity between the probe and target DNA molecule (Fig. 9.9). After a few washing steps (to remove non-hybridised probe), any hybridised double-stranded DNA molecules are digested with a restriction enzyme. If a DNA heteroduplex is present (i.e. specific hybridisation of the labelled probe to the target DNA molecule has occurred), then the double stranded heteroduplex fragment will be digested. Dependent on the experimental design either the mutant or the wild type complex is digested. The presence of any mutation(s) may then be determined by gel electrophoresis and the presence of different RFLP patterns, where the presence of wild type, mutant or both types of polymorphism may be determined.

The second *oligomer restriction* variant utilizes the same steps of PCR amplification and hybridisation as outlined above, but a second restriction enzyme is also used that cuts the heteroduplex only if a mutation event has occurred. In this case not only wild type but also mutant alleles are cut, albeit by different restriction enzymes. This second *oligomer restriction* variant generates an internal control giving an indication of the "digestability" of the heteroduplex fragment and preventing mis-interpretation of the results, which could theoretically occur due to inhibition of activity of one or more of the restriction enzymes within the digestion mix. This

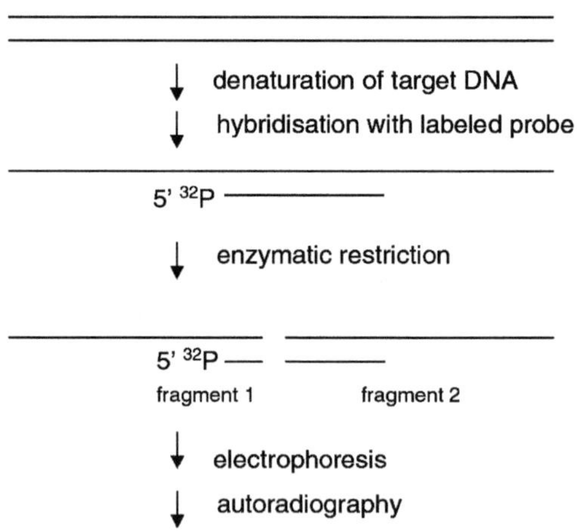

Fig. 9.9 Oligomer restriction analysis. An amplimer is hybridised with a labelled DNA probe which overlaps a particular restriction enzyme recognition site. If the restriction site is not disrupted by the presence of one or more nucleotide changes (mutations), then the probe will be cut into two pieces by the corresponding restriction enzyme. The presence/absence of fragments can then be visualized by electrophoresis

strategy has been used to detect mutations leading to sickle cell anaemia: where a mutation leads to the loss of a *Dde*I site but the addition of a *Hinf*I restriction enzyme digestion site [Simsek et al., 2002].

(iv) Allele specific oligonucleotide (ASO) analysis: The ASO technique is especially relevant in clinical situations involving genetic testing and counseling. The method allows the identification of point mutations without the need for changes in restriction enzyme site sequences. However, the technique is not useful for regions of DNA where mutations may exist in close proximity to each other, or where multiple mutations, deletions, insertions, or more complex rearrangements occur. The ASO method *per se* is a probe hybridisation technique, where the target is fixed to a membrane, but where probe hybridisation occurs at higher stringency levels than oligomer restriction analysis. This increase in stringency is achieved by the inclusion of tetramethyl ammonium chloride in the hybridisation wash buffer along with hybridisation conditions at relatively elevated temperatures. For example, using ASO analysis, a 19-meric oligonucleotide probe will hybridize at a temperature of 60°C and a 15-mer probe at a temperature of 55°C. This change in duplex thermostability allows for the detection of point mutations due to the differential stability of binding of specific probes to either wild type or mutant target amplimers. Nucleotide mutations resulting in an adenosine/guanosine, guanosine/adenosine, or guanosine/cytosine switch drastically affect the stability of probe hybridisation under such stringent conditions [Bakkus et al., 2004; Sawyer et al., 2003].

(v) ASO analysis via reverse dot blotting: Essentially, this is an ASO hybridisation methodology incorporating a capture step (Fig. 9.10). Initially, a panel of unlabeled allele-specific oligomers are extended at their 3′-ends by the addition of several thymidine (poly-dT) nucleotides using the enzyme terminal transferase, else allele-specific oligomers are designed with extra poly-dT 3′-ends, which are chemically synthesised along with the oligonucleotide *per se*. All of these poly-dT tailed allele-specific oligomer probes are then covalently bound to a nylon membrane and the PCR amplimers to be investigated (previously labelled with, for instance, biotin during PCR thermocycling) are allowed to hybridize to the membrane under stringent conditions. After washing, amplimers that have hybridised to the array of probes on the membrane may be detected via their biotin label. This technique enables the detection of multiple allele specific mutations in a single hybridisation experiment per PCR product. The amplimers used may be long or short, so long as the hybridisation kinetics of the probe/target DNA is sufficiently stringent. This technique can also be performed as a sandwich type assay where the PCR amplimer product probes do not have to be (biotin) labelled. After initial ASO hybridisation and washing steps, the membrane is hybridised once more, using a labelled probe, which detects a separate conserved region of the amplimer which is not involved in binding to the target DNA on the membrane. These secondary probes can be of a variable length, dependant on the length of the fragment to be identified and the availability of specific sequence motifs. The advantage of this sandwich-type method is that labelling of the PCR amplification products (usually expensive) is not required.

(vi) Reverse line blot hybridisation: Essentially, this is a commercially available variant of reversed dot blotting (see ii) above). A fundamental difference

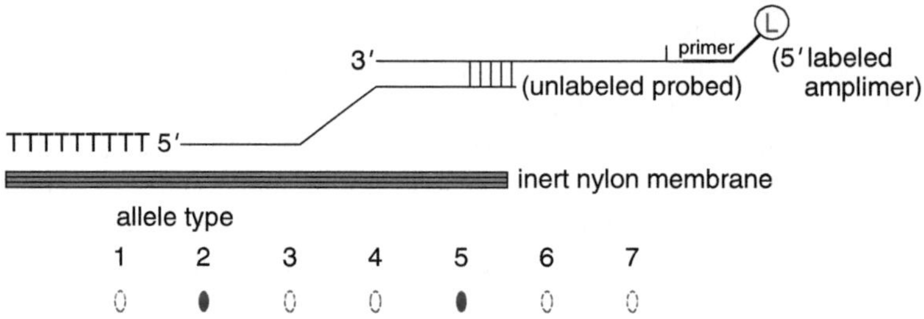

Fig. 9.10 The principle of reversed hybridisation. The scheme on top shows how Oligo-dT extended probes are immobilised on a membrane which is then hybridised with labelled PCR amplification products. Specific hybridisation of the PCR amplification products to seven different probes shown in the bottom section identifies the nature of the PCR amplification products and hence the nature of the target DNA amplified (for example specific genetic alleles)

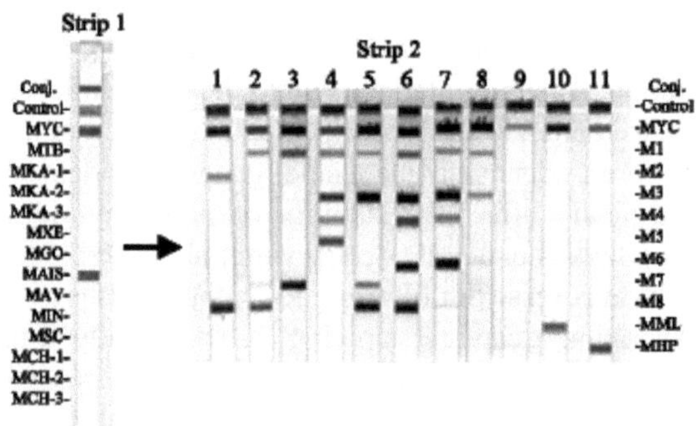

Fig. 9.11 Reversed line blot hybridisation. Mutations in various regions of the RNA polymerase encoding *RpoB* gene in *Mycobacterium tuberculosis* give rise to resistance to the antibiotic rifampin. If various conserved regions of the *RpoB* gene are amplified using PCR, then the resultant amplimers may be screened for the presence of mutations by their ability to hybridize to various DNA probes that contain *RpoB* gene mutations. These DNA probes are attached to a solid matrix (filter) and once the amplimers have been allowed to hybridize, patterns of hybridisation consistent with the presence or absence of specific rifampin resistance encoding mutations may be visualized by ethidium bromide staining. Representative LiPa patterns are visible on the INNO-LiPa MYCOBACTERIA strip [Mijs et al., 2002]. Different subgroups can be distinguished in strip 1, and two species-specific probes are present in strip 2; one for *M. malmoense* (probe MML, row 10) and one for *M. haemophilum* (probe MHP, row 11) (Reproduced with permission from 'The Society for General Microbiology')

however, is that in this system multiple adjacent mutations may be detected. The principle of the test (in this example a genotyping assay for *Mycobacterium tuberculosis* which is the bacterium responsible for tuberculosis) is highlighted in Fig. 9.11. The PCR primers are designed to bind to conserved sequences which enclose regions of DNA prone to mutation (in this example these mutations lead to resistance to a particular antibiotic). Any mutation within the PCR amplimer sequences can then be determined using specifically designed target molecules which contain known sequence mutations and which have been immobilised as lines on a nylon membrane. After hybridisation, hybridised amplimer may be detected using labels attached to one of the PCR primers or via labelled amplimer (added during PCR thermocycling *per se*).

(vii) Discriminant hybridisation: Initially, a variety of unlabeled PCR amplimers are covalently bound to a nylon membrane. These may be quite diverse in nature. Subsequently, 5′-end or 3′-end terminally labelled oligonucleotide probes are hybridised to the unlabelled amplimers under high stringency conditions. Since in this protocol, all of the different PCR products have been hybridised under the same stringency conditions, quantitative differences in the intensity of signal between the different hybridised probes at particular locations on the membrane reflect the presence of complete or partial binding between different probes and targets. It should however be noted, that this technique may vary widely in the results that it delivers, with its reliability being strongly dependent on the accuracy of the stringency conditions used. As a matter of routine, well-characterized positive and negative control samples should be included in every experiment. However, although the results generated by this technique need to be interpreted with care, the methodology has been successfully applied in various scientific fields, including paternity testing, haemophilia testing and HLA-DQ sub-typing [Baird, 1998].

(viii) Amplification refractory mutation system (ARMS): This procedure, (also known as competitive oligonucleotide priming), involves the detection of point mutations through competitive oligonucleotide priming. Each DNA sample to be analysed is amplified in two separate PCRs. In one of the PCRs, the primers are designed to recognize a wild type sequence, whereas in the second PCR the primers are designed to be complementary to a mutant sequence. Moreover, in this second PCR, the primers are designed in such a manner that the nucleotide mutation to be detected is incorporated at the extreme 3′-end of one of the primers. If the mutation is not present, then the mismatch at the 3′-end will inhibit extension of the DNA chain by the thermostable DNA polymerase during thermocycling, yielding an absence of PCR products. In all ARMS methodologies, it is also necessary to design and incorporate PCR primers that amplify a conserved region of DNA to act as an internal positive amplification control (in effect generating a multiplex PCR). This ensures that a lack of amplification products in PCR reactions incorporating mutant primers actually indicates the absence of a mutation in the target and is not the result of inhibition of PCR amplification by PCR inhibitors. It is also important to note that proof-reading thermostable DNA polymerases should not be used in ARMS protocols as they will remove any 3′-end mismatch on primer/target DNA hybrids, thereby allowing amplification to occur in ARMS PCRs where the mutation is not present.

Sometimes, it may be necessary to introduce a second 3′-end mutation, in the ARMS "mutation-detecting" PCR primer. This second mutation acts to further destabilize primer hybridisation, increase the inhibition of mismatching primer amplification, and consequently increases the specificity of ARMS assays. In fact, if wild type and mutant primers are differentially labelled, then both wild type and mutant ARMS primer pairs may be included in the same ARMS PCR reaction mix. A positive amplification signal will then be generated in all of the subsequent ARMS

PCR reactions, with the differentiation between wild type and mutant amplimers being made on the basis of the label incorporated into the amplimer. ARMS analysis may also be successfully adapted to the PCR ELISA format (see (ix) below).

(ix) PCR ELISA and its variants: The enzyme linked immunosorbent test (ELISA) is a robust assay format that has also been adapted for the identification of PCR amplimers. Such PCR ELISA assays can be performed in 96, 192 or 384 ELISA plate formats, allowing the processing of multiple PCR reaction products at the same time without the need for complex blotting or spotting procedures. Furthermore, the interpretation of large numbers of test results is made somewhat easier due to the fact that spectrophotometric absorbance values are recorded, which are expressed as numerical values, allowing semi-quantitative interpretation of the results. However, a cut-off value, used to distinguish a positive signal from a negative signal, must be experimentally determined for each new PCR ELISA assay.

The methodology used in PCR ELISA is basically similar to performing a PCR followed by a Southern blot analysis, except that the "solid phase" onto which the target PCR amplification products are bound is the surface of a microtitre plate well. Figure 9.12 illustrates the various steps involved in a PCR ELISA assay system, with

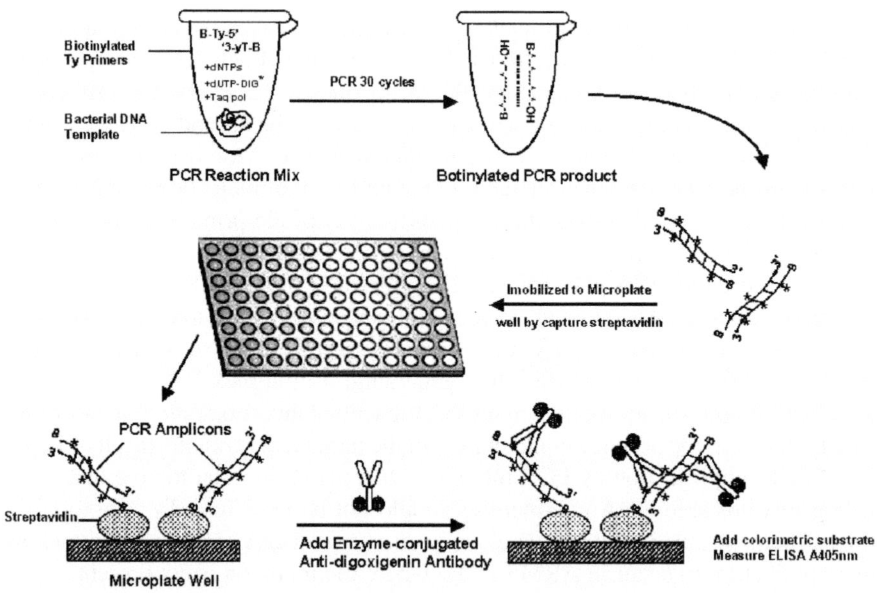

Fig. 9.12 A PCR ELISA system for the identification of PCR amplified DNA fragments. In the example shown, digoxigenin-deoxyuracil triphosphate (DIG-dUTP) is added to the PCR mix alongside dATP, dCTP, dGTP and dTTP and is hence incorporated into the subsequent PCR amplification products. The DIG-labelled amplification products are then hybridised to a complementary biotinylated DNA probe, and the resultant duplexes captured using streptavidin coated microtitre plate wells (biotin-streptavidin binding being extremely efficient and strong). The incorporated DIG molecules can subsequently be identified using anti-DIG antibodies to which peroxidase enzyme has been added (From Luk et al., 1997)

the main requirement being a specific, labelled amplimer that after further processing is able to generate a signal that can be read by a spectrophotometer. Another key requirement is the presence of an anchor mechanism that is required to attach the PCR amplimer to the wells of a microtitre plate. Various amplimer anchoring methodologies are currently commonly used, with many taking advantage of the extremely strong binding that takes place between streptavidin and biotin molecules. These include:

(a) Streptavidin coated microtitre plate wells used in combination with biotinylated PCR amplification products.
(b) Streptavidin coated microtitre plate wells used in combination with single stranded PCR product targets/biotinylated probe hybrid – the double stranded PCR amplimer is first denatured into single strands and then hybridised with specific probe DNA which has been previously labeled with biotin. This mix is then applied to streptavidin coated microtitre plate wells.
(c) α-Digoxygenin, fluorescein or enzyme coating of the microtitre well, leading to binding of an amplimer which has an appropriate specific antibody attached.

Several commercially available PCR ELISA test systems are currently available, including assays for the detection of the sexually transmitted pathogen *Chlamydia trachomatis* [Verkooyen et al., 2002]. Recent variations on the PCR ELISA theme include "Dynamic Allele Specific Hybridisation" (DASH) and "Detection of Immobilised Amplified Products" DIAPOPS! The "dynamic allele specific hybridisation" protocol or DASH (Hybaid), is illustrated in Fig. 9.13. In this protocol, PCR is performed where one of the primers has been 5′-end biotinylated. After PCR thermocycling, the double stranded amplimers are coupled to streptavidin-coated microtitre plate wells. An alkaline wash is then used to remove the non-biotinylated strand of the amplimer and a specific probe is added in combination with an intercalating fluorophore (a dye that inserts between double-stranded DNA heteroduplexes). After washing to remove unbound DNA, a fluorescence measurement is taken and the "base" level of fluorescence recorded. The temperature is then gradually raised towards the melting temperature of the probe/target hybrids, so that the hybrids begin to dissociate and the level of fluorescence gradually decreases (as the trapped fluorescent intercalating dye is released). This process allows the relative stability of the probe/target heteroduplex to be measured, enabling a distinction to be made between any 100% complementary (completely hybridised) PCR products and PCR products containing nucleotide mutations (which generate less stable hybrids). The "Detection of Immobilised Amplified Products" or DIAPOPs (Nunc) technique utilises PCR amplification and post-PCR follow up reactions all within the same 96 or 384 well microtitre plates. Initially, one of the PCR primers (used at a concentration 1/8 of the concentration of the other unbound primer) is chemically attached to the walls of microtitre plate wells and PCR mix and template DNA subsequently added. All ingredients necessary for PCR amplification are then available within each microtitre plate well. After thermocycling using an especially designed PCR thermocycler that can accommodate microtitre plates, the amplimers are separated into single strands, and the unbound single strands washed away. Single stranded PCR

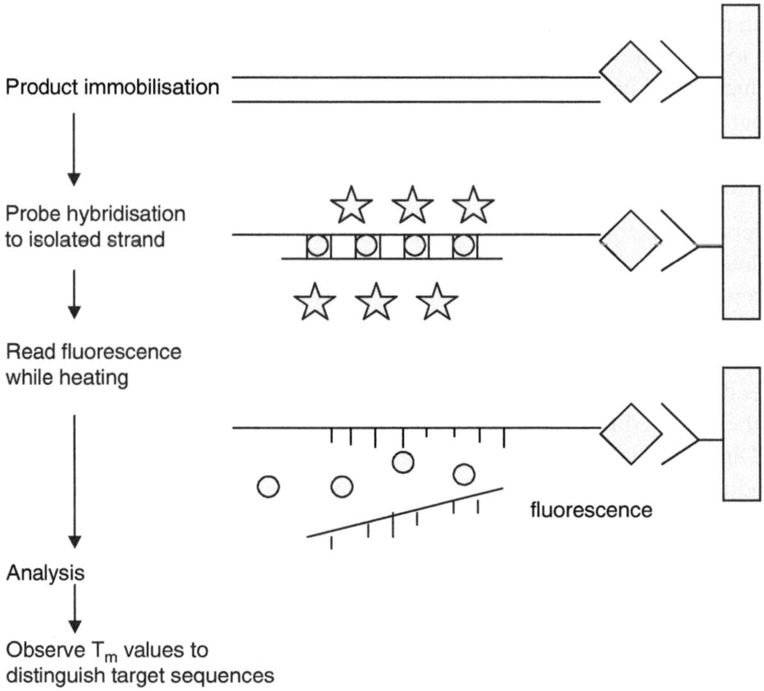

Fig. 9.13 Dynamic allele specific hybridisation. A biotinylated PCR amplification product is bound to a streptavidin coated matrix and the non-biotinylated DNA strand of the PCR product is removed by alkaline treatment. A specific complementary single stranded DNA probe is then added to the single-stranded captured PCR amplimer in the presence of an intercalating fluorochrome, which binds between the two DNA strands (filled circles). After a washing step, the temperature is gradually increased so that the melting temperature of any hybridised heteroduplexes may be determined by measuring the decay of fluorescence (filled stars)

amplimers generated from the primer that was chemically attached to the microtitre plate are then available for hybridization using target specific DNA probes [Said et al., 1994].

(x) Micro-arrays and DNA chips: Over the past 15 years many advances in hybridisation technology have occurred. However, until recently there has been an upper limit on the number of different DNA sequences that may be utilised in a single hybridisation experiment, limited by the number of amplimers or probes that may be attached to the solid phase (e.g. the size of a nylon membrane). Traditionally the number of different PCR amplimers or probes that could be attached to the solid phase ranged from one up to a few dozen oligonucleotides, but recent revolutionary developments in the fields of miniaturization and robotics have significantly increased the technological possibilities. Current "DNA micro-arrays" for example, may contain several hundreds or thousands of probes physically attached to a reasonably sized solid phase membrane, whilst "DNA chips" (composed of a borosilicate glass surface and with dimensions in the square centimetre range), have an even larger capacity for the number of probes which may be immobilised upon their very small surfaces. In fact, the distinction between DNA micro-arrays and DNA chips is quite arbitrary, being based on the number and the density of immobilised probes present on their surface. However, this distinction is not strictly observed and many researchers use both terms interchangeably.

Micro-array technology is generally somewhat more flexible than DNA chip technology [Pollack et al., 1999; Southern et al., 1992] with a diversity of solid phase media being available for use in micro-arrays, including nylon membranes and microscope slides. Also, the range of nucleic acid molecules that may be attached to the solid phase is generally larger as attachment does not rely on "photolithographic technology" (see following paragraph below). The range of available nucleic acid molecules that may be immobilised on micro-arrays includes PCR amplimers, cDNA clones, RNA, genomic clones and also (chemically synthesised) oligonucleotides. The pace of current developments is so rapid that a complete review is almost impossible, however several commercially prepared micro-arrays are available for purchase "off the shelf", including amongst others, several DNA gene micro-arrays available from Invitrogen, whose "*Human Gene Filters* release I to IV" arrays contain over 5,000 human DNA clones (including named genes and expression tagged sequences), and whose *Human Named Genes Genefilter* contains over 4,000 gene specific probes. Other companies offer human tissue specific micro-arrays that harbour cDNA clones specifically expressed in a variety of different human tissues. Rat and mouse micro-arrays are also available which contain over 5,000 rat or mouse specific sequences. Yeast gene micro-arrays contain probes for all 6,000 genes identified covering the complete genome sequence available for this organism [Christie et al., 2004]. However, though commercial micro-arrays are available, laboratories are increasingly operating in a self supporting manner by setting up centralised facilities for "in house" micro-array manufacture with poly-L-lysin or silane coated glass slides being very popular as solid phase materials. Using these materials, DNA samples can be arrayed at distances of between 500 and 700 μm using volumes as small as 10 nl [Battaglia et al., 2000].

DNA chips may also be created via sequential chemical synthesis allowing the ordered immobilisation of nearly half a million probes onto the surface of borosilicate glass via patented photolithographic technology [Jacobs and Fodor, 1994]. The principle of this photolithographic process involves the organo-chemical synthesis of oligonucleotides via chemically reactive intermediates generated by photo-activation. The method is homologous to classical solid phase oligomer (PCR primer) synthesis, though in this case the synthesis of thousands of parallel oligonucleotides containing different sequences is facilitated by miniaturisation. The Affymetrix company holds the strategic patents for this "photolithographic technology" meaning that as such the technology is not widely available. However, custom made chips can be ordered from Affymetrix and several general DNA chips are being sold separately by this company, including DNA mutation detection chips for the human immunodeficiency virus, the p53 tumour suppression gene and cytochrome P450 [Vahey et al., 1999]. Dedicated chips for gene-expression profiling are available as well. These chips contain thousands of gene-specific oligomer targets from mouse, man, rat, yeast, *E. coli* and *Arabidopsis thalania* [Marshall, 2001]. The number of commercially available Affymetrix chips is steadily increasing.

The general technology required to immobilise very large numbers of probes onto small areas is commercially available and the same commercial enterprises can also provide customised and detailed DNA chip or array mediated hybridisation analyses of PCR products, mixtures of PCR products, cDNA preparations and RNA samples upon request. Another very important point is that specialised equipment is required to read the signals produced by such high-density probe arrays, requiring dedicated software for data management after micro-array result visualisation. In many cases, insights into disease processes (e.g. gene expression in healthy versus diseased cells), may be achieved using dedicated computer software which can add or subtract the data obtained from two distinct hybridisation experiments on the same DNA array. If the amplified products from each PCR possess labels that fluoresce at different wavelengths, then combining the micro-array results *in silico* using the specialised computer software, will generate three different colours dependent upon the hybridisation (presence or absence) of PCR amplimer to the individual micro-array oligomers.

Micro-array and DNA chip technology facilitate the probe mediated screening of genetic polymorphisms or expression profile differences using an unprecedented density of probes per hybridisation reaction, effectively allowing the analysis of genome-wide point mutations, deletions and insertions all in a single hybridisation experiment. The current state of affairs in micro-array and DNA chip technology means that what is new technology today is out-of-date by tomorrow. For example, an important new development is that tissue samples can also be attached to micro-arrays, with recent advances allowing more than 1,000 minute tissue sections to be applied to a single glass slide. This facilitates the detection of differential gene expression in various healthy and diseased tissues. In the future, this will enhance our insight into the mechanisms of many pathogenic processes ranging from cancer to infectious diseases.

9.3 Real-Time Analysis of PCR Amplimers

The *in vitro* analysis of real-time PCR amplification products is achieved via the use of fluorescent dyes/probes and is reliant upon the measurement of the presence or absence of fluorescence either during or immediately after PCR thermocycling (the amplimers are not removed from the PCR vessel). Currently, several different protocols have been developed that directly couple the "on-line" measurement of fluorescence to PCR product amplification. All of these protocols require the use of a specific real-time PCR thermocycler (e.g. Fig. 9.14) as well as specific PCR reaction vessels. The essential requirements for real-time amplimer analysis include the use of optically transparent reaction vessels, a source of ultraviolet (UV) illumination and a spectroscopic fluorescence reader (the latter two devices being built into the PCR

Fig. 9.14 The Roche LightCycler – an example of a real-time PCR machine

thermocycler itself). Real-time assays function well over a broad spectrum of initial target concentrations, i.e. ranging from 3 pg to 30 ng target material.

9.3.1 In vitro Analysis Using Intercalating Chemical Dyes

The simplest real-time PCR analysis systems work by measuring the relative intercalation of fluorescent dyes within double stranded PCR amplification products, with an increase in the amount of amplimer synthesised during each PCR cycle

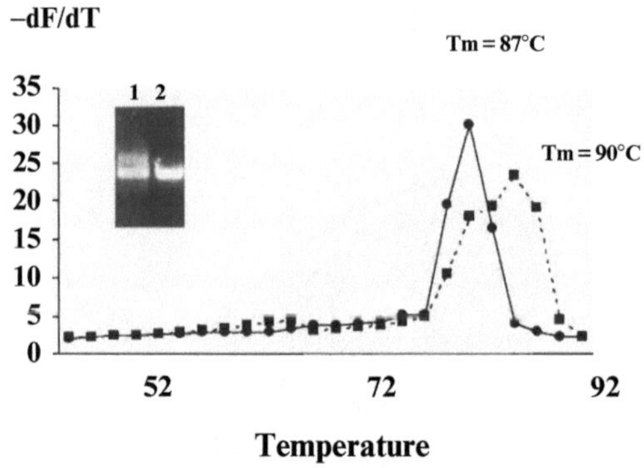

Fig. 9.15 Melting-curve profile of PCR products containing specific and nonspecific amplicons. Melting curve analysis performed after PCR amplification of cDNA from ®TC-3 cells induced 7 h with different cytokines concentration: IL-1, 104 U/L and 105 U/L (■) or 0.3104 U/L and 0.3105 U/L (•), respectively. An agarose gel electrophoresis of PCR products from both reactions is shown in the inset. Lane 1(■), lane 2 (•) (Reproduced from Klein et al., 2000. With permission)

being related to an increase in the amount of intercalated dye and hence an increase in the fluorescent signal obtained. By measuring the amount of dye intercalation and using various amplification standards, an assessment can be made as to the amount of target molecules present at the initiation of PCR amplification whereas melting dynamics also helps to identify the product as such (Section 8.3). Examples of intercalating dyes used in real-time PCRs include SYBR Green I and Hoechst 33258, neither of which interferes with PCR thermocycling (Fig. 9.15). Several companies market complete systems for real-time PCR analysis, with PCR amplification being performed in specially designed, thin, PCR reaction tubes (capillaries), e.g. the *LightCycler* system from Roche.

9.3.2 FRET Quenching Assays

Fluorescence resonance energy transfer or "FRET" is a property that depends upon the molecular interaction between two fluorochromes (a reporter and a quencher), and is determined by the physical distance between the two molecules [Didenko, 2001]. Energy emission by a reporter (donor) fluorochrome molecule is quenched (absorbed) by an acceptor fluorochrome molecule, leading to the extinction of fluorescence or (self)quenching. This mutual quenching effect observed between donor and acceptor fluorochromes is determined by their spectral overlap (i.e. the wavelengths at which they fluoresce or absorb) and the physical distance between them (Fig. 9.16). This critical distance between acceptor and donor fluorochromes can vary from just a few Angstrom units, equivalent to a single nucleotide in size, to multiple nucleotides in length, with the exact distance defining the amount of quenching achieved [Watrob et al., 2003]. Once this limiting distance is exceeded,

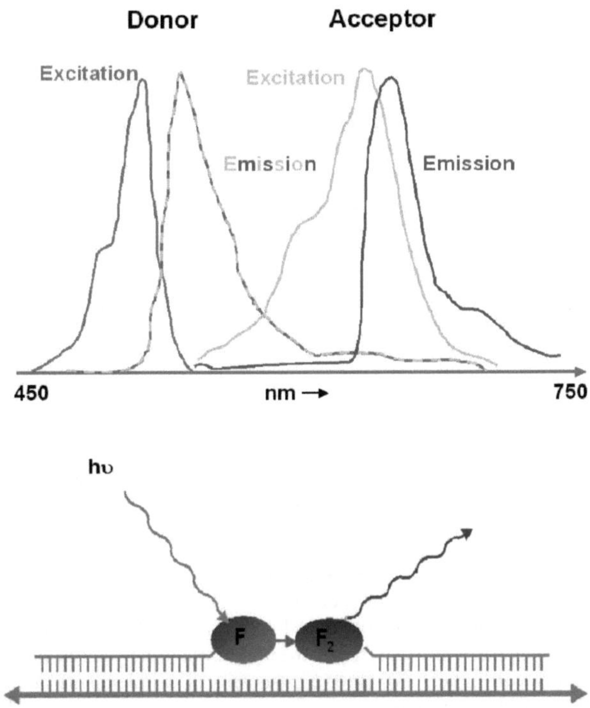

Fig. 9.16 The principle of fluorescence resonance energy transfer (FRET); Top panel: schematic representation of the overlapping spectral characteristics of quencher and donor, essential if adequate quenching of donor excitation by the acceptor is to occur. Bottom panel: Steric position of the quencher and the fluorophore in a fluoresence quenching test (Reproduced from Roche LightCycler Protocols. With permission) (*see Color Plates*)

fluorescence of the reporter molecule is no longer quenched. The FRET principle as such is used in detection systems using "molecular beacons", whereby oligonucleotides are specifically designed to exhibit tertiary structure and form self-binding "hairpin-loops", which position both reporter and quencher fluorochromes in close proximity to each other. Non-hybridised molecular beacons retain their hairpin-loop structure meaning that any fluorescent signal emitted by the reporter molecule is quenched due to its close proximity to the quencher molecule. Molecular beacons hybridised to a complementary region of target DNA however, adopt a linear structure instead of a hairpin-loop with the result that the reporter and quencher are spatially further apart. This very much reduces the absorbing activity of the quencher molecule, allowing fluorescence to escape and be recorded. Special types of molecular beacons have been developed, e.g. "sunrise" and "scorpion" primers [Didenko, 2001] (Fig. 5.4). In practice, molecular beacons have been utilised to quantify and characterise amplified target DNA as well as monitor genetic polymorphism (e.g. single nucleotide polymorphism or SNPs) within organisms [Marras et al., 2003] (Fig. 9.17).

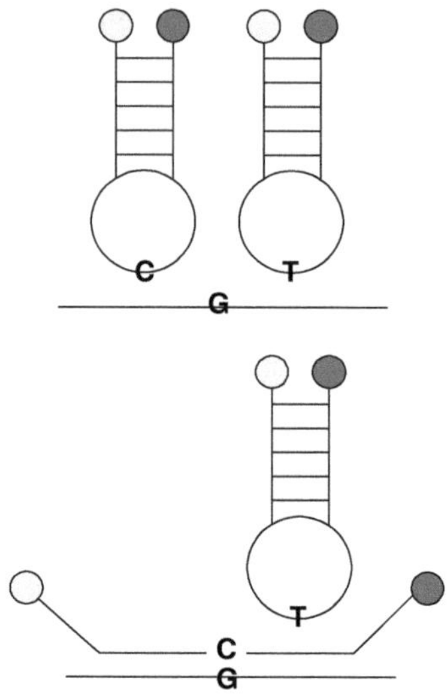

Fig. 9.17 One possible application of FRET and molecular probes in identifying single nucleotide polymorphisms (SNiPs). The two molecular beacons shown only hybridise completely when either a guanosine or an adenosine nucleotide SniP is present. If the donor molecule on each type of molecular beacon fluoresces at a different wavelength, then the fluorescent wavelength in the final PCR reaction mix will indicate whether a guanosine or an adenosine nucleotide is found at the respective position in the amplimer

9.3.3 TaqMan Probes

A variation on molecular beacons is TaqMan technology as developed by Perkin Elmer. This technology uses oligonucleotide probes that are labelled at their 5′-ends and 3′-ends with a quencher and a reporter molecule. Taqman probes are designed to anneal to a central complementary region of the target DNA being amplified during PCR thermocycling. The probe is included in the initial PCR mix and hybridises to single stranded amplification products during the annealing step of each PCR cycle (provided of course that the necessary complementary region of DNA has been amplified). During the subsequent elongation step, the thermostable DNA polymerase will extend the single stranded DNA fragment from the hybridised primer towards the Taqman probe. Upon encountering the Taqman probe, the 5′-3′ exonuclease activity of the polymerase hydrolyses (digests) the Taqman probe into smaller fragments, thereby physically separating the quencher and reporter molecules. The reporter molecules are then free to fluoresce, with the amount of fluorescence recorded being a measure of the amount of specific PCR amplification product formed. Taqman probe PCRs can provide precise, quantitative data on the amount of target DNA present in the initial volume of template DNA added to the initial PCR mix (Section 8.3).

9.3.4 *FRET Enhancement Reactions*

Another application of FRET is not based upon the quenching of reporter fluorescence, but rather on the enhancement of reporter fluorescence. In this methodology, a donor molecule transfers energy to an acceptor molecule, which in turn fluoresces with an increased intensity. In essence, an oligonucleotide labelled with a donor molecule at its 3′-end hybridises precisely upstream of a second oligonucleotide labelled with a 5′-end acceptor fluorochrome (e.g. LightCycler Red 640 [Pals et al., 1999]). Upon co-hybridisation of the two oligonucleotides and the application of a suitable light source, strong fluorescence will occur due to the synergistic effect of the two fluorophores being in close proximity (Fig. 9.16). In real-time PCR assays, mismatches along the length of one or more of these oligonucleotides will slightly reduce the expected annealing and melting temperature (T_m) of that particular FRET probe, generating a melt-curve (–[dF/dT] derivative plot) where the decay of fluorescence (due to dissociation of the FRET probes from the target DNA) occurs at a lower than expected temperature (Fig. 9.18). This method is suited for determining the presence/absence of point mutations within PCR amplimers using real-time technology.

9.4 Nucleic Acid Sequencing

The analysis of PCR amplimer by nucleotide sequencing is the most accurate method for assessing whether the correct amplimer has been amplified or if the amplified region contains any nucleotide changes (mutations). No hybridisation assay or method based on amplimer conformation (structure) can compete with

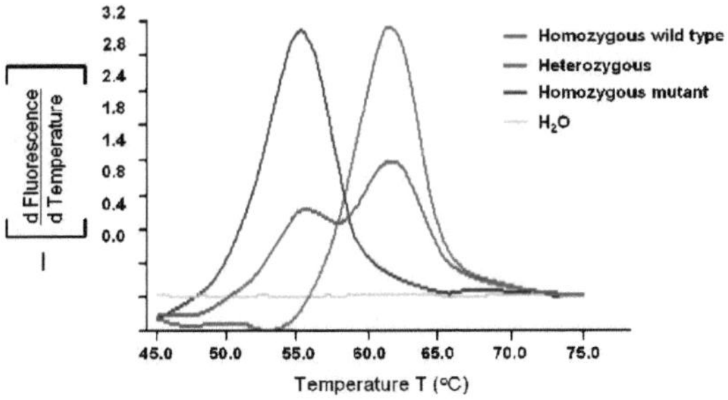

Fig. 9.18 Genomic DNA from a homozygous wild type, a homozygous mutant, and a heterozygous mutant were amplified and analysed on the LightCycler using the hybridisation probes format. The difference in melting temperatures of the samples can easily be visualised by comparing the first negative derivatives (–dF/dT) of the melting curves. The characteristic melting temperatures of the genotypes allow discrimination (Reproduced from Roche LightCycler Protocols. With permission) (*see Color Plates*)

nucleotide sequencing with respect to sequence resolution, meaning that nucleotide sequencing is the current "gold standard" for amplimer characterisation. As such, sequencing of PCR amplimers is frequently employed as a post-PCR follow-up step for (i) mutation screening *per se*, (ii) defining genomic subtypes of bacteria or viruses, (iii) elucidating the molecular basis of (inherited) disease and (iv) for the optimisation and development of novel PCR primer sequences. Although full-scale and expensive nucleotide sequencing machines are available to purchase from commercial sources, for small-scale sequencing projects it is often more cost-effective to post the "raw" or "cleaned" PCR products to companies that have been specifically set up to offer DNA sequencing services and that possess their own custom sequencing facilities. In this case, the raw or purified PCR products (i.e. without or with removal of contaminating thermostable DNA polymerases, unused nucleotides and primers, etc. from the final PCR mix) are sent by mail to the chosen DNA sequencing company, and the sequence data is returned either by mail or e-mail within a period of a few days. In order to increase the reliability and accuracy of the sequencing reaction, it is better to send purified PCR products or have them purified "on site" by the sequencing company, though it is feasible to use unpurified PCR amplification products directly in the sequencing reaction if the PCR protocol is fully optimised and the primers absolutely specific. Two major choices exist for the purification of PCR products: (1) the separation of PCR amplification products by gel electrophoresis, followed by cutting-out of the band of interest and final separation of the DNA amplimer from the agarose matrix, or (2) purification of the amplimer using one of the many commercial kits available which are specifically designed for the purpose. Commercial kits available include amongst others, the Geneclean II/SPIN systems (BIO101), CONCERT kit (Gibco) and the Wizard PCR prep (Promega) kits.

The primer used in any particular PCR sequencing reaction may be specific for the PCR amplification product to be sequenced (usually one of the PCR primers), or in the case of cloned fragments, may be a primer complementary to distinct sequences present within the cloning vector *per se*. Most cloning vectors contain regions of DNA that flank the cloning site that are designed to be used as PCR and/or PCR sequencing primers (the most common flanking sequences being M13 and T7 phage sequences).

With respect to the sequencing protocol *per se*, the method developed by Sanger et al., [1977], although other non-PCR-based techniques may be used in certain circumstances for sequencing short stretches of DNA or detection of SNPs, e.g. pyrosequencing (Biotage AB, Uppsala, Sweden). In the original method, four separate PCR amplification reactions were performed using a single primer, which had been designed to bind to the 5′-end of the DNA template to be sequenced. In each of these four PCR mixes, one of the deoxynucleotides (either dATP, dCTP, dGTP or dTTP) had been partially replaced by its corresponding dideoxynucleotide analogue (either ddATP, ddCTP, ddGTP or ddTTP). These dideoxynucleotides act as premature DNA chain "terminators", effectively preventing any further extension of the DNA chain when incorporated into the growing DNA strand by thermostable DNA polymerases, (in effect the 3′-deoxygenated position in the heptose sugar molecule of DNA is not compatible with the formation of a phosphodiesther bond, a bond required for the addition of further nucleotides to the growing DNA chain by the thermostable DNA polymerase). After PCR amplification, the four PCR sequencing reaction mixes contained a variety of amplification products, all of different lengths, with the 3′-end of each product being that particular dideoxynucleotide terminator which was added to the original PCR sequencing mix. These differentially sized molecules were then separated by denaturing polyacrylamide gel electrophoresis, which was able to separate the single stranded, prematurely terminated, DNA fragments with a resolution of a single nucleotide, yielding sequence data up to a maximum of approximately 800 bp in length (Fig. 9.19). Advances in the original Sanger methodology (and in particular the use of fluorescently labelled dideoxynucleotide terminators), mean that current PCR sequencing protocols are able to utilise a single PCR reaction vessel containing all four dideoxynucleotide terminators (each bound to a different fluorescent dye and emitting light in the green, yellow, red or blue regions of the spectrum). When included in a PCR sequencing reaction, these terminators will be incorporated into the growing DNA chain by thermostable DNA polymerase, generating a series of hundreds of DNA fragment sizes (all containing one of four possible different coloured termini). During gel electrophoresis, the position of each fluorescent peak is measured using a CCD camera as the terminated single-strands of DNA migrate through the gel [Lamture et al., 1994]. Appropriate labels include JOE, FAM, HEX and TAMRA

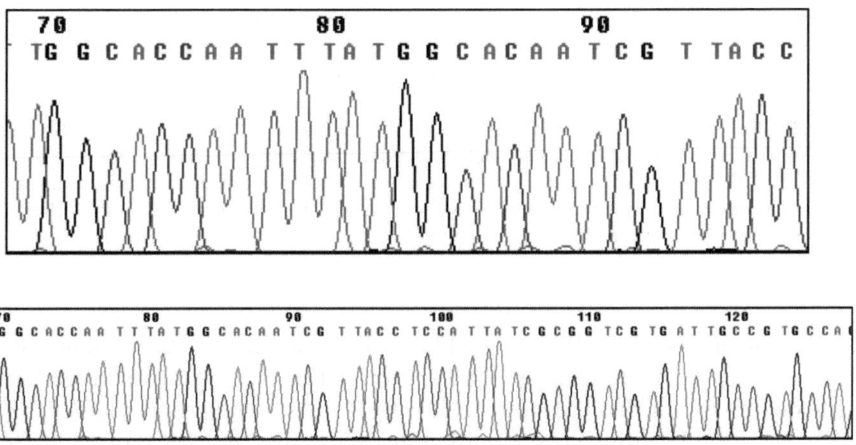

Fig. 9.19 The principle behind the DNA sequencing of method of Sanger et al. [1977]. If a dideoxynucleotide triphosphate (as opposed to the usual deoxynucleotide triphosphate) is incorporated into a growing DNA chain by a thermostable DNA polymerase, then termination of that growing DNA chain will occur at the point where the dideoxynucleotide triphosphate is incorporated. By using a mixture of four dideoxynucleotide triphosphates labelled with different fluorescent dyes in a PCR mix alongside the usual deoxynucleotide triphosphates, then a series of labelled differently sized amplimer molecules are generated which may be separated by denaturing polyacrylamide gel electrophoresis. The sequence of the original target DNA may then be determined based upon the particular fluorescent dye detected for each fragment length (*see Color Plates*)

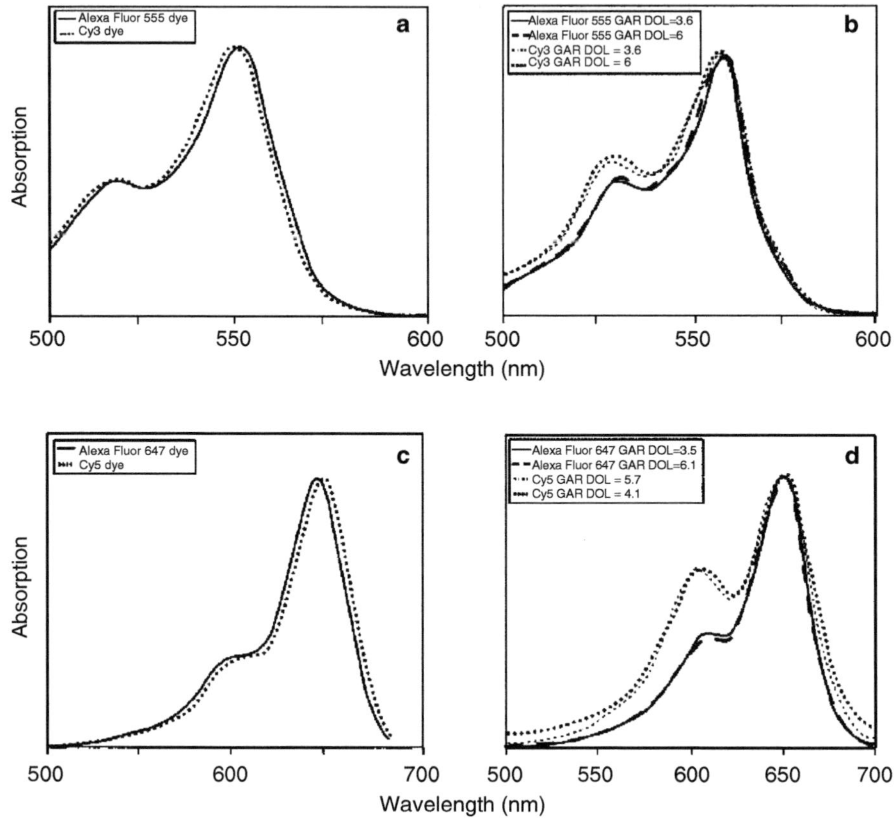

Fig. 9.20 Key characteristics of fluorochromes belonging to the CyDye group (Amersham Biosciences Corp., New Jersey, USA), including emission spectra and maxima for Cy2, FluorX, Cy3, Cy3.5, Cy5 and Cy5.5 and Cy7 fluorochromes. Comparison of normalised absorption spectra of unconjugated and conjugated Alex Fluor and Cy dyes (Reproduced from Berlier et al., 2003. With permission from Dennis G. Bashkin)

(Perkin Elmer) and BODIPY/Alexa Fluor [Torimura et al., 2001]. Some of the characteristics of fluorescent dyes are shown in Fig. 9.20. An alternative Sanger methodology uses a fluorescently labelled PCR sequencing primer and unlabelled dideoxynucleotide terminators, requiring that four different PCR sequencing reaction mixes are used (similar to the original Sanger method).

Unfortunately, nucleotide sequencing is particularly susceptible to problems associated with the presence of stable secondary structure (e.g. self-complementary hairpin-loop structures) within the target DNA. These elements can interfere with DNA strand elongation as they obstruct the thermostable polymerase enzyme, resulting in lowered processivity (reduced total sequence read lengths) due to random DNA chain termination at the position of the secondary structure. Also, target DNA (including PCR amplified DNA) which contains secondary structure may not run uniformly during gel electrophoresis, giving rise to the effect known as "band compression", which complicates the interpretation of sequencing gel results. However, several modified nucleotides are currently available which may help in reducing the effect of secondary structure in PCR sequencing reactions [Motz et al., 2000]. If a fraction of the dGTP nucleotides present in the PCR nucleotide mix are replaced by the modified nucleotide 7-deaza-dGTP then any hairpin-loop present may be destabilised. Alternatively, the nucleotide deoxyinosine tri-phosphate (dITP) may be added to the PCR reaction mix, [Di Mauro et al., 1994]. As alternatives to modified nucleotides, other commercially available PCR sequencing solutions have been developed (e.g. Sequence Rx Enhancer System, Invitrogen), as well as modified enzymes (e.g. SequiTherm Excell II DNA polymerase, Epicentre) which may help to reduce the effect of secondary structure on the sequencing reaction.

Recent advances include the development of various non-nucleotide terminators that possess an a-cyclic chemical structure and are able to be incorporated and terminate growing DNA chains by thermostable DNA polymerases (AcycloTerminators, Perkin Elmer Life Sciences, Boston, USA).

9.4.1 DNA Sequencing Using Non-thermostable DNA Polymerases

Initially, Sanger-based DNA sequencing reactions were performed with thermolabile *E. coli* Klenow fragment or T4 DNA polymerase enzymes, and after polyacrylamide gel electrophoresis, sequence results were recorded by autoradiography or chemiluminescence (dependant on the type of label used). Unfortunately however, both the *E. coli* Klenow fragment and T4 DNA polymerase enzymes are heat labile, which meant that PCR cycle sequencing was very inefficient and wasteful of enzyme, as fresh enzyme had to be added after every PCR sequencing cycle. In order to circumvent this problem, sequencing methods using non-thermostable enzymes traditionally used single stranded DNA as starting material, so that thermocycling of the reaction was less important. In order to make single strands from double-stranded DNA, one of the DNA strands was either digested by specific exonucleases, or alternatively, treated with chemical denaturants. Single strands of DNA complementary to the PCR primer could be captured by attaching the relevant biotin labelled sequencing primer to a streptavidin coated solid phase (e.g. a nylon membrane), allowing the captured single strands to then be purified away from the solid phase and used in a sequencing reaction. This procedure allowed for both strands to be separated from each other and independently sequenced. A more complex method for making single-stranded DNA involved the action of an RNA polymerase and reverse transcriptase enzyme. Firstly, the DNA template to be sequenced was transcribed into single stranded RNA (using for example T7 DNA dependent RNA polymerase), which was then converted into single stranded DNA using a reverse transcription enzyme (e.g. MMLV or AMLV). This sequencing technique (via an

RNA intermediate and reverse transcription) was called "gene amplification with transcript sequencing" (GAWTS) [Lind et al., 1996]. However, this GAWTS protocol is quite expensive, leads to a considerable loss in the quantity of sequencing product, and results in an increased number of unintentional mutations in the final sequence data (introduced by the non-proofreading RNA polymerase and reverse transcriptase enzymes).

9.4.2 PCR Sequencing Using Thermostable DNA Polymerases

The use of thermostable DNA polymerases in nucleic acid sequencing reactions yields several advantages, not least the fact that fresh enzyme does not have to be added to the reaction mix after every sequencing cycle. Further, PCR sequencing protocols help in greatly reducing problems associated with sequencing DNA that contains stable secondary structure, and significantly improves the sensitivity of the sequencing procedure due to a higher number of ddNTP terminated amplification chains being formed. Double stranded DNA sequencing of both the (+) and the (−) strands is also less of a problem, since the DNA duplex may be separated into individual strands by the simple application of heat (one simply needs to add the appropriate complementary (+) or (−) strand primer to the PCR sequencing mix). Increasingly important, PCR sequencing is more adaptable to high throughput applications and is relatively cheaper than the classical procedure which utilised non-thermostable DNA polymerase. The choice of which thermostable enzyme to use is largely dependant on the proofreading ability of the enzyme (Chapter 7), with the most frequently used enzyme currently being AmpliTaq CS, though non-proofreading genetically engineered variants of the Vent and Pfu enzymes, as well as Bst DNA polymerase I, may also be used

9.4.3 The Fidelity of PCR Sequencing Reactions

During Sanger sequencing, the 3′-5′-end proofreading ability of thermolabile *E. coli* Klenow fragment, T4 DNA polymerase, and thermostable Taq type DNA polymerase enzymes may periodically result in the removal of the chain terminating ddNTP nucleotide from DNA strands, allowing previously terminated DNA molecules to act as primers for further amplification and sequencing. This may result in an uneven spread of product lengths upon gel electrophoresis and hence uneven sequence data. Therefore, enzymes (either thermolabile or thermostable) without proofreading ability are preferred for use in Sanger DNA sequencing reactions. The use of such non-proofreading polymerases means that any occasional nucleotide misincorporation error introduced by the DNA polymerase during DNA strand elongation will not be corrected and may appear in the final sequence data. However, the number of nucleotide sequence errors appearing in the final sequence data due to errors introduced during the PCR sequencing reaction will tend to be very low, due to the fact that: (i) the introduction of sequencing errors by non-proofreading DNA polymerases enzymes is essentially random and (ii) the DNA strand being elongated is not amplified, hence any sequence errors are not re-amplified. In contrast, sequence errors resulting from any misincorporated nucleotides introduced during the PCR amplification of target molecules may "contaminate" the final sequence data as DNA molecules containing sequence errors may act as templates for further amplification. However, even in this scenario, only errors introduced during the first few cycles of PCR amplification are likely to be detected in the final sequence data (as these will form a substantial fraction of the final amplimer concentration). Sequence errors introduced during the later stages of PCR amplification will be amplified, but will be outweighed by the number of correctly amplified PCR products present within the final PCR. In order to take into account the possibility of sequence errors in the final data, it is recommended that both strands of the double-stranded DNA target be sequenced at least once (i.e. sequenced in both directions) in order to identify any possible sequence incorporation errors introduced by non-proofreading DNA polymerases. In a recent development, the use of a specialized thermostable DNA polymerase AcycloPol and non-nucleotide-based AcycloTerminators (Perkin Elmer Lifesciences, Boston, USA) has been reported to be more resistant to nucleotide misincorporation when excess target DNA is present in the sequencing reaction as compared to the more traditionally used ThermoSequenase (Amersham Biosciences Corp, New Jersey, USA) and dideoxynucleotide terminators. Further details on the sequencing fidelity of thermostable DNA polymerases are provided in Table 7.1.

References

Baird ML. 1998. Using the AmpliType PM+HLA DQA1 PCR amplification and typing kits for identity testing. Methods Mol Biol 98:261–277.

Bakkus MH, Bouko Y, Samson D, Apperley JF, Thielemans K, van Camp B, Benner A, Goldschmidt H, Moos M, Cremer FW. 2004. Post-transplantation tumour load in bone marrow, as assessed by quantitative ASO-PCR, is a prognostic parameter in multiple myeloma Br J Haematol 126(5):665–674.

Baldino F, Chesselet MF, Lewis ME. 1989. High resolution in situ hybridisation histochemistry. Methods Enzymol 168:761–777.

Battaglia C, Salani G, Consolandi C, Bernardi LR, De Bellis G. 2000. Analysis of DNA microarrays by non-destructive fluorescent staining using SYBR green II. Biotechniques 29:78–81.

Berlier JE, Rothe A, Buller G, Bradford J, Gray DR, Filanoski BJ, Telford WG, Yue S, Liu J, Cheung CY, Chang W, Hirsch JD, Beechem JM, Haugland RP, Haugland RP. 2003. Quantitative comparison of long-wavelength Alexa fluor dyes to Cy dyes: fluorescence of the dyes and their bioconjugates. J Histochem Cytochem 51:1699–1712.

Christie KR, Weng S, Balakrishnan R, Costanzo MC, Dolinski K, Dwight SS, Engel SR, Feierbach B, Fisk DG, Hirschman JE, Hong EL, Issel-Tarver L, Nash R, Sethuraman A, Starr B, Theesfeld CL, Andrada R, Binkley G, Dong Q, Lane C, Schroeder M, Botstein D, Cherry JM. 2004. *Saccharomyces* Genome Database (SGD) provides tools to identify and analyze sequences from *Saccharomyces cerevisiae* and related sequences from other organisms. Nucleic Acids Res 32:311–314.

Dabora SL, Sigalas I, Hall F, Eng C, Vijg J, Kwiatkowski DJ. 1998. Comprehensive mutation analysis of TSC1 using 2D DNA electrophoresis with DGGE. Ann Hum Genet 62:491–504.

De Koning MC, Van der Marel GA, Overhand M. 2003. Synthetic developments towards PNA-peptide conjugates. Curr Opin Chem Biol 7:734–740.

De Sousa Menezes J, De Almeida Drummond Franklin D, Seki H, Rumjanek FD. 2003. SSCP of hypervariable regions of HV1 and HV2 of human mitochondrial DNA: detection by silver staining. Forensic Sci Int 133:242–245.

DiCesare J, Grossman B, Katz E, Picozza E, Racusa R, Woudenberg T. 1993. A high-sensitivity electrochemiluminescence based detection system for automated PCR product quantitation. Biotechniques 15:152–157.

Didenko VV. 2001. DNA probes using fluorescence resonance energy transfer (FRET): designs and applicationsBiotechniques 31:1106–1116.

Di Mauro E, Costanzo G, Negri R. 1994. One-lane chemical sequencing of PCR amplified DNA: the use of terminal transferase and of the base analogue inosine. Nucleic Acids Res 22(18):3811–3812.

Fodde R, Losekoot M. 1994. Mutation detection by DGGE. Hum Mutat 3:83–94.

Gejman PV, Cao Q, Guedi F, Sommer S. 1998. The sensitivity of DGGE: a blinded analysis. Mutat Res 382:109–114.

Fregel R, Maca-Meyer N, Cabrera VM, Gonzalez AM, Larruga JM. 2005. Description of a simple multiplex PCR-SSCP method for AB0 genotyping and its application to the peopling of the Canary Islands. Immunogenetics 57(8):572–578.

Holmila R, Husgafvel-Pursiainen K. 2006. Analysis of TP53 gene mutations in human lung cancer: comparison of capillary electrophoresis single strand conformation polymorphism assay with denaturing gradient gel electrophoresis and direct sequencing. Cancer Detect Prev 30(1):1–6.

Hovig E, Smith Sorensen B, Uitterlinden AG, Borresen AL. 1992. Detection of DNA variation in cancer. Pharmacogenetics 2:317–328.

Ikeda H, Beauchamp RL, Yoshimoto T, Yandell DW. 1995. Detection of Heterozygous Mutation in the Retinoblastoma Gene in a Human Pituitary Adenoma Using PCR-SSCP Analysis and Direct Sequencing. Endocr Pathol 6(3):189–196.

Jacobs JA, Fodor SP. 1994. Combinatorial chemistry – applications of light directed chemical synthesis. Trends Biotechnol 12:19–26.

Jiang-Baucom P, Girard JE, Butler J, Belgrader P. 1997. DNA typing of human leukocyte antigen sequence polymorphisms by peptide nucleic acid probes and MALDI-TOF mass spectrometry. Anal Chem 69:4894–4898.

Klein D, Denis M, Ricordi C, Pastori RL. 2000. Assessment of ribozyme cleavage efficiency using reverse transcriptase real-time PCR. Mol Biotechnol 14(3):189–195.

Kleter B, Van Doorn LJ, Schrauwen L, Molijn A, Sastrowijoto S, Ter Schegget J, Lindeman J, Ter Harmsel B, Burger M, Quint W. 1999. Development and clinical characterization of a highly sensitive PCR-reverse hybridisation line probe assay for detection and identification of anogenital human papillomavirus. J Clin Microbiol 37:2508–2517.

Kutyavin IV, Afonina AI, Mills A, Gorn VV, Lukhatanov EA, Belousov ES, Singer MJ, Walburger DK, Lokhov SG, Gall AA, Dempcy R, Reed MW, Meyer RB, Hedgpeth J. 2000. 3′-Minor groove binder DNA probes increase sequence specificity at PCR extension temperatures. Nucleic Acids Res 28:655–661.

Lamture JB, Beattie KL, Burke BE, Eggers MD, Ehrlich DJ, Fowler R, Hollis MA, Kosicki BB, Reich RK, Smith SR. 1994. Direct detection of nucleic acid hybridisation on the surface of a charge coupled device. Nucleic Acids Res 22:2121–2125.

Lind T, Thorland EC, Sommer SS. 1996. Genomic amplification with transcript sequencing (GAWTS). Methods Mol Biol 65:193–200.

Luk JM, Kongmuang U, Tsang RSW, Lindbergh AA. 1997. An enzyme linked immunosorbent assay to detect PCR products of the *rfbS* gene from serogroup D salmonellae: a rapid screening prototype. J Clin Microbiol 35:714–718.

Marras SA, Kramer FR, Tyagi S. 2003. Genotyping SNPs with molecular beacons. Methods Mol Biol 212:111–128.

Marshall E. 2001. DNA arrays: Affymetrix settles suit, fixes mouse chip. Science 291:2535.

Marsh I, Whittington R, Millar D. 2000. Qualitative control and optimized procedure of hybridisation capture PCR for the identification of *Mycobacterium avium* subsp. *Paratuberculosis* in faeces. Mol Cell Probes 14:219–232.

Martin C, Roberts D, Van der Weide M, Rossau R, Jannes G, Smith T, Maher M. 2000. Development of a PCR-based line probe assay for identification of fungal pathogens. J Clin Microbiol 38:3735–3742.

Mijs W, De Vreese K, Devos A, Pottel H, Valgaeren A, Evans C, Norton J, Parker D, Rigouts L, Portaels F, Reischl U, Watterson S, Pfyffer G, Rossau R. 2002. Evaluation of a commercial line probe assay for identification of mycobacterium species from liquid and solid culture. Eur J Clin Microbiol Infect Dis 21(11):794–802.

Motz M, Paabo S, Kilger C. 2000. Improved cycle sequencing of GC rich templates by a combination of nucleotide analogs. Biotechniques 29:268–270.

Pals G, Pindolia K, Worsham MJ. 1999. A rapid and sensitive approach to mutation detection using real-time PCR and melting curve analysis using BRCA1 as an example. Mol Diagn 4:241–246.

Pollack JR, Perou CM, Alizadeh AA, Eisen MB, Pergamenschlikov A, Williams CF, Jeffrey SS, Botstein D, Brown PO. 1999. Genome wide analysis of DNA copy number changes using cDNA micro-arrays. Nat Genet 23:41–46.Riesner D, Steger G, Zimmat R, Owens RA, Wagenhofer M, Hillen W, Vollbach S, Henco K. 1989. Temperature gradient gel electrophoresis of nucleic acids: analysis of conformational transitions, sequence variations, and protein nucleic acid interactions. Electrophoresis 10:377–389.

Sambrook J, Fritsch EF, Maniatis T.1990. Molecular cloning: a laboratory manual. Cold Spring Harbor, New York.

Said B, Scotland SM, Rowe B. 1994. The use of gene probes immunoassays and tissue culture for the detection of toxin in *Vibrio cholerae* non-O1. J Med Microbiol 40:31–36.

Sanger F, Nicklen S, Coulson AR. 1977. DNA sequencing with chain terminating inhibitors. Proc Natl Acad Sci USA 74:5463–5467.

Sawyer SL, Howell WM, Brookes AJ. 2003. Scoring insertion deletion polymorphisms by dynamic allele specific hybridisation. Biotechniques 35:292–296.

Sharma SK, Ferreira JL, Eblen BS, Whiting RC. 2006. Detection of type A, B, E, and F Clostridium botulinum neurotoxins in foods by using an amplified enzyme-linked immunosorbent assay with digoxigenin-labeled antibodies Appl Environ Microbiol 72(2):1231–1238.

Simsek M, Al-Wardy N, Al-Khayat A, Al-Khabory M. 2002 A PCR-RFLP test for simultaneous detection of two single-nucleotide insertions in the Connexin-26 gene promoter Genet Test 6(3):225–228.

Singer VL, Lawlor TE, Yve S. 1999. Comparison of SYBR Green I nucleic acid gel stain mutagenicity and ethidium bromide mutagenicity in the *Salmonella*/mammalian microsome reverse mutation assay (Ames test). Mutation Res 439(1):37–47.

Southern EM, Maskos U, Elder JK. 1992. Analyzing and comparing nucleic acid sequences by hybridisation to arrays of oligonucleotides: evaluation using experimental models. Genomics 13:1008–1017.

Southern EM. 1975. Detection of specific sequences among DNA fragments separated by gel electrophoresis. J Mol Biol 98:503–517.

Srinivasan K, Morris SC, Girard JE, Kline MC, Reeder DJ. 1993. Enhanced detection of PCR products through use of TOTO and YOYO intercalating dyes with laser-induced fluorescence – capillary electrophoresis. Appl Theor Electrophor 3:235–239.

Tackett AJ, Corey DR, Raney KD. 2002. Non Watson-Crick interactions between PNA and DNA inhibit the ATPase activity of bacteriophage T4 Dda helicase. Nucleic Acids Res 30:950–957.

Torimura M, Kurata S, Yamada K, Yokomaku T, Kamagata Y, Kanagawa T, Kurane R. 2001. Fluorescence quenching phenomenon by photoinduced electron transfer between a fluorescent dye and a nucleotide base. Anal Sci 17:166–160.

Uitterlinden A, Vijg J. 1991. Denaturing gel electrophoretic analysis of minisatellite alleles. Electrophoresis 12:12–16.

Vahey M, Nau ME, Barricks S, Cooley JD, Sawyer R, Sleeker AA, Vickerman P, Bloor S, Larder B, Michael NL, Wegner SA. 1999. Performance of the Affymetrix GeneChip HIV PRT 440 platform for antiretroviral drug resistance genotyping of HIV type 1 clades and viral isolates with length polymorphisms. J Clin Microbiol 37:2533–2537.

Van Gijlswijk RP, Zijlmans HJ, Wiegant J, Bobrov MN, Erickson TJ, Adler KE, Tanke HJ, Raap AK. 1997. Fluorochrome labeled tyramides: use in immunocytochemistry and fluorescence in situ hybridisation. Histochem Cytochem 45:375–382.

Verkooyen R, Peeters MF, Van Rijsoort-Vos JH, Van der Meijden WI, Mouton JW. 2002. Sensitivity and specificity of three new commercially available *Chlamydia trachomatis* tests. Int J STD AIDS 13:23–25.

Watrob HM, Pan CP, Barkley MD. 2003. Two step FRET as a structural tool. J Am Chem Soc 125:7336–7343.

Wickham CL, Lynek C, Ellard S. 2000. Detection of clonal T cell populations by high resolution PCR using fluorescently labelled nucleotides; evaluation using conventional LIS-SSCP. Mol Pathol 53:150–154.

Wolffs P, Knutsson R, Sjoback R, Radstrom P. 2001. PNA based light-up probes for real-time detection of sequence specific PCR products. Biotechniques 31:766–771.

Chapter 10

Ensuring PCR Quality – Laboratory Organization, PCR Optimization and Controls

Three distinct levels of quality control should be considered when setting up PCR applications for use in the clinical laboratory. The primary level of quality control involves the organization of the laboratory and relevant logistical aspects, which are areas of major importance with respect to the avoidance of PCR contamination and the prevention of false positive clinical results. The secondary level of quality control involves PCR design, optimization and assay quality control *per se*. Finally, there may be individual quality considerations that are specifically related to individual types of PCR methodologies (especially reverse transcription PCR methodologies).

10.1 The Primary Level of Quality Control – Laboratory Organization and the Prevention of PCR Contamination

Performing PCR experiments accurately and reproducibly requires a careful and considerate mode of working. Ideally, every laboratory worker should understand what steps and precautions need to be taken in order to prevent PCR contamination, as well as knowing what PCR contamination looks like in the final PCR results. The implementation of Good Laboratory Practice (GLP), attention to detail, and the quality of the personnel involved are the major keys in preventing PCR contamination.

Perhaps the most important PCR contamination control strategy for PCR facilities involves the physical separation of the key PCR related activities into separate laboratories, as outlined in Fig. 10.1. Within the PCR facility, all equipment for specific PCR activities should be dedicated to that particular task and should be restricted to the laboratory where it was originally intended to be used. If primer synthesis is carried out "on site", then this too should be performed in a separate laboratory and preferably by dedicated personnel, (though the current low costs of commercially available oligomers means that PCR primers can be manufactured and shipped to the customer at very competitive rates). For the utmost protection against PCR carry-over contamination, all "pre-PCR" laboratories (i.e. laboratories used for specimen processing, PCR mix preparation and primer synthesis) should be maintained at positive atmospheric pressure to prevent the entry of contaminants and aerosols.

E. Van Pelt-Verkuil et al., *Principles and Technical Aspects of PCR Amplification,*

In contrast, the PCR thermocycling and post-PCR gel electrophoresis laboratories should ideally be maintained at negative atmospheric air pressure in order to prevent the escape of PCR product contaminated aerosols into the general working environment. Further, the PCR thermocycling and post-PCR laboratories should also preferably maintain direct contact with the outside air e.g. via ventilation systems. For those PCR protocols that make use of in vitro methods of PCR amplimer analysis (e.g. when using a LightCycler PCR machine coupled to TaqMan technology), the gel electrophoresis laboratory may not be required, as PCR product analysis is an integral part of PCR thermocycling *per se* and the amplimers remain contained within the PCR reaction vessel (capillary) at all times.

The flow of material and personnel during PCR processing should always be unidirectional, i.e. each step in the PCR process from PCR mix preparation, to specimen

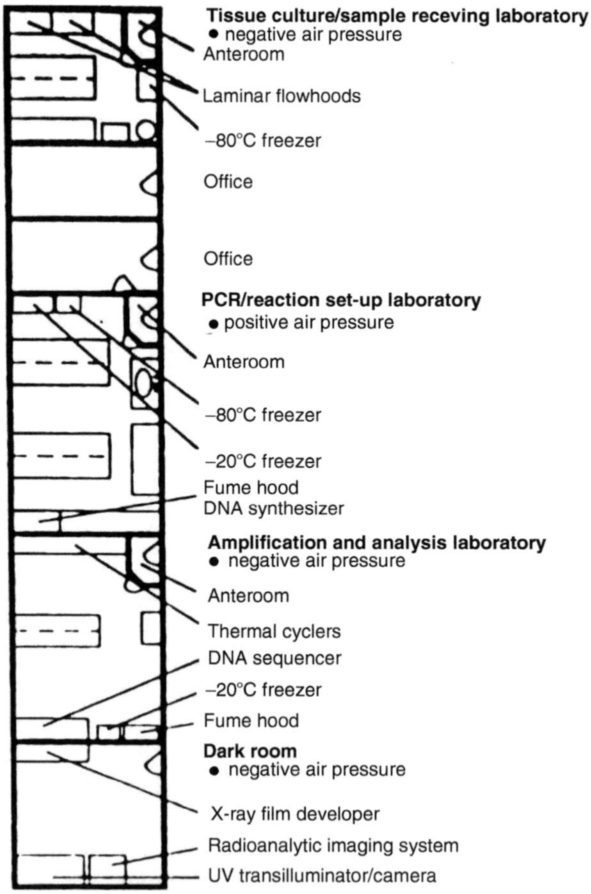

Fig. 10.1 Idealized layout of a routine PCR laboratory suited for high-throughput molecular diagnostics. The electrophoresis equipment is generally used in a distinct, separate room. It may also be used in the negative pressure dark room though (From Sirko and Ehrlich, 1994)

processing, to thermocycling and finally post-PCR gel electrophoresis, should take place in that particular order. Clinical specimens and PCR amplification products are ready sources of contamination, and should not "back-track" into any of the previous PCR laboratories. Ideally, the same unidirectional flow should also apply to laboratory personnel, i.e. a person who has worked in the gel electrophoresis laboratory should not be allowed to work in the specimen isolation laboratory on the same day (because of the increased risk of contamination from PCR amplimer fomites that may be present on clothing, hair, hands, etc.). However, this last recommendation is often unrealistic for a busy clinical laboratory, hence the minimum requirements include the compulsory wearing of separate laboratory coats and disposable latex or neoprene gloves for each PCR laboratory, such that laboratory coats are removed and gloves discarded when moving from one PCR processing laboratory to the other. Laboratory coats should be clearly labelled, regularly cleaned/autoclaved and their use restricted to the dedicated PCR laboratory where they are intended to be used. The wearing of disposable latex or neoprene gloves also offers the added advantage that nucleases (DNases and/or RNases) present on human skin and in sweat are much less likely to contaminate clinical specimens or PCR mixes.

Day to day contamination prevention strategies include keeping all bottles and reaction vessels sealed for as long as possible, with Eppendorf tubes containing clinical specimens, extracted nucleic acid, PCR mix ingredients, or PCR amplimers, all being briefly centrifuged prior to opening (thereby decreasing the risk of aerosol-mediated contamination upon opening the tubes). Aliquoting of key-reagents into single-use (i.e. disposable) aliquots is generally recommended [Kofler and Klausegger, 1999]. Autoclaving or incineration of contaminated laboratory utensils, including laboratory glassware, disposable Eppendorf tubes, pipette tips, buffers and other solutions, is also recommended. Ideally, laboratory bench surfaces should be regularly cleaned with 1 N hydrochloric acid, which leads to the rapid depurination and hydrolysis of contaminant nucleic acid molecules; alternatively overnight exposure to ultraviolet light (or very rarely used γ-radiation), may be helpful in destroying possible contaminants present in any aerosols or fomites which have been created during the working day. Several commercial suppliers (e.g. Q-biogene, Herolab GmbH) sell "PCR workstations", which are essentially enclosed plastic cabinets in which PCR manipulations may be performed, and which may be "sterilized" of contaminating nucleic acid via an in-built ultraviolet (UV) light. The laboratory worker simply places his/her hands inside the cabinet to perform the manipulation and then switches on the UV-light once the work is finished. It has also been documented that UV-light may be used to reduce low level PCR contamination from within Eppendorf tubes by a factor of 1,000 [Ou et al., 1991]. However, it should be borne in mind that UV-light also degrades any DNA controls or size markers left on laboratory work surfaces, and many plastics become brittle after prolonged UV irradiation. If problems with work surface contamination persists after UV-light irradiation or cleaning with 1 N hydrochloric acid, then treatment of the surfaces with psoralene may be tried [Jinno et al., 1990]. However, the psoralene reagent, despite being a chemotherapeutic for phototherapy, is potentially toxic and requires special handling and disposal considerations (Fig. 10.2).

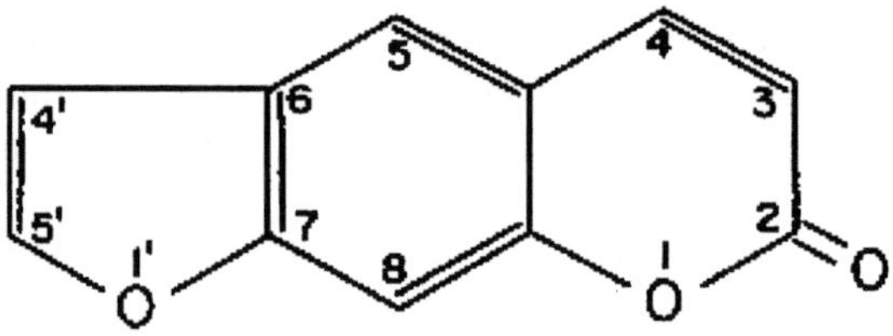

Fig. 10.2 The structure of psoralene. Psoralene intercalates into the DNA double helix and is a fixative that can be used for DNA decontamination

10.1.1 Sources and Routes of Contamination

Several sources of contamination in PCR facilities have been identified, with one of the most important source being the generation and spread of aerosols. Essentially, under the correct circumstances, vigorous pipetting, vortexing, etc., it is possible for prokaryotic and eukaryotic nucleic acid molecules, plasmids, amplification products from previous PCRs, or even complete virus particles or bacteria, to become airborne or aerosolized. These aerosols may persist in the air for many hours, which serves to facilitate their spread via currents of air or draughts, e.g. the opening and closing of laboratory doors, throughout the entire PCR facility. In this way, contaminating material may also come into contact with work surfaces, clothing, hair, hands, or even PCR mixes being freshly prepared. Further, contaminating material present on hands, clothing or hair may be indirectly introduced into "virgin" PCR mixes being prepared in another laboratory. Another important source of PCR contamination worth mentioning is the fomite. Here, excessive vortexing or pipetting leads to the spillage of small quantities of contaminated material onto laboratory coats, gloves, etc. also facilitating dispersal of the contaminant material throughout the PCR facility. Fomite contamination is likely to be especially prevalent in facilities where laboratory coats, gloves, vortexing apparatus, etc. are transferred between the separate PCR laboratories, especially if the "unidirectional workflow" rule is not observed and material/protective clothing etc from the gel electrophoresis laboratory is transferred to one of the other PCR preparation laboratories (i.e. PCR mix preparation, specimen preparation or PCR thermocycling laboratories).

Unfortunately, the procedures that give rise to aerosols and fomites occur in all of the separate PCR laboratories (i.e. PCR mix preparation, specimen preparation, PCR thermocycling and gel electrophoresis laboratories), though the production and spread of such contamination can be somewhat reduced by incorporating the simple procedures outlined in Section 10.1 into the PCR work routine. In the disconcerting event of PCR contamination occurring within the PCR facility, then laboratory coats, protective gloves, centrifuges, pipettes, pipette accessories, blotting equipment, scalpels, microtome knives, dessiccators, electrophoresis supplies, ultra-violet-trays, ice, water/oil baths, glassware, PCR vessels, stocks of PCR ingredients, waste containers, and large volumes of used gel electrophoresis buffer (often many litres containing millions of copies of various PCR amplification products) should all be considered as potential sources. This usually means that laboratory equipment, laboratory coats and work surfaces all have to be thoroughly cleaned, whilst remaining stocks of PCR ingredients (especially polymerase enzymes, dNTPs, water and buffer solutions) should ideally be replaced.

10.1.2 PCR Contamination Issues Within Individual PCR Laboratories

1. PCR mix preparation (set-up) laboratory. This is a "clean" room where PCR mixes are prepared and where a DNA oligomer (PCR primer) synthesizer (if available) may also be situated. The main feature of this laboratory is the presence of PCR stock solutions, dedicated pipettes, pipette tips, reaction tubes, centrifuges, etc., which are used to prepare PCR mixes. All specimens, plasmids, or PCR amplification reactions/products are completely prohibited from this laboratory. For batches of PCRs, it is advisable to prepare a "master-mix" of PCR ingredients, which contain multiple volumes, e.g. multiples of 5 or 10, of PCR mixes. Such master-mixes may be stored frozen at −20°C or −80°C, and when required, thawed and aliquoted into individual PCR tubes prior to PCR thermocycling (and before addition of any nucleic acid template). When preparing large volume master-mixes, it is usual to omit one of the key ingredients (usually the thermostable DNA polymerase) and add this immediately prior to aliquoting and/or addition of the specimen nucleic acid. The use of master mixes helps in reducing the number of liquid manipulations performed as well as increasing the accuracy of pipetting manipulations. Commercially available "ready-to-use" master mix solutions may be purchased either as large volumes or as pre-aliquoted individual PCR mix volumes. A variation on commercial master-mixes is the availability of dried beads containing PCR master-mixes in single volume quantities. Several combinations of dried PCR mix are available, some requiring the addition of nuclease free water and the specimen nucleic acid only. As long as the quality of the beads (i.e. the concentration and activity of ingredients, lack of contaminants, etc.) remains guaranteed by the supplier, then reliable high-throughput diagnostic PCR assays using these beads are possible, with the advantage that there is a decreased chance of contamination occurring due to serial pipetting manipulations.

2. Specimen receipt and/or nucleic acid preparation laboratories. This is a second (separate) "clean" room, which like the PCR mix preparation laboratory, should not be used to store PCR amplimers or plasmids. Sample-to-sample contamination, and an increased risk of false positive results being obtained, is a very real risk in this area, especially when large numbers of specimens containing high concentrations of target molecules are being processed. Therefore, all specimens should be kept physically apart from each other, with specimens not being opened simultaneously but rather each specimen container being opened and then immediately closed before processing the following specimen.

The formation of aerosols and fomites may be avoided and actively prevented by: (i) using sealed storage and PCR vessels, (ii) using as few manipulations as possible, (iii) keeping containers and reaction vessel lids closed as much as possible during specimen processing, and (iv) working within an easily cleaned safety cabinet or PCR workstation. Special care has to be taken with potentially infectious or otherwise health-endangering specimens (oncogenic cells, known HIV positive sera, etc.). Working with viable viruses/bacteria/cells poses a real health risk and the proper protective clothing should be worn throughout, with viable cells preferably being killed and lysed early in the nucleic acid extraction protocol. Paraffin embedded tissue sections (e.g. for DNA extraction, liquid PCR or *in situ* PCR protocols) should be obtained using thoroughly cleaned microtomes, knives, brushes, etc., which if re-used, must be cleaned between samplings (a procedure which may be laborious but reduces the probability of generating false positive results). Tissue sections should be stored on trays cleaned using 1 N hydrochloric acid to remove contaminating DNA.

After extraction, aliquots of the extracted nucleic acid (usually 1–5 μl) should be added to their respective PCR mixes, attempting to ensure that no aerosols or fomites contaminate any surrounding reaction mixes. The PCR vessels should then be transferred to the respective PCR thermocycler (housed in a separate laboratory) for thermocycling. For some older generation PCR thermocyclers (which do not possess a heated lid), a drop of sterile mineral oil should also be added to the top of the PCR reaction mix prior to thermocycling in order to prevent evaporation and condensation of the reaction mix into the top of the PCR tube during thermocycling.

3. The PCR laboratory. Essentially, this is the simplest area within the PCR laboratory. This is the area reserved for PCR thermocyclers and no manipulations of PCR samples should occur here. It is acceptable however, to house PCR thermocyclers in the same laboratory as other types of biotechnological machinery such as DNA sequencers (but not nucleic acid extraction robots, etc.).

4. Laboratory for post-PCR analysis and gel electrophoresis of PCR products. In this laboratory, PCR reaction vessels are opened post-PCR thermocycling, and aliquots of the PCR reaction mix (possibly containing many millions of copies of amplification products) are pipetted into (agarose) gels for electrophoresis (unless real time PCRs involving in vitro fluorescent detection methodologies are being used, e.g. LightCycler and TaqMan assays). This process carries with it a very high risk of aerosol and/or fomite formation. Therefore, this particular laboratory should be regarded as an exceptionally high risk source of PCR product contaminants, which necessarily means that cross-contamination protocols should be strictly observed, and ideally the room should be maintained at negative atmospheric pressure in order to prevent the escape of amplimer containing aerosols into surrounding laboratory areas. Recognized quality control guidelines for preventing PCR contamination (e.g. dedicated laboratory equipment, changing laboratory coats between laboratories, etc.) should be observed most stringently in this laboratory [Kunakorn et al., 2000; Klaschik et al., 1999].

10.1.3 Detecting and Preventing PCR Contamination

It is almost inevitable that the PCR diagnostician will be confronted with contamination problems at some time. Circulating PCR amplification products (aerosols or fomites), positive control specimens and DNA clones used for the synthesis of positive control nucleic acid material, are all possible sources of contamination in PCR facilities. In the worst case scenario, there may be no other choice than to shut down the offending PCR laboratory, throw away all working solutions, thoroughly decontaminate all work surfaces, laboratory coats, pipettes etc and order new supplies of PCR reagents. As "prevention is better than cure" (and often less costly with regard to the amount of time and money spent!), several PCR-specific methods have been developed to help prevent PCR contamination problems arising in the first place. These include:

1. Prevention of PCR product carry-over contamination. PCR product carry-over contamination (via aerosols or fomites) is perhaps the most important issue of PCR contamination that needs to be addressed in a PCR facility, as the numbers of amplification products generated during PCR is extremely high and each individual amplimer can possibly serve as template for future PCR assays. An elegant strategy to prevent such amplimer mediated carry-over contamination uses a specific enzyme to digest contaminating PCR amplimers, ensuring that such amplimers cannot function as templates in subsequent PCR reactions. The commercially available "Gene-Amp PCR carry-over prevention kit" (Applied Biosystems) is one well-known example (Fig. 10.3). The methodology behind this kit involves the substitution of the nucleotide base deoxythymidine (dTTP) in the PCR mix by the nucleotide base deoxyuracil (dUTP), which is subsequently incorporated into all PCR products. This results in the synthesis of "meta-stable" amplification products, which have the same properties as dTTP containing amplimers (i.e. they do not interfere with post-PCR electrophoresis or hybridization reactions). However, the amplification product modifications introduced, result in the dUTP containing amplimers becoming susceptible to digestion by the *E. coli* derived enzyme uracil-N-glycosylase. After addition of a quantity of uracil-N-glycosylase enzyme to freshly prepared PCR mixes (prior to the addition of newly extracted template nucleic acid), and incubation at 18–50°C for 30 minutes, any contaminating dUTP amplification products from previous PCRs will be digested and therefore unable to act as template in any subsequent PCRs. The uracil-N-glycosylase enzyme is then removed from the PCR reaction mix by heating at 95°C for 10 minutes (denaturing the enzyme), a process that also cleaves the a-basic dU-containing PCR products. The final result is that both the contaminating amplification products and uracil-N-glycosylase enzyme are inactivated in the fresh PCR mixes, so that contamination free PCR amplification can be subsequently performed [Kofler and Klausegger, 1999]. Of course this method only works once dUTP PCR protocols are routinely incorporated into the PCR schedule. An alternative anti-contamination method involves the use of gamma- or UV-irradiation, which may be employed to irradiate freshly prepared PCR mixes (prepared without dNTPs or polymerase as these ingredients are

themselves inactivated by gamma- or UV-irradiation), thereby denaturing any contaminating amplimers. After irradiation, dNTPs, thermostable polymerase and template nucleic acid may be added to the irradiated mix and PCR amplification performed. Finally, if PCR is performed in the presence of certain psoralene compounds, then cyclobutane adducts (i.e. new combinations of organic molecules) are formed within the pyrimidine (thymidine and cytosine) nucleotides present within the PCR products. These adducts may subsequently be degraded by photo-inactivation, with pre-PCR illumination inactivating any contaminants present in the reaction mix [Jinno et al., 1990].

It should be noted at this point that some PCR protocols are particularly susceptible to carry-over contamination, namely "nested-PCR" and "consensus-PCR" protocols. Nested PCR protocols increase the specificity and sensitivity of a PCR amplification reaction by re-amplifying a small proportion of the *primary* PCR amplification product using a separate pair of *secondary* PCR primers. These *secondary* PCR primers are designed to hybridize internally to the PCR product generated in the *primary* PCR reaction [Linssen et al., 2000] (Fig. 10.4). The *secondary* PCR amplification step acts to increase the concentration of specific product only, as only specific PCR product from the *primary* PCR amplification reaction will contain complementary sequences to the secondary (nested) set of internal PCR primers. Nested PCR protocols are particularly useful when: (1) specimens contain only very low copy numbers of target molecules or (2) the PCR equilibrium favours the generation of non-specific amplification products [Sellon, 2003]. "Nested consensus-PCR" is somewhat similar to nested PCR in that an initial *primary* PCR is performed, but this time the primary primer pair is designed to amplify relatively non-conserved target sequences that are shared between different individuals, or species, etc. The product from this *primary* PCR is then re-amplified in a *secondary* PCR using a species-specific primer pair [Sonneveld et al., 2003].

One particular problem associated with both nested and consensus PCR methodologies, is that the *primary* PCR reaction vessels (containing many millions of amplified cDNA amplimer copies) have to be opened in order to pipette amplimer from the completed *primary PCR* to the *secondary* "virgin" PCR mix. This fact, along with the increased sensitivity of such PCR protocols, means that the risk of aerosol and fomite contamination is greatly increased when compared to single protocol PCRs.

2. "Anti-contamination" primers. The contamination of PCRs with positive control material (e.g. plasmids containing viral or bacterial genes) can be problematic. Although there is not a precise strategy to prevent this sort of contamination (except strict laboratory "hygiene" and GLP), the presence of contaminating plasmid-based positive control material in PCR reaction mixes may be determined using PCR primers which have been designed to amplify sequences from within the contaminating vector itself, as well as primers designed to amplify the target DNA [Van den Brule et al., 1989]. These primers are often referred to as "anti-contamination" primers, which is not completely correct since contamination is detected rather than prevented.

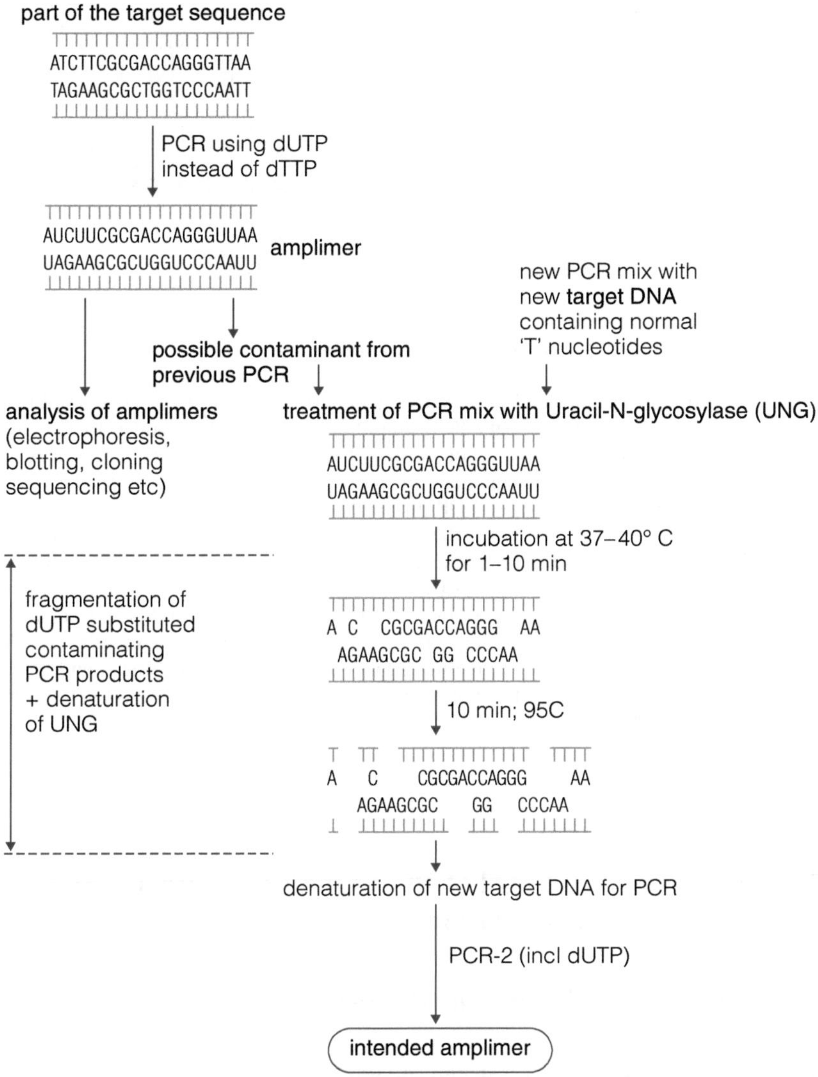

Fig. 10.3 The GeneAmp PCR carry-over contamination prevention kit. Deoxyuracil-triphosphate (dUTP) replaces deoxythymidine-triphosphate (dTTP) in the PCR mix, which means that any PCR products generated may be digested by the enzyme uracil-N-glycosylase. The addition of uracil-N-glycosylase enzyme to freshly prepared PCR mixes, followed by two incubation steps, efficiently removes any cross-contaminating PCR products which may have been "carried-over" by aerosols or on clothing etc from the post-PCR laboratory

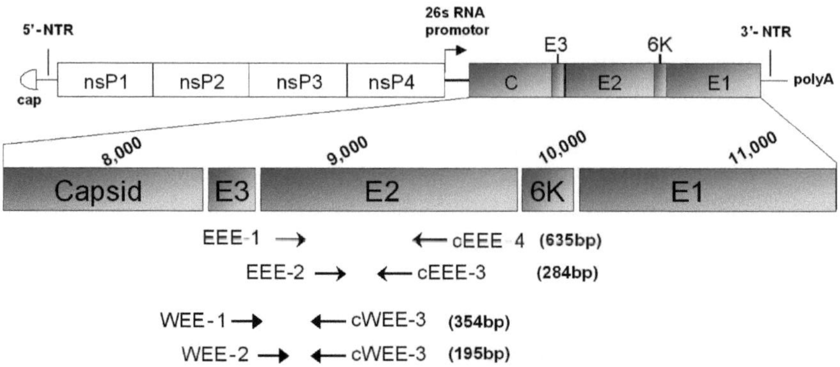

Fig. 10.4 The principle of "nested" PCR. After completion of a primary PCR amplification reaction, a small fraction of the primary PCR amplification products is added to a secondary PCR reaction mix containing a second set of "nested" PCR primers. These primers are designed to specifically hybridize to internal sequences present within the amplification products formed in the primary PCR reaction. The nested PCR protocol increases both sensitivity and specificity of PCR amplification, but also increases the risk of carry-over contamination from previous nested PCR experiments. In this figure, a schematic drawing of the alphavirus genome is shown. Primer pairs EEE-1/cEEE-4 and EEE-2/cEEE-3 comprise a nested set of primers yielding a PCR product of 635 bp (first round of amplification) and a PCR product of 284 bp (second round of amplification), respectively. Alternatively, primer pairs WEE-1/cWEE-3 and WEE-2/CWEE-3 represent a semi-nested set of primers yielding a PCR product of 354 bp (first round of amplification) and a PCR product of 195 bp (second round of amplification), respectively (Adapted from Linssen et al., 2000)

10.2 The Secondary Level of PCR Quality Control – PCR Design and Optimization

10.2.1 Extrinsic and Intrinsic Factors

In theory, many different physical and chemical variables need to be taken into account in order to achieve complete optimization, and hence high quality/reliable results, in PCR assays. However, exact optimization of all these variables would require a prohibitive number of tests to be performed, such that a choice usually has to be made balancing in the degree of optimization, time spent, and laboratory costs when performing PCR optimization experiments. The most important variables affecting PCR accuracy and reproducibility may be grouped into (i) extrinsic factors – the components external to the PCR mix that need to be considered during the development and implementation of PCR methodologies within clinical laboratories, and (ii) intrinsic factors – the range of values best suited for use with respect to physical and reactant parameters. Extrinsic factors affecting PCR optimization include the set up of the laboratory, types of clinical samples and PCR target molecules, the PCR reactants to be used, etc. (Table 10.1). Intrinsic factors include the range of values associated with ion concentration, pH, the type of buffer used, the thermocycling temperature parameters, etc. (Table 10.2). To save time and costs, it is possible to perform optimization experiments where titrations of two or more of PCR ingredients are combined in a "checkerboard" format. Several commercial companies sell "PCR optimization kits" that include the key PCR ingredients as checkerboard dilutions. Roche for example, sell a PCR optimization kit that combines different $MgCl_2$ concentrations with different pH values, whilst Stratagene's Opti-Prime PCR optimization kit offers pH, $MgCl_2$ and KCl optimization. A mathematical approach to PCR optimization was undertaken by Cobb and Clarkson (1994), who used modified Taguchi methods to reduce the number of optimization steps required for both a typical and RAPD PCRs. More detailed information about extrinsic and intrinsic PCR factors may be found in the relevant chapters within this book.

Table 10.1 Overview of the major extrinsic factors for consideration during the development and implementation of PCR assays in order to deliver high quality and reliable results

Parameter	Consideration	Additional information
Separate PCR areas	PCR preparation laboratory Specimen extraction laboratory Thermocycler laboratory Electrophoresis laboratory	Radioactive materials?
Chemicals	Purity, costs, amounts, stability	Ready to use solutions, repeated freezing and thawing
Thermocyclers	Quality, flexibility, reliability, costs, space requirements	
Type of enzyme	Exonuclease activity, reverse transcriptase activity, half-life, storage, buffer requirements	
Type of target	Source, nature of material, isolation protocol, secondary structure, positive/ negative controls	DNA/RNA, native, RNase free, digested, heated
Primer design	Sequence specificity, GC content, secondary structure	
Primer synthesis	Scale, purity, costs, modification	
Primer purification	Precipitation, HPLC	
Special primers	Labels, e.g. biotin, fluorescent dyes, special nucleotides, e.g. inosine, protein nucleic acids	

Table 10.2 Intrinsic factors associated with high quality and reliable PCR results

Physical parameters
Number of cycles – 20–50 cycles
Pre-thermocycling – 95°C for 10 minutes (Hot-start)
Denaturation – 92–98°C
Annealing – 40–70°C
Elongation – 55–75°C
Extra steps – reverse transcription, uracil-N-glycosylase
Special protocols – booster PCR, touch down/up PCR, nested PCR, RT-PCR
Reactant parameters
Magnesium ion conc.– 0.5–5 mM
Primer conc. – 0.1–1 µM
dNTP conc. – 20–200 µM
Enzyme conc. – 0.2–2 units/reaction
pH – 8.5–10
Type of buffer – sodium, potassium or ammonium ions
Additives – DMSO, betain
Templates – RNA/DNA, (un)diluted

10.2.2 *The Developmental Steps Needed to Achieve High Quality PCR Results*

The following scheme (Fig. 10.5) describes the six major steps that needed in order to optimize PCR assays to yield high quality results.

Step 1: Purified or Cloned Target Nucleic Acid in Water/Extraction Buffer

During the first step of optimizing a new PCR protocol, either purified or cloned target nucleic acid, in water or nucleic acid extraction buffer, is used as the target material (and not nucleic acid isolated from clinical material). The availability of such target material is very important when setting up a new assay or when converting from, for example, a classical PCR format to a real-time PCR format (Fig. 10.6). The use of artificial targets (in water or extraction buffer), facilitates and simplifies the initial process of PCR validation (specificity, sensitivity etc) of novel diagnostic tests, and facilitates "agarose gel" assessment of PCR product quality before conversion to a real-time PCR format.

At this stage, a standard annealing temperature may be chosen based on the theoretical annealing temperature of the primer pair to be used (i.e. calculated T_m minus 5°C), and a standard thermocycling assay implemented using melting, annealing and extension times based on the length of the amplimer expected to be generated. Commercially available PCR optimization kits may be used to accelerate optimization of the PCR using this purified or cloned target material, though a range of template concentrations of the purified/cloned template DNA (which should ideally range from the lower limit to the upper limit of the target nucleic acid concentration likely to be found in clinical samples to be tested) should be used.

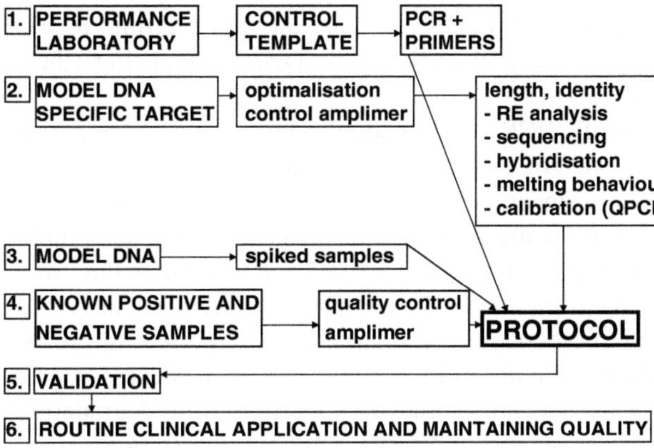

Fig. 10.5 Algorithm for the development of novel PCR protocols. Arrows indicate the logical flow of events

a

CYCLE	AMOUNT OF DNA	AMOUNT OF DNA	AMOUNT OF DNA	AMOUNT OF DNA
	100% EFFICIENCY	90% EFFICIENCY	80% EFFICIENCY	70% EFFICIENCY
0	1	1	1	1
1	2	2	2	2
2	4	4	3	3
3	8	7	6	5
4	16	13	10	8
5	32	25	19	14
6	64	47	34	24
7	128	89	61	41
8	256	170	110	70
9	512	323	198	119
10	1,024	613	357	202
11	2,048	1,165	643	343
12	4,096	2,213	1,157	583
13	8,192	4,205	2,082	990
14	16,384	7,990	3,748	1,684
15	32,768	15,181	6,747	2,862
16	65,356	28,844	12,144	4,866
17	131,072	54,804	21,359	8,272
18	262,144	104,127	39,346	14,063
19	524,288	197,342	70,324	23,907
20	1,048,576	375,900	127,482	40,642
21	2,097,152	714,209	229,468	69,092
22	4,194,304	1,356,998	413,043	117,456
23	8,388,603	2,578,296	743,477	199,676
24	16,777,216	4,898,763	1,338,259	339,449
25	33,554,432	9,307,650	2,408,866	577,063
26	67,108,864	17,684,534	4,335,959	981,007
27	134,217,728	33,600,615	7,804,726	1,667,711
28	268,435,456	63,814,168	14,048,506	2,835,109
29	536,370,912	121,293,220	25,287,311	4,819,686
30	1073,741,324	230,466,618	45,517,160	8,193,466

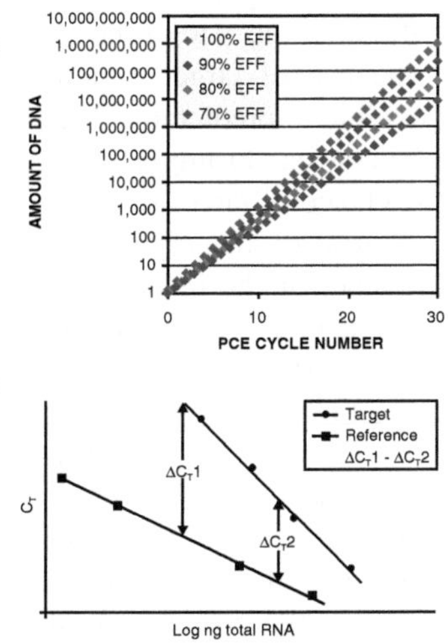

Fig. 10.6 PCR efficiency assessment (A, B). The graphs can be used to deduce the numerical replication value of each and every individual thermocycle and an average value for the entire PCR, based on a presumed replication efficiency value (from 100% to 70% from left to right in A). B shows a graphical version of the Table under A. The parallel assay in (C) allows for a comparison of the amplification efficiency of standard sample C1 with that of test sample C2 (Reproduced from Richard C. Hunt M.A. Ph.D. University of South Carolina School of Medicine, 2000. With permission) (*see Color Plates*)

It is also essential not to forget to include negative controls in this (and indeed all) PCR optimization protocols. Agarose gel electrophoresis, or melt-curve analysis (real-time PCR), may then be used to verify that the amplified product has the same size as that theoretically expected. The exact nature of the amplimer should at this stage be verified using techniques such as Southern blotting, restriction fragment length polymorphism (RFLP) analysis, or DNA sequencing (the most accurate confirmatory test).

Several non-specific PCR amplification problems may arise during this first step of PCR optimization. For example, if several non-specific PCR products are co-amplified (generating smears of DNA fragments in the worst case scenario), then the stringency of the amplification reaction may need to be increased, either by chemical or physical means, e.g. by increasing the annealing temperature, increasing primer length or by altering primer sequences or primer concentrations (Fig. 10.7). For long and accurate PCR protocols, or even simpler PCR protocols which generate relatively long amplimers, the use of PCR additives may be necessary in order to ensure that full-length amplimers are synthesized and to prevent degradation of (DNA and RNA) precursor molecules. If at this stage of PCR optimization a single PCR product of an unexpected length is generated (i.e. a single band that does not appear to be the intended target sequence), then problems could exist with the actual target DNA sequence to be amplified *per se*. These problems may be due to:

1. A high frequency of gene rearrangement within the target DNA – possibly affecting the size of the resultant amplimer.
2. The use of target regions that contain repeat sequences and therefore may exhibit unpredictable behaviour, via the addition or loss of repetitive DNA sequences, during cellular replication (Fig. 10.8).
3. A choice of target sequence that contains stable hairpin structures that hinder the progress of the polymerase during PCR amplification – this phenomenon causes a product of the incorrect size to be amplified, and is especially prevalent when inverted DNA repeat regions are present within the target DNA (N.B. is possible to check for the probability of hairpin structures forming in a particular sequence of target DNA via freely available software) [Shlyakhtenko et al., 1998] (Fig. 10.8).
4. In some cases, alleles may differ because of insertion or deletions of nucleotides or because of heterozygosity in diploid (or polyploid) organisms, so that the PCR is skewed towards the "wrong allele" resulting in the amplification of PCR products of differing sizes.
5. Finally, the presence of an unexpected PCR amplification product of approximately 100 basepairs or less in size may be attributable to primer dimer formation. This is a particular problem if the PCR primer pair share 3′-end sequences that are complementary.

Step 2: Purified or Cloned Target Nucleic Acid in PCR Negative Clinical Material

Once the PCR conditions have been optimized using purified or cloned target nucleic acid in water/extraction buffer, the sampling and extraction assays for the target nucleic acid may be optimized. This is achieved by initially adding various concentrations of purified or cloned target nucleic acid (as used in step 1 above), to a sample or samples of clinical material, e.g. blood, faces, urine, soil samples, forensic substrates etc which have been previously characterized as being negative for the intended PCR target. Using these "spiked" specimens and a negative

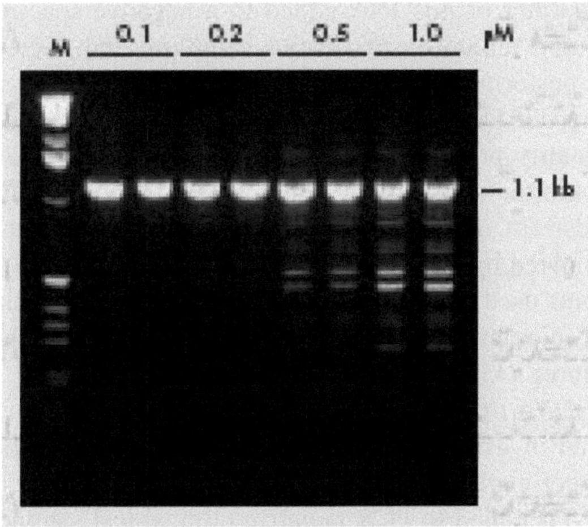

Fig. 10.7 Effect of primer concentration on PCR specificity and quality. Note that when the primer concentration exceeds a certain threshold, non-specific amplification products become more prominent. The test shows the amplification of a 425 bp fragment of the HIV pol gene, starting with ten copies of template DNA in 1 μg of human "background" DNA (Reproduced from Qiagen Critical Factors for Successful PCR, 2003. With permission from Qiagen)

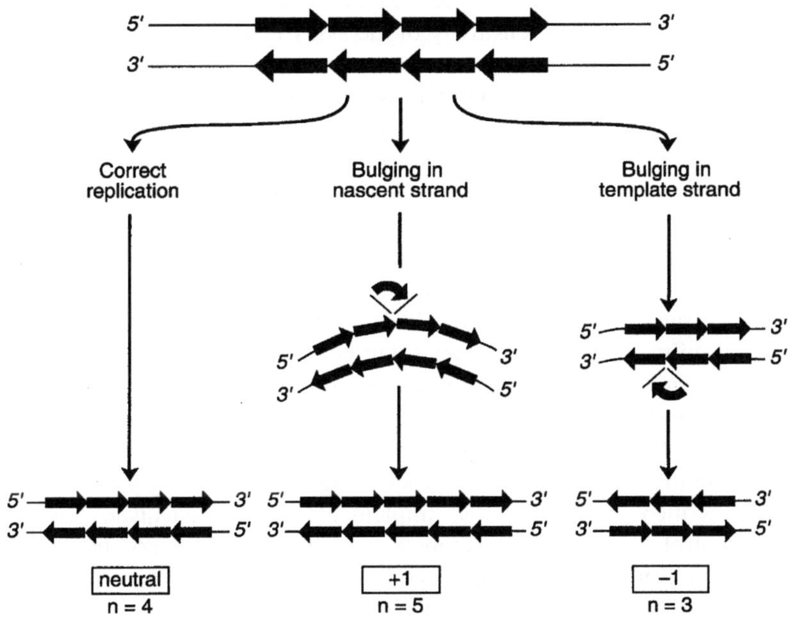

Fig. 10.8 Repeats can cause major problems in the successful amplification of both cDNA and normal PCR targets. Secondary structure formation during the PCR process may lead to deletion or addition of repeat units. Also, hairpin structures may lead to snap-back DNA synthesis (From Van Belkum et al., 1995)

specimen control (water or no DNA added to a negative specimen), experiments into the accuracy and reliability of various sampling and nucleic acid extraction protocols may be performed. If the PCR then fails to generate specific amplimer over a similar range of target nucleic acid concentrations to those used in step 1 above, then the quality problem probably lies with the extraction protocol being used, with PCR inhibitors most likely being present in the final nucleic acid extract [Al Soud and Radstrom, 1998; Al Soud et al., 2000]. In this case, the extraction assay may be optimized by altering, for example, the incubation times, volumes of specimen or reagents used, centrifugation parameters, use of additives, inclusion of another enzyme species, etc. Alternatively, a totally different nucleic acid extraction assay or manufacturer's kit may be tried.

Step 3: Spiking of DNA in PCR Negative Clinical Material

In this step, instead of using purified or cloned nucleic acid, the actual cell type harbouring the target nucleic acid sequence to be detected (e.g. bacteria, viruses, diseased cells, etc.) is added to PCR negative clinical test samples and the nucleic acid is extracted using the extraction assay implemented in step 2. As with step 2, if the PCR fails to generate specific amplimer over a similar range of target nucleic acid concentrations, then the nucleic acid extraction assay may need to be modified or replaced. After successful completion of this step in the PCR optimization procedure, pre-clinical evaluation of the PCR can be initiated.

Step 4: Pre-clinical Validation of the Optimized PCR Assay

During this stage the sensitivity, specificity, and reproducibility of the optimized PCR assay is assessed. The sensitivity of the assay is assessed by using a series of spiked samples containing various concentrations of target DNA as used in step 3 above, whilst the specificity of the PCR assay is assessed using a series of related cell types, e.g. viruses/bacteria of similar species, cell lines exhibiting similar oncogenic properties, cells containing different alleles, etc., in order to ensure that no "false positive" amplification occurs. The reproducibility of the test is assessed by repeating nucleic acid extraction and PCR amplification on duplicate or triplicate aliquots of the same sample. At this stage, two positive control samples are also usually developed. One control being a "strong signal" positive control which generates a large quantity of specific PCR amplimer using the PCR assay being optimized, whilst the other is a "weak, but consistently positive signal" positive control that generates only a low concentration of specific PCR amplimers under the same optimized PCR conditions.

Step 5: Routine Clinical Application

At this step in the PCR optimization process, standard operating procedures (SOP) should be written, authorized and implemented. Also, adequate positive and negative controls should be developed and included in every batch of PCR amplifications. Routine clinical application of the PCR assay may now be undertaken. Commercial test systems usually require stringent testing by independent validation agencies such as the Federal Drug Agency (FDA) and suitable guidelines detailing the possible clinical applications of the newly developed PCR assay are often required as well. These validation procedures are especially essential for laboratories operating under ISO9000 conditions.

Step 6: Maintaining PCR Assay Quality Levels
In order to ensure and maintain the highest level of quality for the optimized PCR assay, it is advisable to regularly compare PCR sensitivity and specificity data with positivity and negativity data obtained using an alternative detection technique (viral/bacterial culture, ELISA, etc.), irrespective of any possible sensitivity differences associated with the alternative technique used. It is also recommended that the assay be regularly controlled using (inter)national approved standards, as for example provided by external quality control schemes [Lewis, 1995; Wallace, 2003]. Using these standards allows statistical comparisons with other (inter)national laboratories and performance criteria of the PCR assay as performed in a particular laboratory to be closely and regularly monitored (Section 11.2).

10.2.3 The Use of Positive and Negative Controls in PCR Quality

The inclusion of positive and negative control reactions during PCR thermocycling helps to identify the presence/absence of false positive and false negative PCR results respectively, results that have a direct bearing on the quality of the PCR assay being performed. Positive controls usually comprise "spiked" PCR negative clinical specimens/water/extraction buffer, i.e. specimens/water/buffer to which various quantities (usually dilutions) of target nucleic acid molecules have been added [Noordhoek et al., 1996]. Alternatively, if an "internal positive control" is required, where the positive control is included in every single PCR to be performed (generating a multiplex PCR protocol where two different sets of primer pairs per reaction mix are used), then the positive control primers are usually designed to target one of the ubiquitous "house-keeping" genes present in tissues e.g. β-globin, GAPDH or β-actin, or else bacterial 16S ribosomal RNA genes (Sections 8.2 and 8.3). Most researchers prefer control targets that are expressed at an equal level as the diagnostic target [Radonic et al., 2004]. On the other hand, negative controls usually comprise PCR reaction mixes to which water or buffer has been added instead of extracted nucleic acid.

Positive controls: Every batch of PCRs should be validated by incorporating at least one positive control into the PCR batch to be amplified. The use of such positive controls helps to exclude the possibility of "false negative" PCR amplification results (i.e. when no amplification signal is achieved in the positive control reaction), which may be brought about by either the poor quality of one or more of the reaction ingredients (e.g. reaction ingredients that are "out of date"), or by erroneous omission of one of the ingredients of the PCR mix. Positive control targets should possess the same type of nucleic acid as that used in the PCR assay itself (e.g. DNA, or RNA), and should be present at similar copy numbers to the average number of target sequences expected to be extracted from clinical specimens. If a broad range of target sequences are likely to be present in clinical specimens (e.g. a range of between 100 and 10,000 copies per PCR), then a range of positive controls incorporating a minimum, mean and maximum number of DNA copies should be included. Positive controls may be either "external" (amplified in a separate PCR), or "internal" (present in every PCR reaction mix and requiring the use of multiplex PCR protocols) components of a PCR batch. External positive controls may comprise nucleic acid extracted from tissues, cells or organisms (in which case the nucleic acid extraction procedure *per se* is also controlled), or dilutions of previously amplified target DNA (not recommended for clinically relevant PCR assays). Alternatively, external PCR positive controls may comprise PCR negative clinical specimens "spiked" with target nucleic acids, or thoroughly tested clinical specimens which have been proven to contain known quantities of target nucleic acid. Internal positive controls may comprise artificially modified target DNA sequences that share the same primer target sequences at their 5′- and 3′-ends as the intended target sequence. These modified target DNA sequences necessarily contain small inserts, deletions or extra restriction enzyme digestion sites, so that the amplimers arising from the positive control sequence may be easily distinguished from target sequences originating from the clinical specimen *per se*. Alternatively, internal positive controls may actually be endogenous to the clinical specimen to be tested, i.e. a non-target DNA sequence that is always found in the genome of the clinical specimen to be PCR amplified (Sections 8.2 and 8.3). In this case, separate (multiplex) PCR primer pairs will have to be developed in order to co-amplify both target and positive control sequences in the same PCR.

Negative controls: Every batch of PCR assays should be validated by incorporating at least one negative control into the PCR batch prior to amplification. This negative control helps to verify that any specific PCR product amplified within the batch of PCRs is not the result of contamination (either from externally derived PCR amplification products or contaminating target DNA), which may be introduced into the PCR batch via contamination of the laboratory environment or contamination of "stock" PCR mix ingredients (i.e. ingredients not aliquoted into single-use containers). Negative controls comprise a PCR mix prepared with the same volume and concentration of ingredients as the PCR mix being used to test specimens, but with the addition of a volume of sterile nuclease free water or sterile nuclease free buffer instead of extracted template nucleic acid. For many research applications, a single negative control per batch of PCRs is sufficient and is usually placed at or near the end of the test PCR mixes within the PCR thermocycler. However, for PCR applications that influence clinical treatment or are involved in for example drug development, a negative control should be included after approximately every five PCR reactions. After completion of thermocycling, all negative controls should contain no amplification products (apart from perhaps primer dimers) and the finding of PCR amplification products within any negative control means that the whole test batch should be repeated (even if only a single negative control is positive). That said, however, performing PCR amplification in the absence of nucleic acid templates may give rise to more frequent primer dimer and other non-specific PCR artefacts, mainly due to changes in the equilibrium of the amplification reaction. In most cases, these artefacts are readily recognizable as small fragments (less than 100 bp in length) or smears upon gel electrophoresis, if any other size of PCR amplification product is obtained within a single negative control, then it is recommended that the whole batch of PCRs be repeated. Negative controls help to check for primer/non-target mis-hybridization, primer dimer formation, random DNA synthesis and even DNA repair during PCR amplification [Tzanakaki et al., 2003], all of which can affect the specificity and sensitivity of a given PCR assay. Again it should be emphasized that negative control samples are absolutely essential for PCR assays performed in the molecular diagnostic clinical laboratory.

10.2.4 Causes and Solutions for False Positive and False Negative PCR Results

False positive results in PCR assays: False positive PCR results can be recognized by either; multiple amplification products (not multiplex PCR protocols), an incorrectly sized amplification product, or an amplification product of the correct size but originating from a non-specific DNA sequence, being observed in one or more of the PCR negative control reactions. In extreme cases, non-specific PCR amplification products may be observed as a smear of amplification products upon gel electrophoresis (Section 9.2.1). Several factors are implicated in generating false positive PCR results, namely (1) one or both of the PCR primer pair is not sufficiently selective (e.g. a small primer length, the presence of repeat regions within the primer, an incompletely optimized PCR), (2) the PCR primer annealing temperature is not sufficiently stringent (i.e. the calculated T_m is too low), or (3) too many PCR amplification cycles are being used (the concentration of PCR amplimers becomes so great that non-specific primer/amplimer and non-specific amplimer/amplimer hybridization occurs. All of these factors allow non-specific primer hybridization (either complete or partial) to non-target nucleic acid sequences to occur. To solve these problems, attempts may be made to re-optimize the PCR reaction mix, re-design the PCR primers (possibly to a new target sequence), increase the annealing temperature upward in 2–3°C increments, reduce the number of PCR cycles performed, or even reduce the amount of initial extracted sample DNA added to the PCR reaction mix (e.g. by diluting the template). Finally, if the primers used are partially degraded, then their average length will be shorter than intended and their specificity of hybridization will be correspondingly reduced. This situation may arise if primers have been frequently freeze-thawed, contaminated with DNases (e.g. fungal contamination), or stored under acidic storage conditions (which facilitates degradation of the primers).

For *in situ* PCRs, "nicking" of the template DNA may result in non-specific DNA synthesis in both specimen and control PCRs. Nicking of double stranded DNA is a frequent occurrence during the histological conservation of tissues and cells, where enzyme nuclease activity cleaves one of the two DNA strands, yielding possible 3′-end termini which may act as non-specific primers in PCR amplification reactions. Another cause of non-specific DNA synthesis occurs if the terminus of a single stranded DNA fragment folds back on itself as a loop, with the folded DNA acting as a primer for DNA polymerase amplification. This event will not completely inhibit effective PCR amplification occurring, though both polymerase and nucleotides will be depleted from the PCR mix and the efficiency of the PCR will decrease.

Finally, in certain PCR methodologies (e.g. reverse transcription or RT-PCR protocols), it may be possible to inhibit the amplification of non-specific PCR products in both specimen and control PCRs by using one of the PCR primers in the reverse transcription step instead of random hexamers or poly-dT primers (which both generate a very diverse range of cDNA templates).

False negative results in PCR assays: If a PCR positive control, or clinical sample known to contain an adequate number of PCR specific target molecules for visualization after gel electrophoresis/real-time PCR, generates no amplimer, then the test result is regarded as a "false negative" PCR result. False negativity in PCR assays may be facilitated by a number of mechanisms, involving problems associated with the target nucleic acid, the PCR reaction mix, primer design, the thermocycling protocol (too high an annealing temperature, insufficient number of cycles performed, etc.), or the sensitivity of the detection system used.

1. Insufficient amount of target DNA in the sample. In some cases the quantities of a particular target molecule may be so low that even an optimized PCR assay does not generate observable amplification products within the programmed number of PCR cycles normally performed. In this case, an increased number of PCR cycles may be necessary to generate products, it being advisable to repeat the PCR but increase the number of cycles in steps of three up to a maximum number of 50 cycles. Alternatively, a larger volume of the nucleic acid extract may be added to the PCR mix prior to amplification, or the nucleic acid extraction protocol itself may be further optimized. The addition of glycogen, which serves as a co-precipitant for nucleic acids, may be helpful in increasing the yield of nucleic acid during the extraction procedure. Sometimes, false negative results that arise from an insufficient amount of target DNA may be compensated for via the use of post-PCR hybridization reactions, or even a separate post-PCR amplification step. This post-PCR amplification methodology utilizes a second "nested" PCR amplification step and a separate pair of PCR primers that hybridize internally to the primary target sequence amplified.

2. Taq Polymerase inhibition or denaturation. Several tissues and specimens contain substances that inhibit DNA polymerase activity, though not all inhibitors have been identified or their mechanism of inhibition elucidated down to the molecular level. Therefore, though it is possible to add a small volume of untreated tissue or specimen directly to a PCR mix (e.g. as used in bacterial colony PCR reactions), the use of extensive nucleic acid extraction and purification prior to PCR amplification of template nucleic acid is strongly recommended. Some of the better known PCR inhibitors present within clinical materials include, hemoglobin and its degradation products, heparin, and bile salts, though not all thermostable DNA polymerase enzymes are equally susceptible to all PCR inhibitors [Al Soud and Radstrom, 2001]. For example, Tth polymerase (in contrast to some other thermostable DNA polymerases), is not sensitive to myoglobin contamination [Belec et al., 1998; Al Soud and Radstrom, 2000;]. Also, it has been found that Tth and

Tfl enzymes are both sensitive to some (as yet unknown) PCR inhibitor present in tears [Wiedbrauk et al., 1995].

If magnesium ions are sequestered by contaminants or by an excess of dNTPs, then the effective concentration of magnesium ions may fall below the optimum value required by the thermostable DNA polymerase to generate PCR amplification products. Thermostable DNA polymerases may also be inhibited by "template shielding" of the target DNA by contaminants present within the nucleic acid extract. This type of amplification inhibition may be detected by adding (also known as "spiking") a known quantity of amplifiable target DNA to the nucleic acid extract prior to PCR amplification. This spiked target DNA may comprise a plasmid or DNA containing a cloned housekeeping gene, or DNA derived from cells, bacteria or viruses. The use of such "exogenous" internal control templates in quality controlling PCR amplifications and/or nucleic acid extraction methodologies is definitely recommended, especially for PCR assays performed in the molecular diagnostic clinical laboratory where patient material is being examined.

Some extraction assays use proteinase K, a proteolytic enzyme which if not sufficiently removed or inactivated after nucleic acid extraction (e.g. by prolonged boiling for at least 10 minutes), will act to degrade the thermostable DNA polymerase added to the PCR reaction mix. It should be noted however, that it may still be possible to generate visible amplification products even if small quantities of proteinase K remain active in the nucleic acid extraction eluate.

3. Incorrect pH and/or buffer systems. Not every PCR assay amplifies equally efficiently at the same pH and in the same buffering system. The thermostable DNA polymerase used may require a specific buffer or pH in order to efficiently amplify target DNA. Further, it has been shown that different buffering systems affect the sensitivity of thermostable DNA polymerases to certain PCR inhibitors [Belec et al., 1998].

4. Insufficient sensitivity of the amplimer detection procedure. If only a small amount of PCR amplification product is generated, post-PCR detection methods (e.g. agarose gel electrophoresis) may not detect the small amount of amplimer generated, as the detection of a PCR fragment by gel electrophoresis and ethidium bromide staining requires more than 1ng of amplification product to be present in order to be visible within the gel. If low concentrations of amplimers are still observed after PCR optimization then increasing the number of PCR cycles, increasing the amount of extracted nucleic acid added, or the development of a nested PCR protocol may help. Alternatively, more sensitive detection methods (other than agarose gel electrophoresis), such as Southern hybridization, immunochemical PCR ELISA, the use of fluorochrome labelled primers, etc., may solve the problem.

5. Problems with gel electrophoresis. On rare occasions, gel electrophoresis of >1 ng of PCR amplification product does not result in visible detection of amplification product after staining. In this situation, the loading buffer (the buffer that contains a visible dye and a component to help stop any PCR products floating out of the gel wells), may not have been correctly prepared. Commercial loading buffer concentrates are available ready for immediate use and are widely employed. Alternatively, the wells in the gel may be too small to allow for the pipetting of an

adequate volume of final PCR mix to allow for the visualization of the PCR amplification product. In this case, the use of gel wells with a larger volume may be beneficial, or alternatively "cleaning and concentrating" the PCR amplimer prior to gel electrophoresis may also be considered (e.g. using the *DNA Clean and Concentrator-5* kit, Zymo Research).

6. Mutations in the template or the oligomers. A 3′-end mismatch between primer and target DNA will tend to completely abrogate PCR amplification. Although such problems should have been detected during PCR optimization, transcriptional errors when ordering PCR primers (e.g. an incorrectly copied nucleotide), impure primers, or errors made during primer synthesis, may all give rise to unexpected amplification problems. Your primer supplier should provide information and quality control data regarding all of these problems. If however, problems with PCR are related to the presence of natural mutations within the target DNA, then the problem may be solved by the introduction of an inosine nucleotide at the relevant position of the corresponding primer. Alternatively, the primer may be shortened or lengthened by a few nucleotides or redesigned to hybridize to a region upstream of the original annealing site. Also, a "degenerate" primer may be designed by ordering primers containing all four possible nucleotides at the position where mutations occur. However, in order to circumvent these problems, it is always recommended to design PCR primers that hybridize to evolutionary conserved target regions, for example genes that are essential for the viability of an organism/cell.

7. Depurination. Repeated heating of a nucleic acid solution leads to degradation of purine (i.e. adenosine and guanosine) nucleotides. This degradation effect may be observed at typical PCR extension temperatures of approximately 72°C and is even more pronounced at PCR DNA denaturation temperatures of approximately 94°C. The longer the time at which the target DNA is exposed to these elevated temperatures the more degradation will occur. Therefore, when very long amplimers requiring prolonged extension times are being synthesized, it may be worthwhile considering reducing the PCR extension step temperature to 68°C.

8. Problems with the primer hybridization/DNA denaturation. Primer hybridization may be very inefficient or even non-existent if the PCR annealing temperature is too high for the PCR primer pair used. This may be especially important with AT-rich primers or where one of the PCR primers has an appreciably higher annealing temperature than the other. Sometimes, increasing or decreasing the ramp rate (i.e. the rate of heating or cooling) during PCR thermocycling may lead to significant improvements in the kinetics of primer hybridization. Problems may also arise if the temperature within the PCR vessel does not correspond to the recorded temperature of the heating block of the PCR machine, possibly leaving the DNA template molecules incompletely denatured. This effect may be caused by delayed heat transfer due to the insulating capacities of the particular PCR reaction vessels used, or to the presence of air between the reaction vessel and the heating block (air acts as an insulator). In this case, the uniformity and type of plastic used in PCR reaction vessel construction are critical. It is also true that the temperature across the surface of a Peltier heating block is not uniform, but tends to become

lower nearer to the edge of the block. Therefore, PCR reaction vessels should be preferably placed at the centre of the heating block. Equipment and software is available for regular quality control testing the performance of PCR machines (e.g. DRIFTCON, CYCLERtest B.V., Landgraaf, the Netherlands).

An improvement in the yield of PCR amplification may also be facilitated by programming the denaturation step of the first few PCR cycles to be somewhat longer (15–30 seconds) than usual. The PCR buffer system may also be rendered more denaturing by for instance, lowering the magnesium ion concentration, reducing the concentration of potassium ions or the addition of DMSO, formamide, glycerol or betain to the PCR mix.

9. Problems with PCR thermocycling. The efficiency of PCR amplification may be influenced by problems associated with PCR thermocycling *per se*. Most "hot-start" thermostable DNA polymerases require an extensive denaturation step prior to initiation of PCR thermocycling. If this initial denaturation step is not adequately performed, for example by not allowing enough time for denaturation, or by having too low a denaturation temperature, then the quantity of active DNA polymerase available for PCR amplification may be much less than expected, resulting in a reduction in the yield of PCR amplification products (even though most of the inactive DNA polymerase will become activated as PCR thermocycling continues). Problems related to the final quantity of PCR amplification product may also be related to the length of time allowed in the elongation step of the PCR cycle. Too short an elongation time may result in the favorable generation of incomplete amplification products, which are not themselves able to be exponentially PCR amplified (as one of the PCR primer pair will not be able to hybridize to the shortened DNA fragment). This will lead to a reduction in the quantity of amplimer produced. Standard PCR thermocycling parameters for the many different PCR methodologies described are available on the Internet.

10. Loss of nucleic acid during nucleic acid extraction. The goal of the nucleic acid extraction procedure is to separate and remove unwanted tissue and cellular components, including proteins, lipids and carbohydrates, away from the nucleic acid to be (RT)PCR amplified whilst inhibiting any enzymatic nuclease activity that may be present within the specimen (nucleic acids are digested very efficiently by exonuclease and endonuclease enzymes liberated from their cellular compartments during nucleic acid extraction). RNA molecules are particularly sensitive to nuclease activity as all bases contain a reactive 2′-hydroxyl group, which increase the chance of a chemical "elimination" reaction occurring, producing water and thereby destroying the sugar ring. This renders RNA particularly vulnerable to (enzymatic) hydrolysis, with the U-A phosphor-ester bond being the most sensitive to hydrolysis. Most important to remember however, is that nucleases (and especially RNases) are ubiquitous in the environment and are extremely stable, sometimes even able to survive phenol extraction and heat denaturation procedures (see Section 4.4).

During nucleic acid extraction, the activity of any released nucleases may be decreased by; (1) maintaining the temperature of incubation and any centrifugation steps below the optimum temperature for nuclease activity of 37°C (e.g. by keeping the extraction mix as much as possible at 4°C or on ice), (2) inactivation of nucleases on glassware, water and other consumables by chemical treatment, e.g. by using diethylpyrocarbonate (DEPC), (3) chemical inactivation and/or inhibition of the nucleases using (hazardous) chemicals, e.g. phenol or guanidinium salts, or (4) removal of essential nuclease metal ion co-factors by chelation (chemical removal).

If required, the quality of extracted total DNA or total RNA may be assessed using gel electrophoresis. Extracted DNA should migrate as a thick band at a position with a molecular weight greater than 20 kb by agarose gel electrophoresis. However, if the DNA has become (partially) degraded by the action of nucleases, then a smear will be observed in the gel. Dependant on the degree of DNA degradation and the size of the PCR product to be amplified, it may still be possible to use the digested DNA in a PCR assay, so long as the average length of the degraded DNA is larger than the size of the expected amplimer [Kleter et al., 1998]. In general, the shorter the fragment of DNA or RNA to be PCR amplified, the lower the likelihood of amplification problems occurring due to degraded nucleic acid [Buffone et al., 1991; Greer, 1991; Soukup et al., 2003]. In the case of severely degraded DNA (frequently found in nucleic acid extractions from archived tissues for example), an intact template may be partially restored using "reconstructive polymerization" [Golenberg et al., 1996]. This re-constructive polymerization assay depends on the concatemerization of several smaller PCR products derived from the (fragmented) target region. Success using this method is not however guaranteed. The quality control of extracted RNA may best be performed using denaturing agarose gel electrophoresis, achieved by adding formamide to the gel in order to remove any secondary structure present within the RNA. In such a system, crude RNA preparations should show at least two distinct bands (1,500 and 2,300 bases for bacteria, 1,800 and 2,800 bases for eukaryotes), consistent with the presence of the larger ribosomal RNA (rRNA) molecules and the smaller messenger RNA (mRNA) molecules. Of course, if an extraction methodology is used which is specifically designed to isolate mRNA only, then the rRNA bands should not be observed upon denaturing agarose gel electrophoresis.

Storage of extracted nucleic acids may be particularly problematic. "Working" solutions of DNA (i.e. solutions in regular use every day or week), may be stored at 4°C for a period of up to several months, though DNA quality may deteriorate at 4°C. Such deterioration is usually due to the presence of residual nucleases remaining after nucleic acid extraction, or autolytic acidic degradation (DNA dissolved in water is a weak acid). Storage of DNA in TRIS-EDTA (TE) buffer helps to circumvent DNA degradation due to acidity, though EDTA "chelates" (removes) magnesium ions from solution, a property which may cause problems if the DNA/TE buffer solution is added direct to the PCR mix (magnesium ions are essential for the activity of thermostable DNA polymerases). As a further note, the degradation effect of acid on DNA may be particularly relevant for certain acid fixed tissues (picric acid containing fixatives being particularly notorious), where the yield of intact DNA after extraction may be very low.

RNA solutions and DNA stock solutions should be stored in manageable quantities at –20°C or even –70°C in order to prevent any damage which may occur by repeated freeze-thawing cycles. Freezing or storage in aqueous ethanol helps stop any residual nuclease activity, especially if frozen in liquid nitrogen at –196°C (where the activity of enzymes is extremely low).

11. Non-extractable nucleic acids. In the past, during certain *in situ* hybridization assays, it was observed that some proteins irreversibly bound to the DNA or RNA nucleic acid being tested, leading to an inability to dissociate the proteins from the nucleic acids in these tissues. Though protein/nucleic acid interactions are not uncommon, for example the binding of chromatin proteins to eukaryotic nuclear DNA, this particular problem was found to be associated with an extremely strong cross-linking reaction facilitated by the use of certain histological fixatives (e.g. non-buffered formalin). This cross-linking often rendered the bound proteins completely resistant to proteases, preventing the extraction of pure nucleic acid and ultimately leading to false negative PCR amplification results. Nowadays, these particular histological fixatives are much less likely to be used, though it is wise to be aware of this problem if using *in situ* PCR methodologies.

12. Secondary structure of the target nucleic acid. Stable hairpin loops or other forms of internal complementarity may prevent complete denaturation of the target nucleic acid and hence hinder hybridization of PCR primers. Several compounds, e.g. DMSO, betain, urea and others, may be added to the PCR mix to act as destabilisers of DNA secondary structure [Kang et al., 2005; Rees et al., 1993]. Another method for helping to unwind DNA secondary structure is to simply increase the denaturation temperature of PCR thermocycling, or simply change the type of thermostable DNA polymerase being used. In particularly difficult cases, restriction digestion of the template DNA with enzymes (that do not have recognition sites within the target DNA), may help to solve the problem [Baldrick et al., 1999]. Since the secondary structure found in RNA molecules are usually more stable than in DNA molecules, removal of secondary structure during RT-PCR processing is even more important than during simple DNA PCR processing. The problem here lies with the thermolabile nature of MMLV and AMV reverse transcriptases, with each having its own particular merits with respect to reverse transcription. Some thermostable DNA polymerases also show (inefficient) reverse transcriptase activity, though are much less frequently used than MMLV and AMV enzymes in RT-PCR protocols (Section 7.5).

10.3 Quality Considerations Specific for RT-PCR Methodologies

Reverse transcription-PCR (RT-PCR) assays are basically identical to DNA PCR assays except that RNA is used as the intended target material, and this must first be reversed transcribed into cDNA prior to PCR amplification. It is this reverse transcription step which is specific for RT-PCR assays and which brings with it several problems related to false positive and false negative PCR results.

10.3.1 Problems Likely to Cause False Positive Results in RT-PCR Assays

In general, false negative results are more likely to occur than false positive results in RT-PCR assays. However, there are several scenarios which may lead to the generation of false positive RT-PCR results.

1. Amplification of pseudogene sequences. Pseudogenes are genes that were once active, but through the course of evolution have lost their function (i.e. are not translated and transcribed), due to either (i) the accumulation of mutations, (ii) immuno-selective pressure, or (iii) the fact that the original gene has been duplicated and the duplicate is no longer required. In some cases, pseudogenes may also arise by reverse transcription of mRNA and insertion of the cDNA into the host's chromosome. Pseudogenes may still show some sequence homology to their original functional gene, a feature referred to as "molecular mimicry". In eukaryotes, pseudogenes are frequently found which lack the particular intron sequences (introns are non-protein coding DNA sequences that are spliced out of RNA molecules in order to make fully functional mRNA) found in the functional DNA gene sequence. Such intron-free pseudogenes may give rise to false positive results during RT-PCR amplification if residual DNA is present in the reverse transcription mix, and if the reverse transcription primer-annealing site is not affected by sequence mutations. However, it is common practice and indeed highly recommended that a DNA digestion step be included in all RT-PCR protocols (see below) [Finke et al., 1993; Foss et al., 1995]. (N.B. Intron carrying pseudogenes may be recognized in normal PCRs by their larger size, as intron sequences are still present in the corresponding functional gene.)

2. Contamination with genomic DNA. Contaminating genomic DNA that is not completely removed from the extracted RNA prior to reverse transcription will generate false positive RT-PCR results. Therefore, the removal of contaminating residual DNA is essential, and this is usually achieved by adding an RNase free DNase enzyme to the extracted RNA, followed by incubation for 1 hour at 37°C, and finally heat inactivating the DNase prior to reverse transcription. The problem of false positive RT-PCR results due to contaminating genomic DNA is especially problematic with regard to RT-PCR amplification performed on paraffin embedded tissues, where separation of the RNA away from the DNA is much more difficult.

The actual presence of any contaminating genomic DNA in an isolated RNA eluate may be determined by simply preparing a reverse transcription mix complete with extracted RNA but which does not contain reverse transcriptase enzyme. During the reverse transcription step of the RT-PCR protocol, no RNA will be reverse transcribed in this reaction mix and hence no amplification products should be observed after RT-PCR thermocycling. However, if contaminating DNA is present, a positive signal (specific PCR product) will be observed. Alternatively, a portion of the RNA solution may simply be added to a PCR mix and PCR amplification performed. Again, no amplification products should be observed after PCR amplification [Matthews et al., 2002]. Another alternative control (specific for the majority of eukaryotes) is to design PCR primer(s) that target an intron splice site of the gene of interest. As introns are spliced out of mRNA, an RT-PCR signal will only be obtained if contaminating genomic DNA is present in the RT-PCR. In any case, it is essential that DNA contamination controls be included in all batches of RT-PCRs performed, and that they should always yield negative results [Dakhama et al., 1996; Koopmans et al., 1993].

10.3.2 Problems Likely to Cause False Negative Results in RT-PCR Assays

Common problems specifically associated with false negative RT-PCR results may be divided into three general areas.

1. RNA degradation problems.

(i) RNA is very susceptible to hydrolysis by RNA degrading (RNase) enzymes, which tend to be ubiquitous within the environment and are very stable, resisting prolonged heating and even retaining activity after treatment with phenol. The addition of "carrier" tRNA to RNA extraction methodologies or to purified specimens may provide a "buffer" against the activity of these enzymes, by acting as a substitute target for RNases.

(ii) RNA is susceptible to chemical hydrolysis especially under alkaline conditions. Therefore, extraction procedures using elevated pH values should be avoided if possible. The nucleotides are also particularly vulnerable to degradation during chemical isolation procedures.

(iii) Native reverse transcriptase enzymes (from MMLV and AMV), exhibit RNase H activity, that is they digest the RNA strand of RNA/DNA hybrids which may reduce the yield of cDNA actually reverse transcribed. This may be a particular problem with regard to the reverse transcription of low concentrations of RNA, or the reverse transcription of full-length mRNA transcripts. RNase H negative reverse transcriptase variants are commercially available.

(iv) Low yields of RNA during extraction are often due to limiting capacities of the RNA binding component of the extraction system (e.g. small centrifuge spin columns or special gel matrices). If only limited amounts of RNA are available prior to extraction it may be helpful to add (RNase free) glycogen prior to precipitation, as it can be helpful in facilitating the complete precipitation of small quantities of RNA. Also, addition of a protein "carrier" (e.g. BSA) to the reverse transcription reaction mix may help increase the efficiency of reverse transcription of small quantities of RNA into cDNA.

(v) In general, RNA is easily damaged by freeze-thawing, and extracted RNA should ideally be aliquoted into smaller volumes prior to storage. Commercial RNA storage buffers are now available that allow RNA to be safely stored at room temperature for a limited period of time.

(vi) Tissue or cellular material processed for *in situ* hybridization or RT-PCR may be completely lacking in RNA, due to the fact that RNase mediated enzymatic

RNA degradation may still occur even after a tissue sample has been fixed and embedded in paraffin wax [Tournier et al., 1987].

2. Incomplete reverse transcription of RNA molecules.

During the reverse transcription of eukaryotic mRNAs, the mRNAs are copied from the 3′-end terminal poly-A tract towards the 5′-end. In order to generate full-length cDNA molecules, the reverse transcription reaction must proceed to the end of the mRNA molecule. For shorter length mRNAs of up to 500 basepairs, the process of reverse transcription is relatively efficient. However for longer mRNAs the process becomes increasingly inefficient. Also, the poly-A tract of eukaryotic mRNAs is not particularly stable to chemical degradation and is also liable to degradation via the RNase H activity of native reverse transcriptase enzymes. Several commercial companies now offer specially developed reverse transcription buffer systems specifically designed to increase the efficiency of reverse transcription and cDNA amplification from average sized (<12 kb) to long (12–15 kb) mRNA molecules. If incomplete cDNA transcripts are generated, for example when the buffer system is not balanced and optimized for reverse transcription, then RT-PCR assays will generate low quantities of specific amplification products or may fail altogether due to the reduced number of sites available for annealing of both primers of the PCR pair.

3. Obstruction of cDNA synthesis.

(i) Eukaryotic mRNAs contain a 3′-end poly-A tract, which is often used as a binding site for complementary poly-dT reverse transcription primers. The use of poly-dT primers approximately 20–25 nucleotides in length necessitates the use of low annealing temperatures due to the low number of nucleotides present in the poly-dT primers. However, this low annealing temperature is not always compatible with melting GC-rich target RNA molecules with the result that the complete removal of RNA secondary structure in these molecules is not always achieved. Any remaining RNA secondary structure may block reverse transcription activity and lower the yield of cDNA generated. Therefore, for all types of reverse transcription primer used (poly-dT, random hexanucleotides, PCR primers), it is usual to heat the RNA to approximately 70°C for 2 minutes (to remove any secondary structure), and then immediately place the RNA on ice for 5 minutes (in order to maintain the "open" structure of the RNA before adding reverse transcriptase enzyme and finally performing a reverse transcription incubation at 37–42°C for 1 hour.

(ii) For RNA molecules that do not possess a poly-A tract (e.g. transfer RNA, ribosomal RNA, bacterial mRNA, etc.), random hexanucleotide primers (commercially available mixtures of random nucleotide sequences of six nucleotides in length) may be used for efficient cDNA synthesis. The use of random hexanucleotides means that many thousands/millions of cDNA sequences will be generated of different lengths, with the advantage that multiplex or multiple PCR assays may be performed using a single reverse transcription reaction. For specimens to be used in a single RT-PCR assay, a reverse transcription primer may be designed (under the same conditions as used for PCR primers) which hybridizes to a region of the RNA molecule not predicted to form a stable secondary structure. Alternatively, the complementary (antisense) PCR primer may be used, which should limit the diversity of cDNA products generated during reverse transcription.

References

Al Soud WA, Radstrom P. 1998. Capacity of nine thermostable DNA polymerases to mediate amplification in the presence of PCR inhibiting substances. J Environ Microbiol 64:3748–3753.

Al Soud WA, Radstrom P. 2000. Effects of amplification facilitators on diagnostic PCR in the presence of blood, feces and meat. J Clin Microbiol 38:4463–4470.

Al Soud WA, Radstrom P. 2001. Purification and characterization of PCR-inhibitory components in blood cells. J Clin Microbiol 39(2):485–493.

Baldrick E, Xamena N, Cabre O. 1999. Overcoming false negatives due to the genomic context in PCR amplification. J Biochem Biophys Methods 40:45–48.

Belec L, Authier J, Eliezer-Vanerot MC, Piedouillet C, Mohamed AS, Gherardi RK. 1998. Myoglobin as a PCR inhibitor: a limitation for PCR from skeletal muscle tissue avoided by the use of *Thermus thermophilus* polymerase. Muscle Nerve 21:1064–1067.

Buffone GJ, Demmler GJ, Schimbor CM, Greer J. 1991. Improved amplification of CMV DNA from urine after purification of DNA with glass beads. Clin Chem 37:1945–1949.

Cobb BD, Clarkson JM. 1994. A simple procedure for optimizing the PCR using modified Taguchi methods. Nucleic Acids Res 22:3801–3805.

Dakhama A, Maeck V, Hogg JC, Hegele RG. 1996. Amplification of human beta-actin gene by the reverse transcriptase PCR: implications for assessment of RNA from formalin fixed paraffin embedded material. J Histochem Cytochem 44:1205–1207.

Finke J, Fritzen R, Ternes P, Lange W, Dolken G. 1993. An improved strategy and a useful housekeeping gene for RNA analysis from formalin fixed paraffin embedded tissues by PCR. Biotechniques 14:448–453.

Foss AJ, Guille MJ, Occleston NL, Hykin PG, Hungerford JL, Lightman S. 1995. The detection of melanoma cells in peripheral blood by reverse transcription PCR. Br J Cancer 72:155–159.

Golenberg EM, Bickel A, Weihs P. 1996. Effect of highly fragmented DNA on PCR. Nucleic Acids Res 24:5026–5033.

Greer CE. 1991. PCR amplification from paraffin-embedded tissues: effect of fixative and fixation time. Am J Clin Pathol 95:117–124.

Jinno Y, Yoshiura K, Niikawa N. 1990. Use of psoralen as extinguisher of contaminated DNA in PCR. Nucleic Acids Res 18:6739.

Kang J, Lee MS, Gorenstein DG. 2005. The enhancement of PCR amplification of a random sequence DNA library by DMSO and betaine: application to in vitro combinatorial selection of aptamers. J Biochem Biophys Methods 64(2):147–151.

KlaschikS, Lehman LE, Raadts A, Book M, Hoeft A, Stuber F. 1999. Comparison of different decontamination methods for reagents to detect low concentrations of bacterial 16S DNA by real-time PCR. Mol Biotechnol 3:231–242.

Kleter B, Van Doorn LJ, Ter Schegget J, Schrauwen L, Van Krimpen K, Burger M, Ter Harmsel B, Quint W. 1998. Novel short fragment PCR assay for highly sensitive broad-spectrum detection of anogenital human papillomaviruses. Am J Pathol 153:1731–1739.

Kofler B, Klausegger A. 1999. Simplified PCR set-up using a frozen preformulated mix for the detection of CMV. Diagn Microbiol Infect Dis 34:33–35.

Koopmans M, Monroe SS, Coffield LM, Zaki SR. 1993. Optimization of extraction and PCR amplification of RNA extracts from paraffin embedded tissue in different fixatives. J Virol Methods 43:189–204.

Kunakorn M, Raksakai M, Niemhorn S, Chongtrakool P, Pracharktam R. 2002. Overcoming the errors of in-house PCR used in the clinical laboratory for the diagnosis of extra-pulmonary tuberculosis. Southeast Asian J Trop Med Public Health 30:84–90.

Lewis SM. 1995. Quality assurance programmes in the United Kingdom. Ann Ist Super Sanita 31:53–59.

Linssen B, Kinney RM, Aguilar PM, Russell KL, Watts DM, Kaaden O, Pfeffer M. 2000. Development of reverse transcription PCR assays specific for the detection of equine encephalitis viruses. J Clin Microbiol 38:1527–1535.

Matthews JL, Chung M, Matyas RJ. 2002. Persistent DNA contamination in competitive RT-PCR using cRNA internal standards: identity, quantity and control. Biotechniques 32:1412–1417.

Noordhoek GT, Van Embden JD, Kolk AH. 1996. Reliability of nucleic acid amplification for detection of *Mycobacterium tuberculosis*: and international collaborative quality control study among 30 laboratories. J Clin Microbiol 34:2522–2525.

Ou CY, Moore JL, Schochetman G. 1991. Use of UV irradiation to reduce false positivity in PCR. Biotechniques 10:442–446.

Radonic A, Thulke S, Mackai TM. 2004. Guideline for reference gene selection for quantitative real-time PCR. Biol Bioch Res Commun 313:856–862.

Rees WA, Yager TD, Korte J, Von Hippel PH. 1993. Betaine can eliminate the base pair composition dependence of DNA melting. Biochem 32:137–144.

Sellon RK. 2003. Update on molecular techniques for diagnostic testing of infectious diseases. Vet Clin North Am Small Anim Pract 33:677–693.

Shlyakhtenko LS, Potaman VN, Sinden RR, Lyubchenko YL. 1998. Structure and dynamics of supercoil stabilized DNA cruciforms. J Mol Biol 280:61–72.

Sirko DA, Ehrlich GD. 1994. PCR laboratory facilities, protocols and operations. In: Ehrlich GD, Greenberg SJ (eds) PCR-based diagnostics in infectious disease. Blackwell, Boston, MA, pp. 19–44.

Sonneveld T, Tobutt KR, Robbins TP. 2003. Allele specific PCR detection of sweet cherry self incompatibility (S) alleles S1 to S16 using consensus and allele specific primers. Theor Appl Genet 107:1059–1070.

Soukup J, Krskova L, Hilska I, Kodet R. 2003. Ethanol fixation of lymphoma samples as an alternative approach for preservation of the nucleic acids. Neoplasma 50:300–304.

Tournier I, Bernuau D, Poliard A, Schoevaert D, Feldmann G. 1987. Detection of albumin mRNAs in rat liver by *in situ* hybridization: usefulness of paraffin embedding and comparison of various fixation procedures. J Histochem Cytochem 35:453–459.

Tzanakaki G, Tsolia M, Vlachou V, Theodoridou M, Pangalis A, Foustoukou M, Karpathios T, Blackwell CC, Kremastinou J. 2003. Evaluation of non-culture diagnosis of invasive meningococcal disease by PCR. FEMS Immunol Med Microbiol 39:31–36.

Van Belkum A, Scherer S, Van Alphen L, Verbrugh H. 1998. Short sequence DNA repeats in prokaryotic genomes. Microbiol Mol Biol Rev 62:275–293.

Van den Brule A, Claas ECJ, Du Maine M, Melchers WJ, Helmerhorst T, Quint WG, Lindeman J, Meijer CJ, Walboomers J. 1989. Use of anti-contamination primers in the PCR for the detection of human papillomavirus genotypes in cervical scrapes and biopsies. J Med Virol 29:20–27.

Wallace PS. 2003. Linkage between the journal and the Quality Control Molecular Diagnostics (QCMD). J Clin Virol 27:211–212.

Wiedbrauk DL, Werner JC, Drevon AM. 1995. Inhibition of PCR by aqueous and vitreous fluids. J Clin Microbiol 33:2643–2646.

Chapter 11

Ensuring PCR Quality – Quality Criteria and Quality Assurance

When quality issues are discussed in any manufacturing or result-orientated process, then two key terms are often quoted, namely "quality control" and "quality assurance". These two terms are often used interchangeably, but in fact refer to two different procedures. Quality control refers to a system of routine and consistent controls, which help ensure data integrity, data correctness and data completeness. Moreover, quality control helps identify and correct any errors or omissions that have occurred during processing, as well as recording and archiving processing results and quality control data. On the other hand, quality assurance involves a planned system of review procedures, preferably performed by an independent "third party" laboratory, which monitors quality control procedures and verifies that quality objectives have been (and are being) met.

Over the past few years, PCR applications and protocols have been developed, optimized, and implemented, for use within a wide range of clinical disciplines. As with the development of any "in house" assay whose results influence clinical decision making, quality criteria for these new PCR protocols have had to be developed in order to comply with both good laboratory practice (GLP) and (inter)national guidelines. Quality criteria for "classical" diagnostic tests such as culture-based assays and ELISAs have already been defined and parallel guidelines for quality controlling, as well as quality assuring, novel PCR-based assays are under continuous development and refinement. Efficiency, sensitivity, specificity, accuracy and reproducibility are all extremely important factors with regard to the official acceptance of novel PCR assays within the clinical diagnostic laboratory, especially now that PCR-based techniques are increasingly being used as diagnostic tools and affecting patient treatment regimens. It should be noted however, that not all PCR assays have to comply with strict clinically oriented quality guidelines, a fact often true in relation to PCR assays used within "research-only" environments. This means that the decision over which PCR assays should be strictly quality controlled depends on the use to be made of the results, which ultimately lies in the hands of the laboratory scientists using that particular PCR assay. Some common concepts and terms used in assessing the quality criteria of newly developed or newly implemented PCR assays within the clinical diagnostic laboratory are presented below. For a more detailed discussion, the reader is referred to two particular reviews [Niesters, 2002; Neumaier et al., 1998].

E. Van Pelt-Verkuil et al., *Principles and Technical Aspects of PCR Amplification,*

11.1 Quality Control Criteria and PCR

11.1.1 Sensitivity and Diagnostic Sensitivity

For a particular PCR assay, the sensitivity may be defined as the lowest number of target molecules that can be reproducibly and specifically detected by that particular assay. At the lowest limit, the signal-to-noise ratio should be such that an obvious and distinct signal (i.e. the correct PCR amplification product) is observed. The initial target concentration, nucleic acid extraction protocol, PCR amplification reaction conditions and the post-PCR method used for detecting and identifying the amplification product, all contribute to the sensitivity of the assay as a whole. In fact, it could be argued that the term "efficiency" rather than "sensitivity" should be used with respect to PCR amplification assays. This better highlights the amplification factor achieved per individual PCR cycle. Further, though PCR amplification has in theory an absolute lower sensitivity limit of a single target nucleic acid molecule, in practice, this absolute sensitivity level is never achieved.

Another term, the "diagnostic sensitivity" is expressed as the percentage of cases in which the assay is capable of distinguishing a positive clinical sample from a negative one, the latter having previously been identified on the basis of results obtained using one or more alternative gold-standard diagnostic techniques, e.g. viral cell culture or ELISA. In ideal cases, this value would be 100% and no false-negatives would be obtained. The formula TP/(TP+FN) × 100% (where TP equals the number of true positive samples and FN the number of false negatives), defines the diagnostic sensitivity of a particular assay. In an ideal case, all of the positive samples including weak positive samples, would be detected (FN = 0), resulting in a diagnostic sensitivity of 100%. However, it should be noted that the sensitivity of PCR assays (especially *in situ* PCR assays) are drastically affected by pre-PCR specimen treatment regimens such as histopathological fixation [Sen Gupta et al., 2003].

11.1.2 Specificity and Diagnostic Specificity

The specificity of a particular assay indicates how successful that particular assay is in detecting the target molecule for which the assay was actually intended, which for PCR assays signifies how successfully the designed primers hybridize to their intended target sequence rather than mis-hybridise to another unrelated sequence. The actual definition of specificity in relation to PCR assays has however been a matter of intense debate over the past decade, since different levels of specificity may be defined dependant on the gold-standard assay used as a reference. The most stringent definition of PCR specificity states that an assay only ever amplifies the target region for which that particular PCR assay was intended, without ever amplifying any non-specific product(s). In this scenario, no false negative or false positive results would be attributable to the PCR. However, false positive reactions may still occur due to the carry-over contamination of DNA from specimens, or contamination from previously amplified PCR products amplified using the same PCR assay. Further, the strict definition of PCR specificity does not take into account the fact that false positive and false negative results can never be completely excluded due to the fact that every PCR reaction mix will be subject to slightly different reaction conditions during thermocycling. Also, unexpected DNA sequence variability may occur due to replication errors that have been introduced by cellular/viral DNA polymerases, RNA polymerases and/or reverse transcriptases, a process possibly driven by for example "immune selection-pressure". Such mutations could theoretically facilitate sequence changes in PCR target sequences, (leading to inefficient primer hybridisation), or alternatively, facilitate sequence changes in non-target sequences (allowing PCR primer hybridisation to extraneous DNA sequences). Such non-target DNA mutations may also occur in "foreign" DNA (i.e. DNA from a foreign organism or cell-type contaminating the clinical sample to be tested), also possibly facilitating false positive results.

It should be noted that the definition of PCR specificity also depends upon the type of PCR assay used, with the definition of specificity having to be adapted for PCR assays that make use of

1. A single pair of PCR primers which are designed to hybridize to well defined DNA or reverse transcribed cDNA sequences.
2. A single pair of PCR primers which are designed to amplify closely related, but not necessarily identical DNA sequences (consensus PCR).
3. Two or more selective PCR primer pairs designed to amplify a set of independent targets (multiplex PCR).
4. A single pair of partly degenerate primers designed to amplify related, but not identical DNA sequences (degenerate PCR).
5. A single pair of primers comprising repeat sequence motifs designed to amplify target sequences containing repeat regions of DNA (repeat sequence PCR).

For general, consensus and multiplex PCR types (1, 2 and 3 above), there remains a simplistic definition of specificity because each primer pair is designed to hybridize specifically to a region of DNA without any mis-hybridisation or cross-hybridisation occurring. It has to be noted that the distinction between cross-hybridisation where hybridisation to more than one DNA region is required, and mis-hybridisation where hybridisation to more than one DNA region is undesirable, is an important one to make. For degenerate PCR, the size, sequence and origin of PCR amplification products is unknown. Therefore, sequence analysis in combination with database searching for homologous sequences may be necessary to identify the exact origin of the amplified fragments. Success is not guaranteed using the degenerate PCR methodology and defining the specificity of such assays is virtually impossible. Repeat sequence PCRs hardly fit the context of specificity at all, as the presence of multiple fragments of differing and unknown sequence length and not a single specific or series of specific fragments are expected. Rather, this type of PCR is used to determine the presence of a series of banding patterns (DNA fingerprints), which may be highly specific once a library of such DNA fingerprint banding patterns has been established.

In its most simple form, the diagnostic specificity of a PCR assay defines the percentage of "false positive" results in relation to the number of "true negative" results. This is represented by the following equation: (TN/(TN+FP) × 100%, which means that when FP = 0 then the specificity of the assay is 100%. If the number of false positives (FP) and false negatives (FN) overlap, then it can be stated that the PCR assay is of too poor a quality for the diagnostic laboratory (Fig. 11.1). In practice however, the situation regarding diagnostic PCR specificity is more complex than the simple equation TN/(TN+FP) × 100% suggests. Firstly, the specificity of DNA amplification depends among other things on the nucleotide mis-incorporation (error) rate of the (non-)proofreading thermostable polymerase being used, as well as the annealing temperature (too low an annealing temperature may yield non-specific product whilst a too high annealing temperature yields no product). The specificity of an assay may also be seriously reduced if mis-priming events occur during the initial few cycles of the PCR, particularly during the initial phase of the PCR cycle as the reaction mix

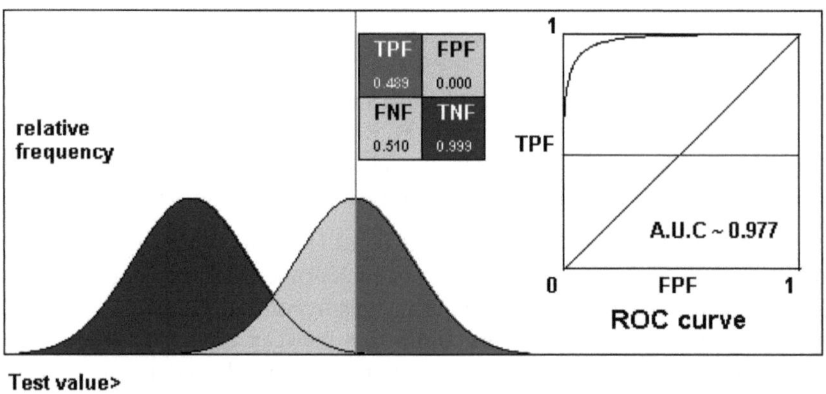

Fig. 11.1 Graph showing changes in the predictive value of an assay related to the number of false-negative or false–positive test results generated by that particular assay. The threshold value distinguishes between true positive (TPF) and true negative (TNF) in 100% of all cases. When the T value is shifted to the right no false-negatives (FNF) are observed, rendering the test 100% sensitive. When the T value is shifted to the left no false-positives (FPF) are observed, rendering the test 100% specific (From http://www.anaesthetist.com/mnm/stats/roc/Findex.htm. With permission from Dr. JM van Schalkwijk)

temperature is being increased from room temperature to the DNA melting temperature of 95°C. As the PCR thermocycler heats up to reach 95°C, at some point the temperature will be high enough to allow partial DNA double strand melting and yet low enough to allow non-specific primer hybridisation to occur. Moreover, any non-specific amplification products generated during this and the first few cycles of PCR will represent a much greater proportion of the final amplification than any specific/non-specific products generated during later PCR cycles. Finally, specificity may vary due to mis-interpretation of gel electrophoresis or during other forms of PCR product detection. For example, ethidium bromide staining is a rather insensitive technique and low concentrations of, or non-specific, amplification products may remain undetected in the gel. Conversely, more sensitive detection methods are useful in visualizing low concentrations of specific amplification products, but they may also allow the visualization of low concentrations of non-specific amplification products too. Finally, it is perhaps important to remember that, although false-positive results should be avoided as much as possible, a certain degree of false-positivity may be tolerated in the PCR assay, dependant on the clinical aspects for which the PCR assay was developed. For example, if we consider infectious diseases, then the administration of a particular antibiotic due to a false-positive PCR result may not always bring about a significant burden being placed on the patient. However, if we consider minimal residual disease (e.g. after treatment for leukaemia), the consequences of further treatment for a patient tested positive via a false-positive result are far more severe (i.e. the possibility of further chemo- or radiotherapy) [Salto-Tellez et al., 2003].

11.1.3 Reference and Threshold Values

The specificity and sensitivity of a test are also dependent on the reference values. This indicates whether or not a distinct threshold value (T) can distinguish between true positive and true negative specimens as based on a gold-standard technology (Fig. 11.1). Sometimes, it may be difficult to precisely calculate the threshold value of a particular assay due to experimentally observed variation (spread) in the negative assay results generated between healthy and affected populations. In this case, a result in the borderline or threshold range could indicate either a healthy or diseased individual, meaning that the combination of high assay specificity coupled to high assay sensitivity is almost impossible to achieve (because of the large number of potentially doubtful results). Threshold values for a particular assay may be determined by referring test results to internally (or externally) verified "control reference" material, i.e. samples with a pre-determined number of target sequences which have been repeatedly tested, and internally (and/or externally) controlled.

11.1.4 The Predictive Value

The predictive value relates assay results to actual clinical disease, and is used to indicate how far a positive or negative test result predicts the presence or absence of actual disease within a given population. The number of false negatives or false positives that can be expected using a particular assay system actually depends on the "prevalence" of illness within the study population (i.e. the number of clinically relevant diseased individuals expressed as a percentage of the study population). Positive and negative predictive values, which define the fraction of cases in which a positive test result actually predicts disease or alternatively the fraction of cases in which a negative result assures the absence of disease, are calculated using the equations: TP/(TP+FP) × 100% and TN/(TN+FN) × 100% respectively [Ferreira and De Avila, 1995]. On the basis of the number of clinically relevant disease cases, the percentage of the population that is truly diseased at a given point in time may be calculated. A variety of web sites facilitating the calculation of NPV or PPV values and the desired population size are available (Fig. 11.2).

It is generally understood that the predictive value of an assay increases as the percentage of clinically diseased individuals within a study population increases. Hence, selecting a particularly disease-prone subgroup or population may artificially increase the accuracy of the predictive value for a particular assay. The concept of predictive value affects the general applicability of PCR assays. For instance, imagine a new population-based PCR screening assay for a low prevalence disease such as cervical cancer due to human papillomavirus infection (prevalence = 2–5%) [Zhang et al., 2004]. If the assay performed has a specificity and sensitivity of 95%, then the predictive value of the assay will lie between 28% and 50%. This is too low to usefully associate a positive test result to clinical disease. The 28–50% positive predictive value indicates that an accurate test result will be expected in only 3–5/10 of cases, which means that linking a positive test result to clinical disease, or assay detection of actual disease, is too low to warrant the assay's use in a population screening strategy. When screening blood donors for the presence of human immunodeficiency virus, only a low percentage of patents screened will actually be positive

Test performance: Population A (n=10 000, incidence 1/1000)		
	D+	*D-*
T+	10	100
T-	0	9890
PPV = 9.1% NPV = almost 100%		

Test performance: Population B (n=10 000, incidence 300/1000)		
	D+	*D-*
T+	2970	70
T-	30	6930
PPV = 97.7% NPV = 99.5%		

Fig. 11.2 Relationship between incidence and predictive values. Note the difference in test performance when disease incidence changes (PPV: positive predictive value; NPV: negative predictive value) With permission from Dr. JM van Schalkwijk

for the virus (i.e. there is a low prevalence of HIV infection in the blood donor population) [Leroy et al., 1998]. Because the assay used has to be 100% sensitive (as false-negative results could result in healthy people being transfused with HIV infected blood), it only makes sense to use the assay on high-risk populations (where the prevalence of HIV is expected to be much higher).

11.1.5 Efficiency

In PCR-related assays, the term "efficiency" defines the smallest number of initial template DNA molecules that can be reproducibly amplified in a "standard" clinical sample using the complete PCR test system including nucleic acid extraction and amplimer detection. As such, PCR efficiency is closely related to PCR sensitivity and specificity. However, the term efficiency utilizes actual quantities of specific versus non-specific amplification products generated, as opposed to frequencies of true negatives and false positive results, and takes into account the presence of background noise as a result from for instance interference by non-specific amplification products. At low specificity, the efficiency of DNA synthesis is lower because PCR reagents are being used to amplify unintended non-specific amplification products, decreasing the amount of specific product that is amplified, as well as decreasing the "signal-to-noise" ratio. This suggests that maximum PCR efficiency does not actually have to equal the maximum yield of PCR amplified DNA, especially if large quantities of non-specific amplification products as well as large quantities of specific amplification products are being amplified. Numerically, the concept of assay "efficiency" may be expressed using the following equation:

$$\text{Efficiency} = \{[\text{object signal}] - [\text{background signal}]\}/[\text{background signal}] \times 100\%,$$

which for PCR-based assays translates to:

$$\text{Efficiency} = \{(\text{specific product concentration} - \text{non-specific product concentration})/\text{specific product concentration}\} \times 100\%$$

The optimal efficiency (signal-to-noise ratio) in PCR assays may be determined with the help of optimization experiments and simple agarose gel electrophoresis to detect the presence/absence of specific and non-specific amplimers under varying PCR conditions.

11.1.6 Error and Accuracy

Assay error and accuracy can be expressed in a simple formula: Error = Observed Value – True Value. In the optimal scenario (0% error, 100% accuracy), the observed value "O" would equal the true value "T", and no false positive or false negative results would be obtained. For quantitative PCR assays, in which an exact value is measured, e.g. copy number of retroviral particles per sample of blood, a high accuracy assay will always result in the same numerical value for a particular internal standard/calibrated sample/negative control, even after repeated testing. However, with respect to qualitative PCR assays (where positive or negative results are obtained), the concept of accuracy is slightly more complex. Here one needs to determine whether the correct result has been obtained by comparison of a number of qualitative process controls (e.g. low concentrations of target DNA) using methodologies other than the PCR assay itself [Connolly et al., 1995]. Without these essential process controls all PCR tests are essentially downgraded to qualitative assays only (yes or no answer provided, nothing more, nothing less).

11.1.7 Precision and Correctness

The accuracy of an assay is dependent on two related concepts, "precision" and "correctness" (see Fig. 11.3 for a graphical presentation). When the precision and correctness of an assay are high, then the assay accuracy will be correspondingly high.

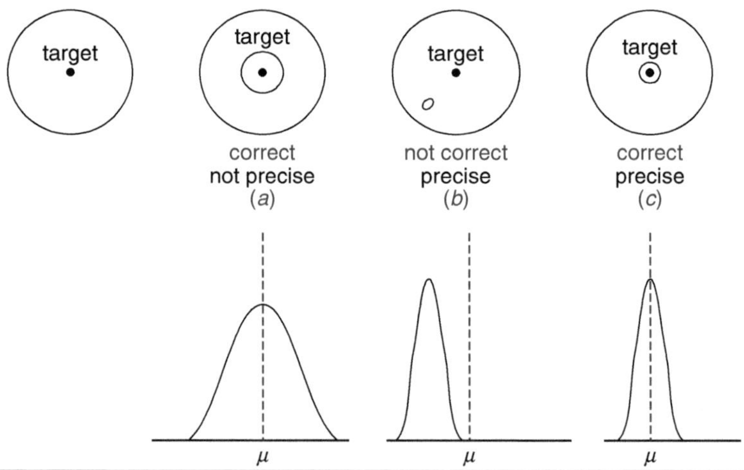

standard deviation (μ)	large	small	small
accuracy	poor	poor	good
systematic deviation	small	large	small
incidental deviation	large	small	small

Fig. 11.3 The accuracy of a particular protocol depends on the definition of the goal to be achieved. Only under optimal conditions can the goal be reached (situation c). Both precision and standard deviation affect the accuracy of a particular assay protocol.

The precision of a test indicates how close an observation is to reality. In relation to PCR-based assays, the precision describes the (lack of) variability among repeated measurements. The most important factor affecting the precision is accidental deviation, which may be generated by a variety of mechanisms including: sample processing artefacts (heterogeneity, the presence of PCR inhibitors, etc.; Fig. 11.4), variation in the amount or quality of ingredients used in the detection system, laboratory handling errors, pipetting errors, imprecise reading of measurements, etc. All observations (results) may be plotted in a single graph and the mathematical spread visualized in order to obtain the statistical "standard deviation" value (Fig. 11.5). An assay is precise when the standard deviation is low. When the accidental deviation is large, the standard deviation becomes correspondingly larger and the assay will exhibit a correspondingly lower precision. In most PCR assays precision indicates the variability of repeated observations, and except for quantitative PCR assays, this concept almost always remains limited to an estimation or a semi-quantitative determination.

The "correctness" of an assay (also described as the "systematic error"), indicates whether or not a series of test results deviate in a single direction away from the "true" results (the true "results" being established using an alternative assay protocol, e.g. ELISA, Southern blotting). For PCR assays, correctness assures that a fragment visualized on a gel is really is the specific fragment one is looking for. In the case of real time PCR, this quality requirement can be met by occasional sequencing of the PCR product. Five different categories of systematic error can affect the correctness value namely: (1) methodological problems, e.g. insufficient specificity of a primer pair, (2) instrumental deviations, (3) changes to the original (optimized) PCR protocol, (4) deviations in any follow-up analysis reactions and (5) inappropriate laboratory operating instructions, e.g. wrongly formulated protocols or poor laboratory technique.

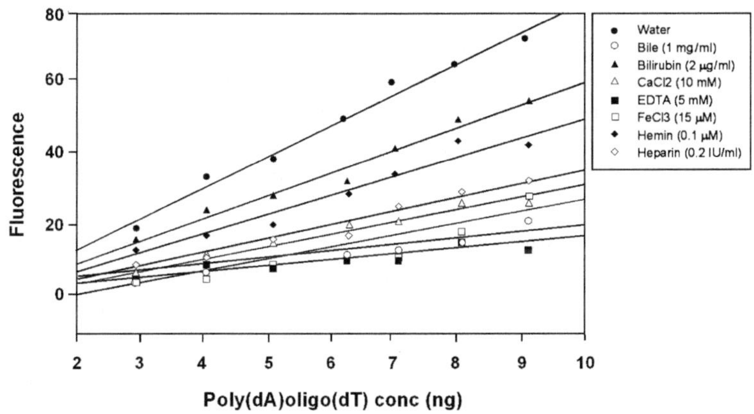

Fig. 11.4 The effect of inhibiting substances on the yield of a real time PCR assay. The inhibitory activity of water, bile, bilirubin, calcium chloride, EDTA, ferric chloride, hemin and heparin on the activity of Tth polymerase was evaluated [Al-Soud and Radstrom. 2001]

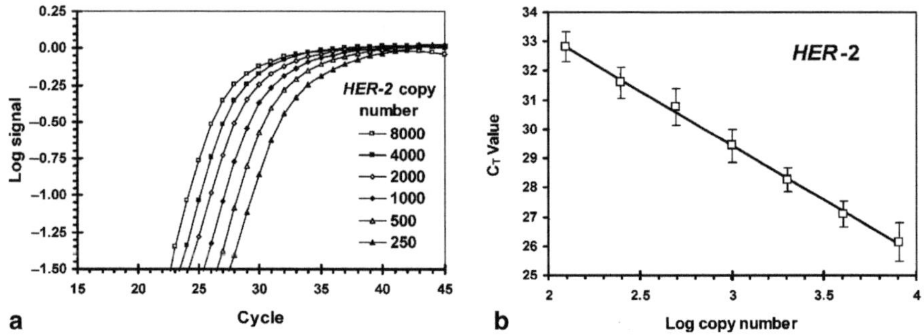

Fig. 11.5 Amplification profiles and mean calibration curves for HER-2/neu. (A) amplification profiles for HER-2/neu – the DNA was serially diluted 1/2 with water from 8,000 to approximately 250 genome equivalents, with 3 ng equalling 1,000 haploid genome equivalents. The negative controls without target DNA showed no signal increase and yielded no visible product. (B) mean calibration curve for six independent quantification experiments for HER-2/neu (two-fold dilutions of human genomic DNA). The calibration curve function includes Standard Deviations: y = −3.7(±0.33) × + 40.4 ± 1.05) [Konigshoff et al., 2003]

11.1.8 Defining the Analytical or Quantification Range and Sensitivity

Defining the analytical range of a PCR assay, a process known as "range finding", involves establishing the concentration range over which the required specificity and reproducibility may be obtained. Range finding experiments involve using the PCR assay to amplify different concentrations of target molecule using a selected PCR protocol for which the entire assay, including nucleic acid isolation and amplimer identification, displays sufficient specificity to generate the correct amplification products. Initial determination of the analytical range involves conventional PCR experimentation using a range of "spiked" control samples covering the low and high range of sensitivity values for that particular PCR assay protocol. Next, a variety of assay positive clinical samples are used whose target titres have been determined to fall within the analytical range (using alternative "gold-standard" techniques), the results tabulated, and finally compared with the "gold-standard" values. The smallest number of target molecules that can still be detected reliably and reproducibly in comparison with gold-standard techniques using clinical specimens is known as the "analytical sensitivity" of the assay and may be as low as one or two genome copies per PCR [Ferreira-Gonzalez et al., 2000]. During the determination of analytical range, all of the different types of materials that are likely to be encountered in the clinical situation need to be tested with a range of likely target concentrations. This is especially important in detecting the presence of false negative results, where the use of an internal positive control target (e.g. the β-globin gene for human tissues and cells) may be especially useful [Niesters, 2002]. At the same time, clinical samples known to be negative for the target nucleic acid must also be included in order to detect any possible false positive PCR results (e.g. by mis-priming or contamination). For PCR assays, it is also important that the measurements all fall within the analytical range, and that for quantitative PCR assays, the standard line for reference values comprises a straight line (N.B. for quantitative PCR assays absolutely homogeneous temperature profiles are required [Wilhelm et al., 2000]). In fact, the analytical range for clinical specimens may cover several orders of magnitude or may be very narrow! Virology related PCR protocols for example, may need to reliably detect and quantify clinically relevant numbers of viral particles, which span a range of six orders of magnitude. However, other generally applicable PCR assays may be required to define an analytical range covering only subtle changes in gene copy number (e.g. a doubling or tripling in the number of targets) [Koningshoff et al., 2003]. In some cases, rather than presenting an analytical sensitivity as the number of target molecules *per se*, it may be necessary to use more convenient units, e.g. units of infectivity or inclusion forming units [Madico et al., 2000]. For some assays, exact quantification is problematic [Ferreira-Gonzalez et al., 2000], as the definition of analytical sensitivity in a system does not fully exclude any false negative results obtained which may have been facilitated by PCR inhibitors and carried through from the original clinical sample [Mahony et al., 1998].

The inclusion of two randomly inserted positive control samples allows for the direct and continuous validation of the analytical range between different batches of clinical specimens. One positive control should generate a weak but consistently positive signal, with a target titre set at the lower end of the analytical range. When this sample is positive in all PCR batches tested, then this indicates that the sensitivity at this particular analytical range point is similar between the batches. In addition, this sample is also a measure of batch to batch reproducibility. The second positive control should generate a strong positive signal, with a target titre set at the upper end of the analytical range. This is an internal control primarily aimed at defining the quality of the reaction ingredients. In addition to the two positive controls mentioned, batch to batch comparisons of analytical range should always include a negative control specimen as well as a few spiked control specimens, with target titres set around the middle of the analytical range, in order to provide information on the absence/presence of PCR inhibitors.

It is very important to know the threshold value of a test above which the target titre becomes clinically relevant. This should be within the analytical range. Novel PCR protocols have been instrumental in defining threshold values for various clinical syndromes (e.g. minimal residual disease in leukaemia). To fall within the limitations of the analytical range some clinical materials may need to be diluted prior to PCR testing. At the other extreme, some clinical specimens must be concentrated by increasing the amount of nucleic acid extract added to the PCR reaction mix. In some clinical laboratories, diagnostic PCR tests are performed in triplicate; with the original clinical specimen being diluted 100-fold and 10,000-fold, and all specimens and dilutions being processed in parallel. This helps to ensure that in at least one of the samples the target concentration lies within the analytical range [Pas et al., 2002] (Fig. 11.6). Moreover, since more than one of the three sample dilutions is frequently positive, a positive result within a particular PCR batch is often verified by at least one other dilution of the same specimen. Increasing the target concentration (by concentrating the extracted nucleic acid prior to use, or by adding an increased volume of nucleic acid extract to the PCR reaction mix), is particularly relevant in cases where the PCR protocol is used primarily to indicate the absence of target sequences within clinical material, e.g. in the measuring of the microbiological "quality" of sewage or outdoor swimming waters, or the detection of genetically modified organisms in foodstuffs.

11.1.9 Recovery, Reproducibility and Quality Assurance

Recovery and reproducibility values indicate the likelihood of obtaining a positive result in a sample that is known to contain sufficient amounts of specific target nucleic acid for visible PCR amplification products to be observed after the appropriate

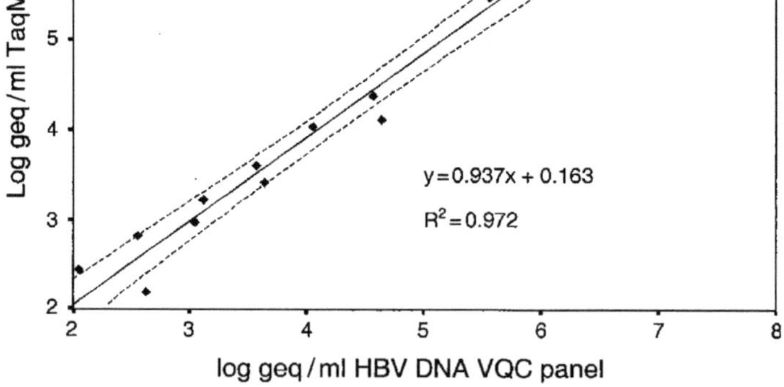

Fig. 11.6 Linear comparison of a real time HBV DNA PCR detection assay and the DNA concentration expected from the viral quality control (VQC) panel. Plotted are the calculated regression line (solid) of these values and the 95% confidence intervals of the mean (broken line) (Reproduced from Pas et al., 2000. With permission from 'American Society for Microbiology')

number of amplification cycles. The assessment of recovery and reproducibility parameters involves repeated testing of multiple samples containing specific quantities of target nucleic acid, all by different persons, using different equipment, in different institutions and at different periods of time. Recovery rates can be assessed by inclusion of positive control samples containing strong and borderline quantities of target. Essentially, in this way, systematic errors (i.e. intra- and inter-laboratory errors) may be detected and laboratory quality criteria defined on the basis of these results (Fig. 11.7) [Bustin, 2002]. Suggestions for the improvement of poorly performing laboratories may then be formulated, enacted and controlled. Such studies are absolutely essential for novel PCR protocols that are to be utilized in the clinical laboratory. For example, comparative recovery and reproducibility studies on a PCR protocol designed to detect the microbial pathogen *C. trachomatis*, revealed that differences in DNA extraction protocols gave rise to major differences in assay sensitivity between various laboratories [Apfalter et al., 2002].

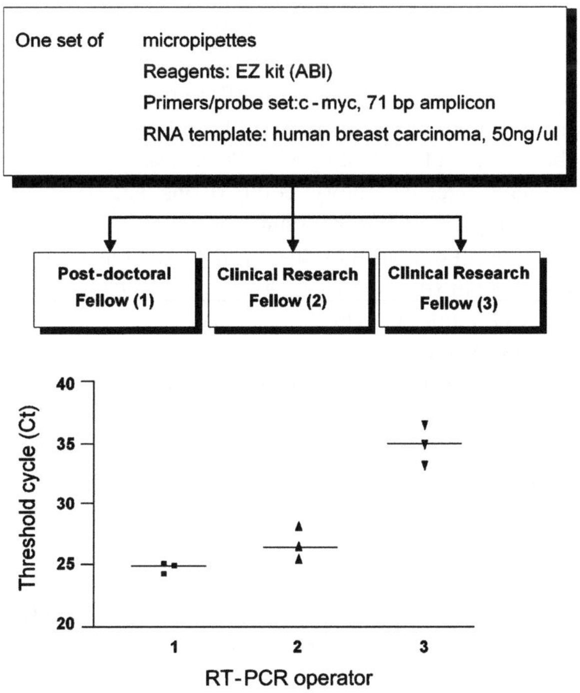

Fig. 11.7 Operator variability. Three researchers used identical micropipettes, mastermix and template to quantify the same target on an ABI Prism 7700 machine using a real-time PCR methodology. Individual 1, was an experienced researcher, whereas individuals 2, and 3, were non-experts who had limited experience with dispensing reagents. The scatterplot shows the replicate C_t values obtained by the three individuals. The variation in results translates into significantly different copy numbers per microgram of total RNA recorded by the individuals (From Bustin, 2002)

11.2 Quality Assurance and Multicenter Studies

PCR mediated molecular diagnostics is a dynamic field of research where new and novel assays and protocols are constantly being developed in the search for greater sensitivity, specificity, reliability, etc. over currently established "gold-standard" detection methods. However, before a new PCR mediated molecular diagnostic assay is accepted into general clinical use, it is important to perform studies comparing the new PCR assay against the "gold-standard" assay currently being used in that particular clinical situation (Fig. 11.8). This validation process may initially be performed using archived or spiked clinical samples, which then must be repeated using fresh clinical material. The positive and negative samples used for both PCR and gold-standard assays should be identical so that a direct comparison of the two tests can be made with regard to their comparative sensitivities and specificities. If the new PCR assay generates an increased number of positive results compared to the gold-standard assay, then the new PCR assay may be either more sensitive (and hence more useful as a diagnostic tool than the current gold-standard) (Fig. 11.8) [Coutlee et al., 2002], or alternatively the PCR may simply generate a greater number of false positive results. In both cases, further comparative testing may be necessary. Even if the new PCR assay does not provide a distinct advantage with regard to sensitivity, specificity, reliability etc over the current gold-standard assay, the new assay may still offer advantages with regard to lower costs, less hands-on time required, increased throughput, etc.

Essential to the process of evaluating new PCR assays (as well as other types of assays) are intra- and inter-laboratory comparisons and the concept of quality assurance. The development of internationally accepted quality control criteria and quality control specimens (similar to those currently developed for diagnostic ELISA assays, immunohistochemistry, etc.), as well as standard operating procedures (SOPs), are particularly important to this process. A range of viral and microbiological proficiency panels have been made available for assessing quality control criteria in participating laboratories. A number of key concepts and conditions related to quality assurance, quality control and new PCR protocols are listed below:

Quality assurance (QA). Quality assurance is a means of monitoring the correct functioning of quality control systems running in a particular laboratory.

METHOD B	METHOD A	
	POS	NEG
POS	a	a
NEG	c	d

Observed concordance: (a + d) / (a + b + c + d)

Fig. 11.8 Definition of observed concordance by comparing Method A to Method B. Method A could be a PCR assay for example, compared to Method B, which could be an *in situ* hybridisation (ISH) assay for example

External Quality Assessment (EQA). EQA concerns the assessment of the quality control procedures as enacted in individual laboratories. Identical reagents and/or clinical samples are distributed from a single coordinating centre to all participating laboratories. PCR testing is then performed and the results compiled by the EQA administrators, allowing a comparison of the results obtained within a particular laboratory (for multiple EQA specimens), as well as (anonymous) comparisons with other laboratories to be made. Such EQA samples often include internationally accepted standards, as in such schemes, it is extremely important to have a consensus on what actually constitutes a positive/negative result or value. Most of the initial PCR EQA schemes have focussed on qualitative assays, i.e. presence or absence data. However, newer EQA schemes also allow the inter-laboratory quality assurance of quantitative PCR amplification protocols to be monitored. For such quantitative PCR assays, it is imperative that the positive control samples used contain amounts of template molecules that lie within the analytical range of the methods employed.

Multicenter Studies. Slightly different from EQA schemes, (where inter-laboratory results are compared in order to assess the efficacy of quality control procedures being undertaken within an individual laboratory), multi-centre studies allow comparisons to be made between both internal and external quality control efforts that concentrate on a particular diagnostic PCR assay. Such studies are performed by a number of research institutions sharing a common diagnostic interest in order to assess the overall quality of national and/or international diagnostic PCR testing, an assessment often required by governmental agencies in order to improve patient care strategies or in the development of novel quality control strategies. With specific regard to PCR assays, multi-centre studies can be particularly important in assessing: (1) whether a particular PCR protocol can detect geographically associated target nucleic acid polymorphisms, (2) whether quantitative PCR assay results show inter-laboratory linearity over the range of likely target concentrations to be found in clinical specimens (usually achieved by including dilution series of target DNA in the samples to be tested in a random manner), and (3) whether all the participating laboratories actually comply with (inter)national quality criteria. Figure 11.9 highlights the organizational structure required for such multi-centre studies. Essentially, each laboratory remains anonymous to the others with each being assigned a unique laboratory identifier code. Next, encoded positive and negative samples selected on the basis of previous results obtained using alternative test formats [Noordhoek et al., 1994; Kolk et al., 1994; Quint et al., 1995], as well as accompanying sample documentation, are distributed from the coordinating central laboratory to all of the individual participating laboratories. If available, subtypes of the target organism, tissue etc are included, as well as a range of positive controls containing target DNA within the analytical range of the assay, negative controls containing no nucleic acid at all, and negative controls containing large amounts of non-target nucleic acid. Once the experimental results have been obtained, the data is anonymously transferred to the coordinating central laboratory, where it is statistically examined and arranged into an accessible format. Occasionally, data processing may be performed by a neutral laboratory (i.e. a group of people unconnected to the multicenter study), which then sends the double-blinded data to the organizing committee. Finally, the tables with the cumulative (and still anonymous) data are disseminated to all participating laboratories, which may then identify their own datasets via their own laboratory code. Using this data, the results from other participating laboratories may be compared and conclusions drawn as to the quality control performance of the PCR assay as used in that particular laboratory. Alterations in laboratory practice or assay protocol may then be made in order to improve the quality performance of the participating laboratory (changes in laboratory practice or assay protocol are tolerated so long as quality performance enhancements of the assay and/or laboratory are achieved). Repeat annual testing helps to determine whether or not quality control improvements in the PCR protocol are being achieved, as well as allowing a comparison to be made regarding the progress of other participating laboratories involved in the scheme. The continual assessment of an assay's performance using slightly different methodologies may eventually lead to diagnostic improvements, a feature that is profitable to all study participants.

The measure of success of these multicenter EQA testing schemes can be seen by the fact that during the early 1990s, when the first multicenter studies for PCR assays were organized, up to 60% of the participant laboratories failed with regard to detecting highly diluted/borderline positive control samples. The current situation, however, is much better, with the percentage of laboratories that fail quality assessment having dropped below 10%, even though there has been a threefold increase in the number of laboratories participating in such schemes. However, a large degree of variability in quality performance may still be observed between laboratories, even when they are using the same (e.g. commercially available) PCR test system. In general, laboratories affiliated to blood banks perform very well in multicenter EQA studies when compared to other types of laboratories.

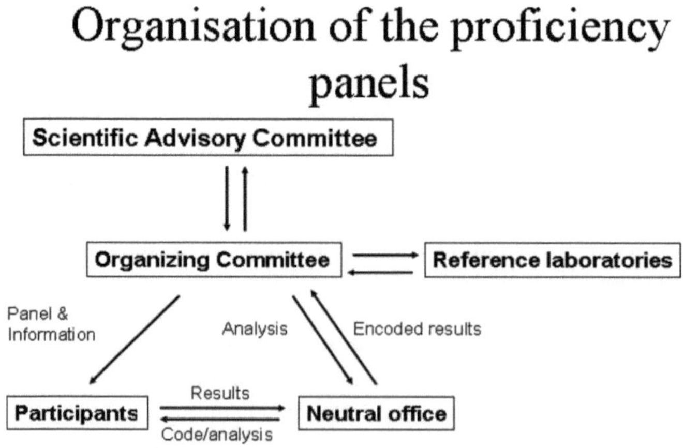

Fig. 11.9 Organizational scheme required for multi-centre investigations in order to determine the accuracy of a particular PCR protocol using quality assurance procedures (From Quint et al., 1995)

References

Al-Soud, W.A. and Radstrom, P. 2001. Purification and characterization of PCR-inhibitory components in blood cells. J Clin Microbiol 39:485-493.

Apfalter P, Assadian O, Blasi F, Boman J, Gaydos CA, Kundi M, Makristathis A, Nehr M, Rotter ML, Hirschl AM. 2002. Reliability of nested PCR for detection of *Chlamydia pneumoniae* DNA in atheromas: results from a multicenter study applying standardized protocols. J Clin Microbiol 40:4428–4434.

Bustin SA. 2002. Quantification of mRNA using real-time reverse transcription PCR (RT PCR): trends and problems. J Mol Endocrinol 29:23–39.

Connolly AR, Cleland LG, Kirkham BW. 1995. Mathematical considerations of competitive polymerase chain reaction. J Immunol Methods 187:201–211.

Coutlee F, Gravitt P, Kornegay J. 2002. Use of P6My primers in L1 consensus PCR improves detection of human papillomavirus DNA in genital samples. J Clin Microbiol 40:902–907.

Ferreira AW, De Avila SD. 1995. Laboratory diagnosis of Chagas' heart disease. Rev Paul Med 113:767–771.

Ferreira-Gonzalez A, Yanovich S, Langley MR, Weymouth LA, Wilkinson DS, Garrett CT. 2000. Enhanced analytical sensitivity of a quantitative PCR for CMV using a modified nucleic acid extraction procedure. J Clin Lab Anal 14:32–37.

Kolk AH, Noordhoek GT, De Leeuw O, Kuijper S, Van Embden J. 1994. *Mycobacterium smegmatis* strain for detection of *Mycobacterium tuberculosis* by PCR used as an internal control for inhibition of amplification and for quantification of bacteria. J Clin Microbiol 32:1354–1356.

Konigshoff M, Rainer JW, Pingoud A, Hahn M. 2003. Her2/Neu gene copy numbers quantified by real time PCR: comparison of gene amplification, heterozygosity and immunohistochemical status in breast cancer tissue. Clin Chem 49:219–229.

Leroy V, Newell ML, Dabis F, Peckham C, Van de Perre P, Bulterys M, Kind C, Simonds RJ, Wiktor S, Msellati P. 1998. International multicentre pooled analysis of late post-natal mother to child transmission of HIV-1 infection. Ghent International Working Group on Mother to Child Transmission of HIV. Lancet 352:597–600.

Madico G, Quinn TC, Gaydos CA. 2000. Touchdown enzyme time release PCR for detection and identification of *Chlamydia trachomatis*, *C. pneumoniae* and *C. psittaci* using the 16S 16S-23S spacer rRNA genes. J Clin Microbiol 38:1085–1093.

Mahony J, Chong S, Jang D, Luinstra K, Faught M, Dalby D, Sellors J, Chernesky M. 1998. Urine specimens from pregnant and nonpregnant women inhibitory to amplification of *Chlamydia trachomatis* nucleic acid by PCR, ligase chain reaction and transcription mediated amplification: identification of urinary substances associated with inhibition and removal of inhibitory activity. J Clin Microbiol 36:3122–3126.

Neumaier M, Braun A, Wagener C. 1998. Fundamentals of quality assessment of molecular amplification methods in clinical diagnostics. International Federation of Clinical Chemistry Scientific Division Committee on Molecular Biology Techniques. Clin Chem 44:12–26.

Niesters HG. 2002. Clinical virology in real time. J Clin Virol 25:3–12.

Noordhoek GT, Kolk AH, Bjune G, Catty D, Dale JW, Fine PE, GodfreyFaussett P, Cho SN, Shinnick T, Svenson SB. 1994. Sensitivity and specificity of PCR for detection of *Mycobacterium tuberculosis*: a blind comparison study among seven laboratories. J Clin Microbiol 32:277–284.

Pas SD, Fries E, De Man R, Osterhaus ADME, Niesters HGM. 2000. Development of quantitative real time detection assays for hepatitis B virus DNA and comparison with two commercial assays. J Clin Microbiol 38:2897–2901.

Quint WGV, Heijtink RA, Schirm J, Gerlich WH, Niesters HG. 1995. Reliability of methods for hepatitis B virus DNA detection. J Clin Microbiol 33:225–228.

Salto-Tellez M, Shelat SG, Benoit B, Rennert H, Carroll M, Leonard DG, Nowell P, Bagg A. 2003. Multiplex RT-PCR for the detection of leukemia associated translocations: validation and application to routine molecular diagnostic practice. J Mol Diagn 5:231–236.

Sen Gupta R, Hillemann D, Kubica T, Zissel G, Muller-Quernheim J, Galle J, Vollmer E, Goldmann T. 2003. HOPE-fixation enables improved PCR-based detection and differentiation of *Mycobacterium tuberculosis* complex in paraffin embedded tissues. Pathol Res Pract 199:619–623.

Wilhelm J, Hahn M, Pingoud A. 2000. Influence of DNA target melting behavior on real-time PCR quantification. Clin Chem 46:1738–1743.

Zhang ZY, Sdek P, Cao J, Chen WT. 2004. Human papillomavirus type 16 and 18 DNA in oral squamous cell carcinoma and normal mucosa. Int J Oral Maxillofac Surg 33:71–74.

Chapter 12

Variants and Adaptations of the Standard PCR Protocol

The standard qualitative PCR protocol (as initially developed) involves the exponential amplification of a specific DNA sequence which is achieved by subjecting an optimized PCR mix to a predetermined number of thermocycles (cyclical heating and cooling steps). Each thermocycle within this standard PCR protocol actually comprises three distinct steps, namely: (i) a nucleic acid denaturation step (to disassociate the double-stranded DNA helix into single strands), followed by (ii) a primer annealing step (to allow the specific hybridization of a 15–25 base pair oligonucleotide primer), followed by (iii) a DNA extension/amplification step (to replicate and amplify the specific target nucleic acid sequence). After the given number of thermocycles are complete, and exponential DNA amplification has been performed, the presence or absence of specific target DNA sequences in the final PCR mix (and hence the presence or absence of specific target DNA sequences in the original sample) is established. Over the last few decades, many useful adaptations/variations of this standard qualitative PCR protocol have been developed. Some of the most important of these variants/adaptations are outlined below.

12.1 Generating Labelled PCR Amplimers for PCR Product Visualization, DNA Probes and Cloning

Several standard PCR protocol adaptations have now been developed which allow for the labelling of PCR amplimers as they are being created during PCR amplification. Some of the possible labels that may be incorporated are shown in Fig. 12.1, and include biotin, digoxygenin, many fluorescent dyes or fluorochromes (e.g. FAM, HEX, JOE, etc.; see also Table 5.4), and even (radioactive γ-^{32}P) phosphate groups. The advantage of attaching specific labels to PCR amplimers, is that the amplimers may be directly visualized or used in other "downstream" post-PCR applications such as solid phase hybridization or reverse line blots (see previous chapters).

Frequently, labelled amplimers are generated by adding specially modified nucleotides (to which labels have been chemically attached) to the PCR mix prior to PCR thermocycling. These modified nucleotides are added to the initial PCR mix alongside the normal, unmodified nucleotides (dATP, dCTP, dGTP and

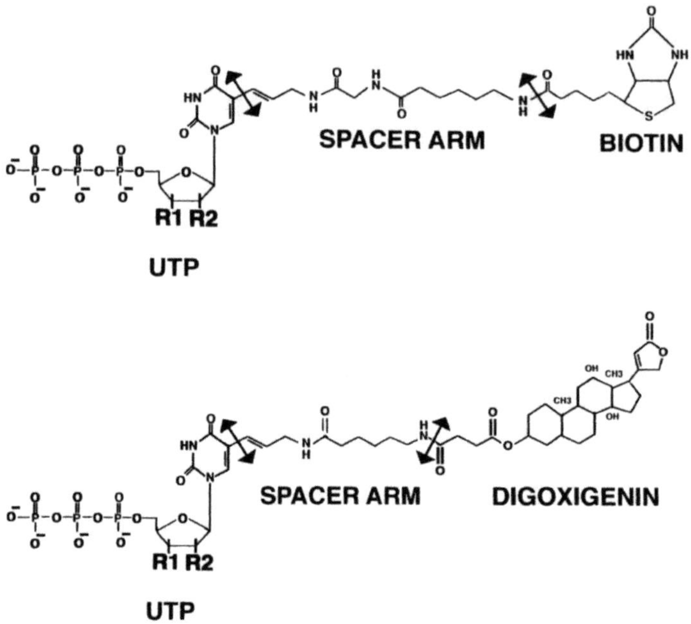

Fig. 12.1 Biotin-11-dUTP and digoxygenin-11-dUTP are two ligands which are frequently used to label PCR amplimers. When included in a PCR mix (along with unlabeled deoxynucleotides), they are incorporated into the growing DNA chain by the thermostable DNA polymerase. The 5′-end position of the pyrimidine nucleotide dUTP (deoxyuracil triphosphate) is most suited as substitution site for the addition of labels. The "11" identifies the number of carbon atoms present in the spacer arm. Radioactive compounds such as α-^{32}P dUTP are becoming increasingly less popular due to safety and disposal problems, though radioactive labelling still represents the most sensitive detection technique currently available. In uridine R1 = OH and R2 = OH, in deoxyuridine R1 = OH and R2 = H and in dideoxyuridine both R1 and R2 are H (Reproduced from Chevalier et al., 1997. With permission from Dennis G. Bashkin, Ph.D.)

dTTP), and become incorporated into the PCR amplimers via the action of DNA polymerase. When using PCR amplimer labelling methodologies, however, further optimization experiments, over and above those performed to optimize the standard PCR protocol, may need to be performed in order to achieve the correct ratio of modified/unmodified nucleotides within the PCR mix. Also, different thermostable DNA polymerases exhibit different incorporation efficiencies with different modified dNTPs, for example, the biotin labeled dNTP "Bio-16-dUTP" is incorporated more efficiently by Pwo (*Pyrococcus woesei*) thermostable DNA polymerase (Roche Diagnostics, Mannhein, Germany) than by Taq (*Thermus aquaticus*) or Tth (*Thermus thermophilus*) polymerases. However, modified compounds such as fluorescein-12-dUTP or digoxygenin-11-dUTP (DIG-11-dUTP) are incorporated equally efficiently by all three of these enzymes, (though the optimum PCR mix ratio of modified/unmodified dNTP may nevertheless vary between 1:3 and 1:20 dependant on the enzyme used).

There are, however, important cost aspects for PCR protocols that use labelled deoxynucleotides in the initial PCR mix. Specifically, a relatively large fraction of the labelled deoxynucleotides will remain unused in the final PCR mix and both DNA strands will be labelled even if post-PCR experiments only require a single-stranded labelled DNA fragment (probe) to be generated. If cost is indeed an important factor, and single-stranded labelled DNA probes are required, then an asymmetric PCR protocol (where one of the PCR primers is present in great excess) using labelled deoxynucleotides may be a useful alternative [Jouquand et al., 1999]. However, asymmetric PCR amplification protocols amplify DNA in a mainly linear fashion, as opposed to the exponential amplification achieved using balanced concentrations of PCR primers, resulting in a significant reduction in amplimer yield. Finally, it should be noted that labelled deoxynucleotides will also be incorporated into any non-specific PCR amplification products generated during PCR amplification.

Alternatives to using *labelled dNTPs* for PCR amplimer visualisation/detection include the use of *end-labelled primers* as well as post-PCR labelling protocols involving either *nick translation*, *random primer labelling*, or *terminal deoxynucleotidyl transferase* (TdT) reactions. Many commercially available kits are currently advertised. The use of end-labelled primers in a PCR does not alter the PCR conditions compared to non-labelled PCR conditions. The use of such kits has advantages: DNA strand specificity can be accomplished and only primers that correctly anneal will be extended and labelled. Most end-labelling PCR mixes employ a single labelled primer, though multiple labelled primers may be useful in differentiating amplimer species in multiplex PCRs. The *nick translation* protocol uses the enzyme pancreatic exonuclease I to generate single nucleotide "nicks" (deletions on one of the strands of double-stranded DNA) in the PCR amplimer, which are then filled in by the 5′ to 3′-end repair mechanism of DNA polymerase I from *E. coli*. Moreover, by including labelled dNTPs in the DNA polymerase I reaction mix, the nicks will be filled in with labelled dNTPs, resulting in a labelled amplimer. *Random primer labelling* protocols work by annealing a random mix of (decanucleotide) primers to single-stranded PCR amplimers and allowing any "gaps" to be filled in using exonuclease-free Klenow fragment DNA polymerase. Figure 12.2 shows two amplimer labelling strategies currently available. The enzyme *terminal deoxynucleotidyl transferase* (TdT) has the ability to add 5–10 (labelled) deoxynucleotides to the 5′-ends of a double-stranded DNA molecule using manganese as a cofactor. If these nucleotides are labelled, then the PCR amplimer itself will also be labelled.

Different labels (e.g. dyes which fluoresce at different wavelengths upon excitation) may be attached to PCR amplimers arising from different PCRs. If closely related (with respect to their sequence identities), then these amplimers may be utilized in competitive hybridization assays, where differential (competitive) hybridization is indicated by the presence or absence of a particular fluorescent dye signal. This procedure is often used to detect differences in gene expression

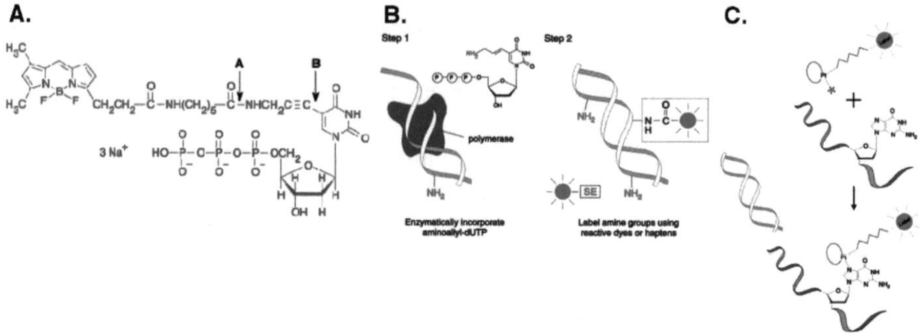

Fig. 12.2 Various important features of nucleic acid labelling technologies. A. The chromatide BODIPY FL-14-dUTP is representative for a variety of other aminoalkynyl dUTP nucleotides. Fluorophore labels are attached through a four-atom aminoalkynyl spacer (between arrows A and B). B. A two-step DNA labelling method in which aminoallyl dUTP is enzymatically incorporated (step 1) after which a reactive fluorophore is chemically coupled (step 2). C. Direct labelling of DNA using ULYSIS reagent. The reagent reacts directly with the N7 position in guanine and generates a stable coordination complex between nucleic acid and label (Copied from "The Handbook" at www.invitrogen.com. Copyright Molecular Probes, Inc.)

between different tissues or organisms, or between induced and non-induced tissues [Verhagen et al., 2000]. Labelled amplimers may be used as DNA probes in for example Southern hybridization reactions, however because single-stranded DNA is used in such protocols, one should ensure that the label is attached to the single-strand of the double-stranded amplimer that is to be used as a probe (see also other chapters).

Other means of labelling PCR amplimers includes the addition of phosphate groups to the 5′-end, allowing the amplimer to be used in cloning and ligation reactions (PCR amplimers are not phosphorylated at their 5′-ends). The enzyme T4 polynucleotide kinase may be used along with ATP (adenosine triphosphate) to phosphorylate the non-phosphorylated 5′-end of double-stranded PCR amplimers, thereby facilitating cloning of the amplimer into (plasmid) vectors using ligase enzymes. Alternatively, 5′-end phosphorylated primers may be utilized in the PCR mix *per se*. 3′-End phosphorylated primers will actually block PCR amplification (inert primer).

12.2 Two-Step PCR Protocol

Unlike the standard qualitative PCR protocol, a two step PCR protocol utilizes two distinct steps in each thermocycle instead of the usual three-steps (of denaturation, annealing and extension). The two-step PCR protocol uses a nucleic acid denaturation step (whereby the reaction mix is heated to a temperature of between 92°C and 97°C), followed by a combined primer annealing and extension/amplification step whereby the reaction mix is heated to a temperature of between 50°C and 70°C. Taqman technology sometimes employs two step PCR protocols as well [Petitjean et al., 2005]. This contrasts with the more usual three-step PCR protocol where the annealing and extension temperatures of each thermocycle are separate. The advantage of using a two-step PCR protocol is that it allows the use of less stringent extension/amplification temperatures, a feature that may be particularly important when short amplimers with low annealing temperatures (e.g. random hexanucleotides or poly-dT primers of approximately 10–15 bp), or degenerate primers where PCR primer sequence mismatches may have to be tolerated, are to be used. During the heating and cooling periods of two-step PCR thermocycling, the ramp rate (i.e. the increase or decrease in temperature per unit time), may be varied in order to facilitate more efficient hybridization and elongation/amplification. A decreased ramp rate may increase the efficiency of a PCR, but unfortunately could also lead to the increased synthesis of non-specific PCR products. In some two-step PCR protocols dependant on the nature of the template (e.g. template molecules that are extremely rich in GC base pairs, it may be necessary to adjust the temperature ramp rate during the latter cycles of the PCR.

12.3 Booster PCR

The *booster* PCR protocol is designed to inhibit the accumulation of non-specific amplimers and primer dimer complexes, which may arise due to non-optimized PCR conditions and/or when target DNA concentrations are initially very low. Essentially, the first 15–20 cycles of booster PCR are performed using dNTP and primer concentrations which are 10–100 times less than in a standard qualitative PCR protocol so that the reaction equilibrium of the booster PCR is significantly shifted towards target-specific amplification, albeit at an initial cost to amplimer yield. After a set number of cycles, when sufficient quantities of specific PCR product have been amplified, thermocycling is paused and an extra quantity of primers and dNTPs (and thermostable DNA polymerase if required) are added to the reaction mix in order to achieve primer and dNTP concentrations similar to a standard PCR protocol. After addition of the extra quantities of reagents, the second *post-booster* phase of PCR amplification is performed using a further 30–50 PCR cycles with the same thermocycling parameters as the initial phase. Alternatives to the *booster PCR* are *nested PCR*, *hot-start PCR* (with or without and the use of thermally activated thermostable DNA polymerases) and *touchdown PCR*. It should be noted that all PCR protocols where the PCR vessel has to be opened after the initiation of thermocycling (including *booster PCR*, some *hot-start PCR*s and *nested PCR* protocols) carry an increased risk of being contaminated by amplified DNA (via aerosols, etc.) from neighbouring reaction tubes.

12.4 Hot-Start and Time-Release PCR Protocols

Non-specific primer hybridization and the formation of primer dimers may occur in all types of PCR protocols, especially if the initial target DNA: non-specific background DNA concentration is very low or multiple primer pairs are used within the same PCR mix (multiplex PCR). If a large concentration of non-specific background DNA is present in a non-optimized PCR mix, the vast majority of mispriming events will occur during the initial pre-PCR heating step, a step required in order to melt the DNA duplex prior to the start of the PCR thermocycling program *per se* [Chou et al., 1992]. During this initial pre-PCR heating step when the temperature is sufficient to dissociate regions of the double stranded DNA template but not sufficient to ensure specific primer hybridization, primers will be able to hybridize to non-target regions of DNA, facilitating the amplification of non-specific PCR products, even at temperatures that are sub-optimal for amplification by the thermostable DNA polymerase. During subsequent PCR thermocycling, the temperature for primer hybridization does not drop below the pre-programmed hybridization temperature used in that particular PCR protocol, so that primer hybridization and amplification during these subsequent PCR cycles is specific. However, if primers have already hybridized to non-target DNA sequences and these non-specific sequences have been extended, then both non-specific DNA and target DNA will be co-amplified, resulting in a reduction in the yield of specific target amplimers and ongoing depletion of reaction components (and essentially a reduction in PCR sensitivity). The result is therefore a non-specific, insensitive PCR protocol, which is not ideal for either the clinical laboratory or research purposes. Several methods have now been developed which help to prevent non-target primer hybridization during the initial pre-PCR heating step. One alternative is the use of various single-stranded DNA "aptamers" that efficiently inhibit the polymerase activity of enzymes from the thermostable *Thermus* genus of bacteria (e.g. *Taq* and *Tth*), as long as the temperature remains below the melting temperature of the hairpin containing aptamer [Lin and Jayasena, 1997] (Fig. 12.3). Another, seldom used procedure involves the addition of "carrier DNA" to the PCR mix which shields all potential primer hybridization sites. Consequently, PCR primer hybridization only occurs at regions of DNA where the primers can successfully compete with the carrier DNA for binding sites, i.e. where the PCR primers bind in a highly specific manner to their specific DNA target sequence. Alternatively, some thermostable DNA polymerases show a much reduced activity at low temperatures, meaning that they are less capable of amplifying non-specific products during the initial PCR pre-heating step. However, neither of these methodologies is widely used in clinical or research laboratories. Instead, "hot-start" and "time release" PCR methodologies have become the most preferred methods used for inhibiting non-specific product amplification during the initial pre-PCR heating step (before the first cycle of thermocycling begins).

(i) Hot-start PCR: The hot-start PCR methodology, involves preparing PCR mixes with one of the essential ingredients (thermostable DNA polymerase, PCR primers,

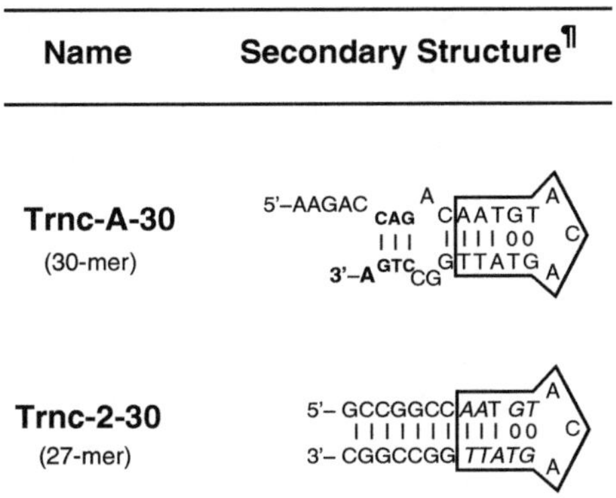

Fig. 12.3 Aptamers are small hairpin forming molecules that bind to Taq polymerase at temperatures below 40°C and exert a sequence specific inhibiting effect on polymerase activity. The complex dissociates upon heating and binding is reversible. Aptamers are used to prevent non-specific amplification (From Lin and Jayasena, 1997. Copyright Elsevier)

magnesium ions, dNTPs or rarely template DNA) missing from the mix. After heating to a temperature of between 92°C and 97°C (the pre-PCR heating step), the missing ingredient is then manually added to the reaction mix and thermocycling is allowed to proceed. Subsequent lowering of the temperature to the pre-programmed annealing temperature will then allow specific primer hybridization to occur and the amplification of specific target DNA during the first and subsequent PCR cycles [Mullis, 1991]. The hot-start PCR protocol helps to ensure that only specific PCR primer annealing occurs and allows the amplification of as few as 1–5 specific target DNA molecules within a background of up to 1 μg of non-specific DNA. A very serious disadvantage of the hot-start methodology involving the manual addition of PCR ingredients is the increased likelihood of carry-over contamination that arises when the PCR tube is opened (particularly if the PCR assay is being performed on multiple samples within the same PCR heating block, as all of the reaction tubes will have to be manually opened and closed, thereby increasing the chance of aerosol mediated contamination). Also, opening of reaction tubes at high temperature allows evaporation of the ingredients to occur, which may lead to small changes in the concentration of key PCR ingredients. To date, several ingenious alternative procedures have been developed to circumvent the contamination problems associated with the manual hot-start procedure. One option involves pipetting mineral oil onto the surface of each PCR mix (prior to any heating and thermocycling steps) and then pipetting the missing ingredient onto the surface of the mineral oil. Upon heating, the viscosity of the mineral oil will decrease and any ingredient pipetted onto the surface of the oil will readily diffuse into the PCR mix itself. N.B: Mineral oil was initially used to prevent evaporation of the PCR mix into the lid of the reaction tube during thermocycling, a necessity for older PCR machines which did not possess heated lids.

In a similar manner, solid DNase-free wax plugs may be used instead of mineral oil. The wax is added to the surface of the PCR mix as a semi-solid material and the reaction tube is then cooled on ice to allow solidification of the wax. The missing PCR reaction mix ingredient is then pipetted onto the surface of the solid wax layer. Subsequent heating of the PCR vessel results in the wax gradually melting, allowing diffusion of the missing ingredient into the PCR mix. In order to save time, many manufacturers now supply DNase-free wax "beads" which have been impregnated with magnesium chloride (an essential PCR ingredient). The wax bead is simply dropped onto the surface of the incomplete PCR mix and upon heating melts releasing the magnesium chloride. Finally, another ingenious hot-start PCR methodology involves the use of primers that contain internal self-complimentary sequences that form a hairpin-like stem-loop ("molecular beacon") structure at low temperatures. Upon heating the PCR mix, thermal energy disrupts the stem-loop structure generating a linear primer, allowing specific hybridization to occur [Kaboev et al., 2000].

(ii) Time Release PCR: Time release PCR has the same objectives as hot-start PCR, namely the reduction of non-specific PCR amplification products, though nowadays the term "hot-start PCR" is usually used for all hot-start and/or time release PCR methodologies. Time release PCR involves the use of novel thermostable DNA polymerases which require an extended period of heat treatment at relatively high temperatures (e.g. ±8 minutes at 95°C) in order to become activated (N.B. various alternative activation protocols have been described [Kebelmann-Betzing et al., 1998]). This means that PCR mixes can be prepared in their entirety prior to PCR thermocycling without the risk of non-specific primer hybridization (as the thermostable DNA polymerase will not be active until the temperature within the PCR mix is above the temperature required for specific primer annealing). The advantages of time release PCRs include the fact that they generate a greater yield of specific amplimer from a given concentration of reagents, are less sensitive to the action of PCR inhibitors, and generate better quality sequence data than non-hot-start PCR protocols [Moretti et al., 1998]. Examples of commercially available *time release PCR* enzymes include; AmpliTaq Gold (PE Biosystems), Thermostart (AbIgene), Hot Start Taq (Qiagen) and Proofstart (Qiagen) (Fig. 12.4). All of these particular enzymes have had their active sites artificially mutated so that their DNA polymerase activity is temporarily inactivated. However, upon heating, the three-dimensional conformation of the active site of the enzymes changes, restoring polymerase activity.

Other commercially available time release PCR-specific thermostable polymerase enzymes facilitate inhibition of amplification via binding of monoclonal antibodies to the enzyme, effectively blocking the active site of DNA synthesis as first described in the mid-1990s [Kellogg et al., 1994]). Upon heating, the monoclonal antibodies are denatured and passively dissociate from the enzyme, de-blocking the active site and restoring DNA polymerase activity. Commercially available examples of such enzymes include; Taq Platinum (Life Technologies), FastStartTaq (Roche), KOD polymerase [Mizuguchi et al., 1999], HotStarTaqDNAPol (Qiagen),

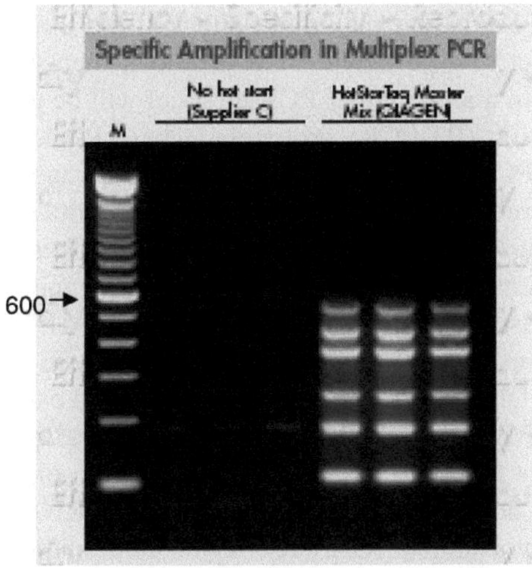

Fig. 12.4 Hotstart procedures involving activatable enzymes prevent the amplification of non-specific PCR procedures. At high DNA dilution there is an obvious clearing effect of the use of Hot Star Taq (five lanes on the right) as compared to non-hot start alternatives. Fragment of the p53 gene were amplified from genomic DNA in multiplex PCR. Parallel reactions were prepared using standard reaction conditions. On the left in the lane marked M, a 100 basepair molecular weight marker is shown with the 600 basepair fragment highlighted (From Critical Factors for Successful PCR at www.qiagen.com)

and RedHot DNA polymerase (ABIgene). It is also possible to purchase certain anti-DNA polymerase monoclonal antibodies from commercial suppliers, e.g. Jumpstart (Sigma) and TaqStart Antibody (ClonTech), which are added along with the non-time release thermostable DNA polymerase to the initial PCR mix. Newer monoclonal antibodies are currently under development which will be able to bind to different regions of specific polymerases. Another example of an antibody mediated time release enzyme is AccuPrime *Taq* DNA polymerase (Invitrogen), an enzyme mixed with accessory proteins of "unknown nature" (i.e. they are a commercial secret), which improve the specificity of primer annealing to target DNA. The protein may be similar to a "minor groove binding protein" previously described [Kutyavin et al., 2000]. The HotPrime polymerase (Qbiogene) is modified in yet another way, in that "heat labile blocking groups" have been coupled to some of the key amino acids of the enzyme. The application of heat (15 minutes at 95°C) dissociates the blocking groups restoring polymerase activity.

Certain classes of short double-stranded DNA fragments are able to inhibit non-specific amplification by *Thermophilus aquaticus* (*Taq*) and *Thermophilus flavus* (*Tfl*) enzymes at temperatures below 60°C. However, this inhibition effectively disappears at temperatures above 60°C [Kainz et al., 2000]. Ideally, the annealing temperature of these short double-stranded DNA fragments should lie between 60°C and 70°C in order to exert their maximum effect. If the titre of these fragments is optimized with respect to a particular PCR protocol, then the quantity of specific amplimers generated in that particular PCR protocol will be greatly increased.

Reverse transcription (RT-PCR) protocols (where RNA is reverse transcribed into cDNA prior to PCR amplification) may also be transformed using either a hot-start or a time release RT-PCR protocol. The DuraScript HSRT PCR kit (Sigma) uses a mutant AMV RT (Durascript), which is robust enough to synthesize cDNA at elevated temperatures (up to 65°C) using random nine-mer primers equipped with 23 base pair oligo-dT extensions. The HSRT kit comprises the Durascript reverse transcriptase enzyme in conjunction with Sigma JumpStartAccu Taq LA thermostable DNA polymerase mix (bringing both exponential amplification and proofreading activity to the PCR). This (and other) adapted RT-PCR hot-start/time release protocols facilitate the reverse transcription of low copy numbers of mRNA and increase both the yield and specificity of RNA amplification. More of these commercially available RT-PCR enzyme combinations are now appearing on the diagnostic market.

12.5 Inverse PCR

PCR amplification proceeds in the 5′–3′ direction using primers that hybridize to known sequences of the target DNA molecule to be amplified. It is however possible to utilize PCR to amplify unknown "flanking" DNA regions that lie immediately upstream or downstream of the normal PCR primers used. This technique is called *inverse PCR* and can be used to amplify flanking regions of upto 3–4 kbp in length. The prerequisites for inverse PCR are that DNA sequences internal to the required flanking regions to be amplified are known, and that "upstream" and "downstream" restriction sites exists within these unknown flanking regions. The known DNA sequences are used to create "inverse" PCR primers, which point in the opposite direction to a standard PCR but still hybridize to complementary DNA. In effect, these inverse PCR primers amplify outwards and away from each other, rather than inwards towards each other (as is the case for a standard PCR).

The inverse PCR methodology *per se* involves; isolation of template DNA followed by restriction digestion using an appropriate enzyme(s), cleaning of the restriction fragments to remove any inhibitory enzyme and buffers, ligation and circularisation of the pooled fragments using DNA ligase, and then final PCR amplification using specific inverse PCR primers. The appropriate restriction enzyme(s) to be used are usually selected by "trial and error" by performing inverse PCR using a range of frequent cutting restriction enzymes, e.g. *Rsa I, EcoR I* etc or suitable frequent cutting isoschizomers (restriction enzymes that recognize the same DNA sequence), or by digesting the template DNA with a range of frequent cutting restriction enzymes and then hybridizing the fragments to PCR products generated from the known (internal) sequences of DNA to check that the restriction enzymes/combinations used do not digest within the known sequence. After digestion of the template DNA, the many fragments are preferably ligated together in varying dilutions. One of these ligated fragment combinations will possibly be a circularized fragment containing the known DNA region of interest with extra upstream and downstream flanking DNA. Subsequent PCR amplification of the total ligation mix using inverse PCR primers, will allow PCR amplification outwards around the circularized DNA forming a linear PCR product. Thereafter, the unknown upstream and downstream flanking DNA sequences may be determined by sequencing using the inverse PCR primers as upstream and downstream sequencing primers (Fig. 12.5). If multiple amplimers are present after inverse PCR, non-contiguous DNA fragments (i.e. digested fragments from another part of the genome) may have erroneously ligated to the required DNA fragment. By electrophoresing, excising, purifying and sequencing each amplimer from an agarose gel, the sequences of all of the multiple inverse PCR amplimers may be determined. By comparing all of the sequences it will be possible to determine a conserved sequence, which will comprise contiguous DNA from the known and flanking DNA regions. The DNA sequences can also be searched for the presence of restriction digestion sequence sites recognized by the enzyme used in the initial inverse PCR digestion reaction. The required PCR product without contaminating DNA insertions should contain only one such restriction site as the ligation of non-contiguous fragments requires at least two restriction sites. Alternatively, restriction digestion of the correct circularized amplimer using the same enzyme as was used in the inverse PCR protocol, should yield only one digestion product. These results also provide information on the length of the upstream and downstream flanking sequences successfully amplified. However, logic dictates that the smallest amplimer will be the amplimer of interest, as the smallest amplimer is the one least likely to contain contaminating non-contiguous DNA.

When successful, inverse PCR can be used to circumvent time consuming and costly cloning and sub-cloning procedures, though the amount of sequence data provided for each flanking region may be somewhat limited (often in the region of 100–500 bp in length) due to restrictions on the size of linear DNA fragments that can circularize and self-ligate. The method is however well suited for improving the amount of sequence information available for short genomic regions, sequencing clones from genomic libraries, generating probes specific for uncharacterized flanking DNA sequences, and determining the presence of genomic translocations and gene fusion events by transposable or randomly integrating sequence elements. Recent advances in this field include PCR-based amplified RFLP [Liu et al., 2004] and long distance inverse PCR [Sonoki et al., 2004].

12.6 Asymmetric PCR

Asymmetric PCR is a PCR methodology where one of the PCR primers is present in a higher concentration than the other (usually at a concentration 1:50 to 1:100 higher) [Gyllenstein, 1989], allowing high concentrations of single-stranded

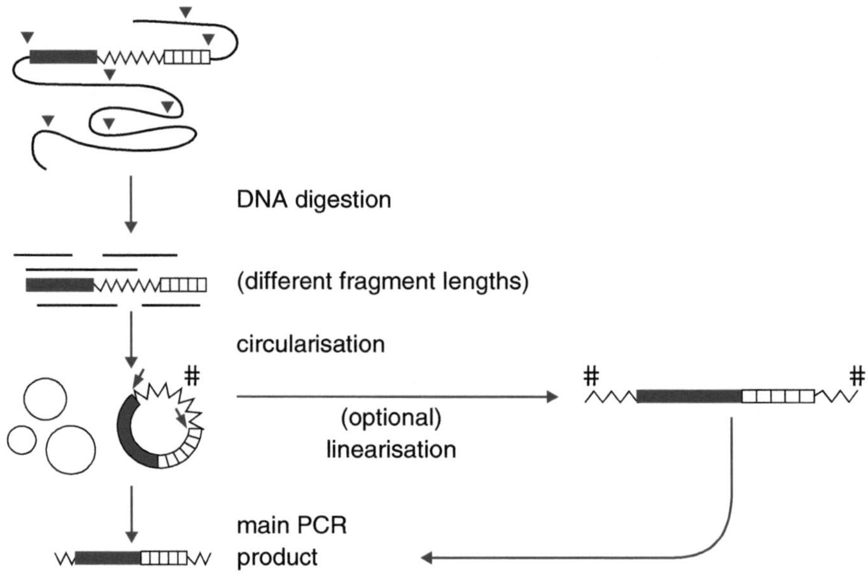

Fig. 12.5 Inverse PCR is a method for identifying unknown regions upstream and downstream (boxed) of a region of known sequence composition (zigzag). The flanking DNA regions are initially digested using an appropriate restriction enzyme (blue) and the resultant linear fragments are circularized by ligation. A PCR is then performed outwards from the known sequence using specifically designed inverse PCR primers. The final PCR product is sequenced to determine the composition of the flanking DNA sequence

DNA molecules up to 1,000 bp in length to be generated. During the first 15–20 cycles of an asymmetric PCR, the target DNA is amplified in a normal exponential fashion, however once the low concentration primer has been almost totally exhausted, exponential PCR amplification fails, as only the high concentration primer remains in the PCR mix. At this point and during the following 30–40 cycles, the vast majority of DNA molecules generated will be single-stranded, generated in a linear fashion. Asymmetric PCR therefore, is essentially a two-phase process, the first involving exponential amplification and the second phase linear amplification. If required for downstream applications, the single-stranded DNA molecules may be separated from any double-stranded amplimers by agarose gel electrophoresis and extracted from the agarose matrix using commercially available kits.

Asymmetric PCR may be utilized for generating single-stranded DNA probes for hybridization reactions, as well as in certain (rarely used) sequencing protocols which require for example the thermolabile *E. coli* DNA polymerase Klenow fragment (see Section 12.7.1) below). Asymmetric PCR also provides a convenient alternative to classical cloning and sequencing strategies for phage λ clones (which are laborious to grow and purify).

12.7 PCR Mediated DNA Sequencing Strategies

Determining the nucleotide sequence of DNA amplimers, plasmids, genomic clones, etc. is an essential course of action for many research and clinical diagnostic laboratories. In order to obtain sequence data, the vast majority of laboratories use PCR mediated sequencing protocols, especially designed nucleotides called dideoxynucleotides and fluorescent labels. The most important advantage of PCR mediated DNA sequencing is the fact that the thermostable DNA polymerases used allow reactions to be performed at high temperatures under stringent conditions, as well as helping reduce any problems which may be encountered with template secondary structure. One particular advance over the past decade, is that PCR sequencing protocols have been extensively developed, simplified and adapted for high throughput applications that use specially designed automated sequencing machines, and (for the vast majority) fluorescent detection using fluorescent dye labels attached to either dideoxynucleotides or one of the PCR sequencing primers. Large advances have also been made with regard to sensitivity via the use of lasers and charge coupled device (CCD) cameras, gel visualization (using several different fluorescent labels in a single PCR sequencing mix), speed (via capillary electrophoresis formats), and sequence data analysis/sequence assembly (using specialized sequence alignment programs, e.g. Megalign, DNAstar; Sci Ed Central, Scientific and Educational Software etc) (see also Section 9.4).

Initially, there existed two major DNA sequencing methodologies, namely the Sanger method [Sanger et al., 1992], which used DNA chain terminating dideoxynucleotides, and the Maxam-Gilbert method [Maxam and Gilbert, 1977] which used chemical processes to sequentially remove nucleotides from a DNA chain. However, the Sanger method has now replaced the Maxam-Gilbert method in both research and clinical laboratories, not least because of the toxic and carcinogenic chemicals required to perform the Maxam-Gilbert method.

In the Sanger method, extension of the DNA strand by DNA dependent DNA polymerase is terminated by incorporation of a specific dideoxynucleotide (a deoxynucleotide analogue) that lacks the necessary 3′-hydroxyl group required for further DNA chain elongation. Moreover, by generating a whole series of prematurely terminated DNA fragments of different lengths whose termini end in one of the four dideoxynucleotide analogues, then gel electrophoresis (to separate the fragments) and detection (to identify which dideoxynucleotide terminates which individual fragment), will allow the sequence of the original template DNA to be determined. It should be noted that all DNA or PCR products for Sanger sequencing reactions should first be purified away from any contaminating DNA, primers, dNTPs, enzymes, etc. which may interfere with the sequencing reaction.

12.7.1 Generating Single-Stranded DNA for Sanger Sequencing Reactions

PCR amplification products and genomic DNA are double-stranded in nature, which makes them unsuitable for use in Sanger sequencing reactions (which require single stranded DNA). However, it is relatively easy to separate double stranded DNA into single strands by heating at 100°C for 5 minutes and then "*snap-cooling* or *snap-freezing*" the resultant single strands by immediately placing them in either ice, a dry-ice/ethanol bath or in liquid nitrogen. An alternative method involves the use of alkaline denaturation (often used for denaturing double-stranded extra-chromosomal DNA or plasmids into single strands), though this method tends to generate poor sequence data when applied to double-stranded PCR products.

One particular method for generating single-stranded DNA from PCR products, involves the use of 5′-end hapten labelled (e.g. biotin) PCR primers. This labelled primer becomes incorporated into subsequent PCR products and can be used to attach the double-stranded PCR products to a solid phase (e.g. streptavidin coated ELISA plate wells, streptavidin coated magnetic beads, etc.). The complementary DNA strand may then be removed (e.g. by alkaline treatment) ready for single-stranded DNA sequencing. Another PCR method for generating single strands involves asymmetric PCR (Section 12.6), which may be used to generate significant amounts (5–10 pmol) of single-stranded DNA up to 1,000 bp in length. However, the main disadvantage of this technique is that asymmetric PCR optimization has to be performed for each new target DNA molecule to be sequenced.

Another method for creating single-stranded DNA for sequencing reactions involves the use of the enzyme lambda exonuclease III [Higuchi and Ochman, 1989]. Basically, lambda exonuclease III is a 5′-3′ exonuclease which is specific for double-stranded DNA and which requires the presence of a phosphate group at the 5′-end position of the double-stranded DNA for its activity. The procedure involves the 5′-end labelling of double-stranded PCR products using T4 polynucleotide kinase and ATP (or alternatively the use of a 5′-end phosphorylated primer), followed by exonuclease degradation of the phosphorylated DNA strand. The remaining single-stranded DNA can then be used as a template in further sequencing reactions without the use of double-stranded denaturation protocols. By far the most and widely used method for generating single stranded DNA for use in Sanger sequencing reactions however, combines Sanger sequencing reactions with thermocycling (*cycle sequencing* methodologies).

12.7.2 Classical Sanger Sequencing of Single-Stranded PCR Products

Classical single-stranded DNA sequencing methodologies comprise two distinct steps and use a heat stable DNA polymerase. The start of the sequencing reaction is performed at 42°C for 5 minutes using a polynucleotide kinase enzyme or DNA polymerase to label either specifically labelled sequencing primers or the nucleotide deoxyadenosinetriphosphate (dATP) with a (radioactive) label, e.g. ^{32}P. In the second step, four new reaction mixes are prepared each containing a normal mix of dNTPs and one of four dideoxynucleotide triphosphates (ddATP, ddCTP, ddGTP or ddTTP). DNA chain extension and chain termination is then performed at the optimum temperature for the polymerase enzyme (70°C) for a short period of time. Chain extension is terminated upon the incorporation of a ddNTP molecule. Products are separated by electrophoresis and visualized on the basis of the specific characteristics of the incorporated label. No amplification is involved in the classical Sanger sequencing methodology.

12.7.3 Direct PCR Sequencing

This sequencing protocol begins with a standard optimized PCR (15–30 cycles) that is used to generate millions of copies of the required DNA for sequencing. After PCR amplification is complete, a small aliquot of the PCR mix is added to a commercially available "sequencing mix" containing a special sequencing thermostable DNA polymerase that exhibits no 5′-end to 3′-end exonuclease (proofreading) activity, an activity which could remove any dideoxynucleotide terminators incorporated into the DNA chain (see Section 7.3). For the following sequencing reaction, one sequencing primer is used along with normal dNTPS, a labelled nucleotide (e.g. α–^{32}P dCTP), and one of the four dideoxynucleotide DNA terminators (either ddATP, ddCTP, ddGTP or ddTTP). The sequencing reaction in itself results in the linear accumulation of single-stranded DNA molecules (as only one primer is used) of different sizes, in which the detection label (e.g. α–^{32}P dCTP) has been incorporated. After PCR cycling, the DNA fragments generated in each reaction mix are analyzed using polyacrylamide gel electrophoresis and visualized by autoradiography or detection of fluorescence (Fig. 12.6). However, this somewhat old fashioned method is labour intensive, requires large quantities of template, and requires the use of very sensitive detection techniques, e.g. radioactive labels and autoradiography, which may have health and safety implications.

12.7.4 Four-Tube Cycle Sequencing

Cycle sequencing is performed using a standard PCR thermocycling machine and a sequencing reaction-specific thermocycling program. Both double-stranded and single-stranded DNA template may be used as starting material, as the temperature achieved during thermocycling is sufficient to thermally disassociate any double-stranded DNA into single strands. Cycle sequencing protocols utilize four standard PCR mixes, with each additionally containing one of four labelled chain terminating dideoxynucleotides (either ddATP, ddCTP, ddGTP or ddTTP). In contrast to a

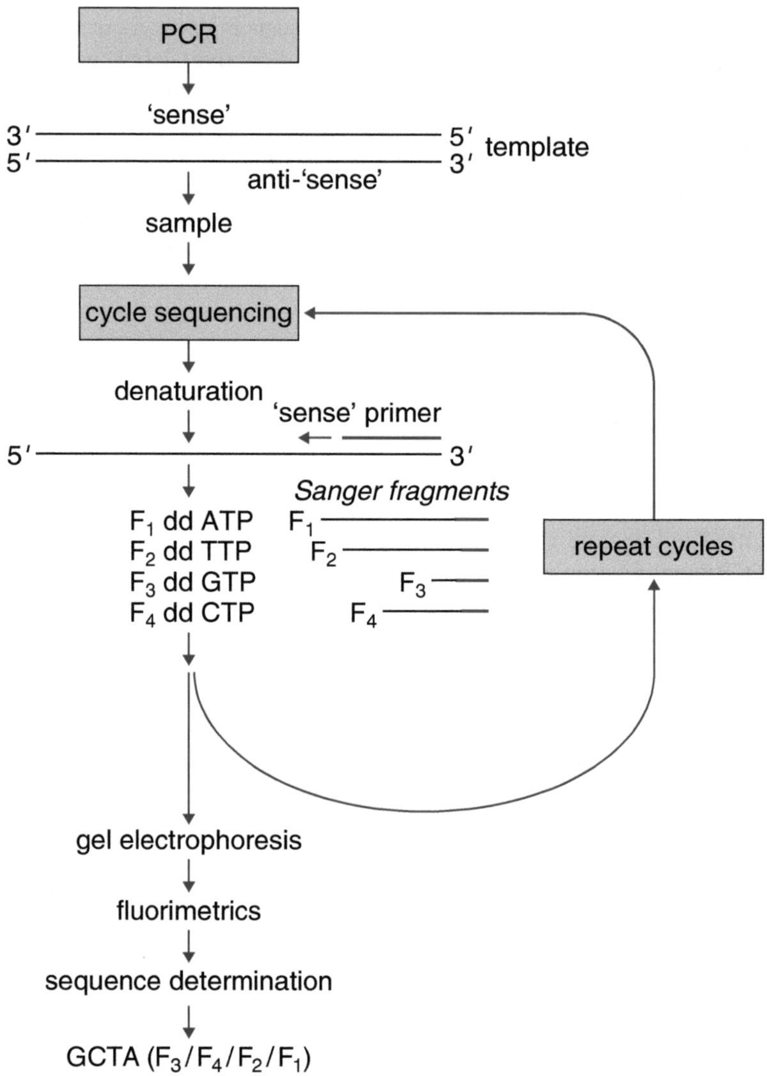

Fig. 12.6 Cycle sequencing using PCR. Cyclical denaturation, annealing and extension of target DNA in the presence of ddNTPs leads to the linear amplification of DNA molecules of different lengths

standard PCR methodology however, only a single (sequencing) primer is added to the sequencing reaction mix, which means that amplification is linear (as opposed to a standard PCR where two primers are used and the amplification is exponential). Various sequencing protocol options (including the use of either labelled dNTPs or labelled primers) are currently available. The terminated fragments in each of the four sequencing reactions are then separated by electrophoresis in a single polyacrylamide gel, and the labeled fragments then detected using (for example) radioactive sensitive film (e.g. Hyperfilm, Amersham International). By examining the developed film, it is possible to associate a particular ddNTP with each terminated fragment length and hence determine the DNA sequence by reading down the gel.

Unfortunately, the fact that DNA synthesis during cycle sequencing is linear, means that this method is relatively insensitive, and requires the use of sensitive labelling/detection systems. Traditionally, radioactive labels have been used because of the high degree of sensitivity that these labels provide. However, developments in the field of fluorescent dye labels which are more able to efficiently transfer laser excitation energy into fluorescence energy (up to six times greater than previous earlier fluorescent dyes, Fig. 12.7), now means that fluorescent cycle sequencing is sensitive enough to generate readable sequence data using a relatively small amount of template and a reduced number of PCR cycles (10 ng for 15 cycles) from DNA fragments of up to 1,000 bp in length.

During the initial stages of the development of cycle-sequencing, it was observed that the available thermostable DNA dependent polymerases had problems in correctly recognizing the respective dideoxynucleotide terminators (ddNTPs), which were consequently incorporated into the growing DNA chain far less efficiently than the normal dNTPs, requiring extensive optimization of the ddNTP

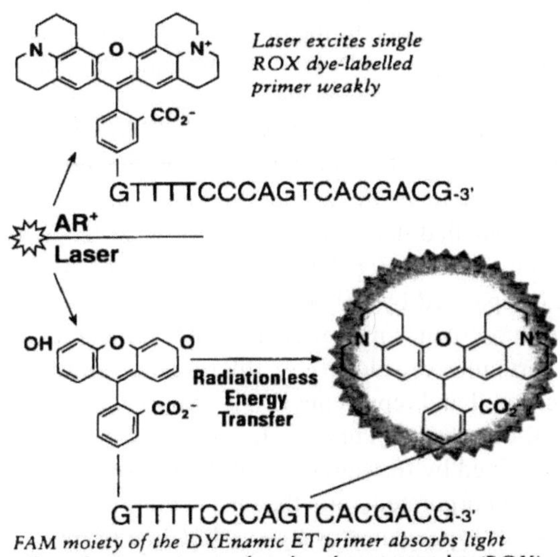

Fig. 12.7 ET fluorochromes are clever combinations of different fluorochromes which show increased fluorescence as compared to the single fluorochrome alone. For example, the excitation of the fluorescent dye ROX (structure on top) by an argon laser induces low levels of fluorescence; however if ROX is excited through transfer of energy from a FAM molecule (structure at bottom), then the resultant fluorescence is six times more intense

versus dNTP concentration. This effect also resulted in the poor sensitivity initially observed using the cycle sequencing method. However, the application of molecular techniques has now allowed the development and manufacture of newer, specifically engineered, thermostable DNA polymerases which show a much improved affinity for ddNTPs, produce uniform band intensities and exhibit a high degree of accuracy. Many companies which supply thermostable DNA polymerases for PCR now also supply variant polymerases specifically designed for sequencing purposes, e.g. Thermosequenase (Amersham International Ltd.), Sequencing Grade *Taq* DNA Polymerase (Promega Corporation) and Amplitaq DNA polymerase FS (Perkin Elmer/ABI).

12.7.5 One-Tube Cycle Sequencing

Sanger cycle sequencing has been completely revolutionized by the development of a broad spectrum of non-radioactive, fluorescent-based, dye labels which can be chemically attached to dideoxynucleotidetriphosphates and which fluoresce with their own specific spectral characteristics. The fact that these fluorescent labels exhibit their own specific fluorescent spectral characteristics (with a minimum of overlap of fluorescent wavelength between each individual dye), means that each of the four dideoxynucleotide terminators used in Sanger cycle sequencing (i.e. ddATP, ddCTP, ddGTP and ddTTP), may be individually tagged with its own specific dye label. Consequently, due to the dissimilar spectral characteristics of each of the four labels, all four tagged ddNTPS may be added to a single cycle sequencing reaction mix, allowing *one-tube cycle sequencing* to be performed. Examples of such fluorescent dye labels include FAM, HEX, TAMRA and ROX.

Basically, the methodology for *one-tube cycle sequencing* is the same as that used for *four-tube cycle sequencing*, except that only one reaction mix (containing all four differentially labelled ddNTPS) is used for each DNA template instead of four separate reaction mixes. After thermocycling, the whole range of terminated DNA fragments are separated by polyacrylamide gel electrophoresis, excited by a static laser, and the fluorescence measured as each fragment migrates down the gel. The quantity and particular wavelength of fluorescence emitted by each fragment is continuously measured and represented on a computer screen as a coloured peak (similar to a normal distribution curve). This allows the terminal ddNTP of each fragment to be determined by measurement of the fluorescent spectrum emitted by the fluorescent dye incorporated into that particular fragment. When collected together, the final result is presented as a wave-like line of different coloured peaks and troughs called a "chromatogram" (Fig. 12.8), with the colour black being associated with the nucleotide guanosine, the colour blue associated with the nucleotide cytosine, the colour red associated with the nucleotide thymidine and the colour green associated with the nucleotide adenosine as standard. Lately, special DNA sequencing machines have been developed which include laser and detection modules (e.g. the Perkin Elmer Cetus 3700 and 7600 series DNA sequencing machines,

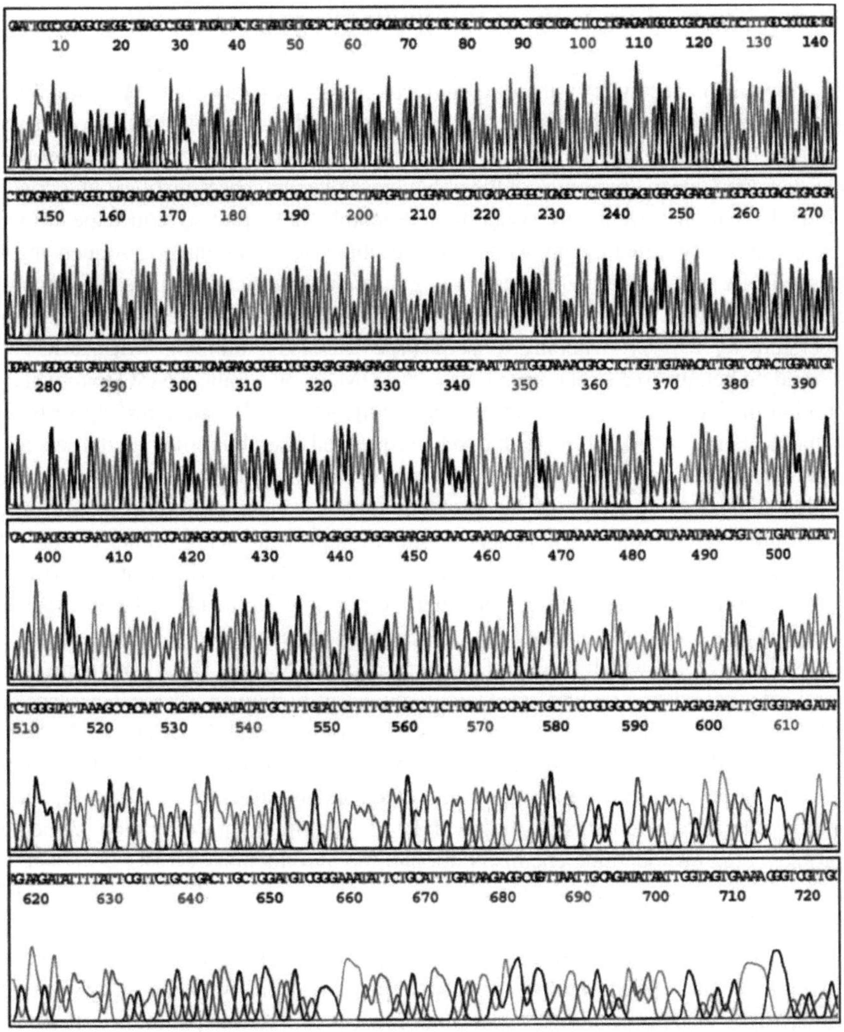

Fig. 12.8 DNA sequencing output: chromatogram showing the consecutive pattern of differentially labelled DNA products. Each peak represents the maximum fluorescence as obtained for each prematurely terminated DNA fragment. The fragments pass through a laser beam during polyacrylamide gel electrophoresis (PAGE)

capillary sequencer models), allowing 700–800 base pairs of sequence data to be generated from a single electrophoresis gel. Pre-prepared sequencing reaction mixes which contain the necessary labelled dideoxynucleotides, sequencing polymerase, dNTPs, magnesium ions etc are now also available for purchase from several commercial sources (e.g. Big Dye Terminator kits, Applied Biosystems).

12.7.6 Difficult to Sequence Templates

The presence of secondary structure in DNA molecules can interfere with Sanger sequencing reactions, as they hinder the progress of thermostable DNA polymerases along the DNA strand and may facilitate "base compressions", where incorrect sequence data is obtained due to "jumping" of the polymerase over any bases involved in secondary structure formation. Several novel options are available for helping to reduce the influence of such secondary structure on sequencing reactions. Options include sequencing the opposite DNA strand, altering the sequencing reaction conditions by increasing the melting temperature (e.g. to 98°C) or by including 40% formamide, increasing the extension temperature by 2–3°C, switching to Sequenase chemistry, utilizing 7-deaza-dGTP or dITP, adding betaine to a final concentration of 1.0M, or by adding DMSO to a final concentration of 5% (vol/vol). Alternatively, the template may be shortened by restriction enzyme digestion outside of the region to be sequenced, thereby removing the secondary structure sequence entirely or generating shorter fragments that are less likely to form complex secondary structure. If the problem involves incomplete sequencing primer hybridization to a region of secondary structure, a new sequencing primer closer to the original sequencing primer may be designed to provide better hybridization properties. Another option is to sequence a set of smaller, overlapping fragments and then align all of the sequences into a single piece of data. Fidelity Systems Inc, offer ThermoFidelase I, a "hyperstable protein" which binds to DNA, and apparently helps DNA polymerases read through any secondary structure present (e.g. regions of high guanosine and cytosine content or repeat regions), as well as protecting DNA from thermal degradation. The protein is simply added to the sequencing reaction mix prior to thermocycling. Finally, it should be noted that problems with the sequencing of secondary structure may arise due to the type of dye-terminator chemistry used, with Big Dye terminators being more susceptible to problems than Rhodamine chemistry, though the introduction of a Big Dye dGTP kit (Applied Biosytems), specifically for sequencing templates with difficult secondary structure, has helped address the problem associated with Big Dye chemistry.

Homopolymeric regions (e.g. long stretches of adenosine nucleotides) can result in "slippage" of the polymerase (exact mechanism unknown) and short repetitive regions (200–500 bp in length) can also cause sequencing problems. It may be necessary in these cases to utilize transposon insertion as a method of introducing specific primer sequences into the regions followed by PCR sequencing in both directions, e.g. EZ-Tn5 and Hyper Mu (Epicentre).

"Next generation" sequencing strategies are currently reaching advanced stages of development and use. The reader is referred to "The year of sequencing", Nature Methods 2008; 5(1):11–21, for more information.

12.8 Touchdown and Touch-Up PCR

The touchdown PCR protocol is particularly suited for: (i) use on highly complex DNA samples containing a limited number of initial template molecules, (ii) for multiplex PCRs which contain sets of primers with very different annealing temperatures, (iii) for the selective amplification of an intended target region where one or more primer sequence is very similar to other sequences present within the template DNA to be used, and perhaps most usefully, (iv) as a general thermocycling protocol for use with many different PCR primer pairs (without having to determine the exact annealing temperature required for each individual PCR primer pair and having to re-program the PCR thermocycling machine).

Touchdown PCR comprises a two-step methodology. In the first step (10–15 thermocycles), the annealing temperature of the very first thermocycle within this step is set at a temperature of approximately 10°C above the calculated T_m of the primer pair being used (e.g. 65–70°C). At every subsequent cycle in the first step the annealing temperature is decreased by a standard increment (usually either 0.5°C or 1°C) until the total number of cycles within the first step is complete and the final cycle annealing temperature is equivalent to the calculated T_M for the primer pair used. In the second step (15–30 thermocycles), thermocycling is performed at a single annealing temperature, the annealing temperature of the final cycle in the first step [Wu et al., 2005]. The advantage of touchdown PCR is that primer hybridization conditions are initially very stringent (and sometimes even too stringent to allow primer annealing to occur!). However, as first step thermocycling progresses, and the annealing temperature gradually decreases, the annealing temperature will begin to approach the T_m of maximum stringency for the primer pair(s) being used, resulting in the amplification of specific target DNA only. As the annealing temperature is gradually lowered, the conditions for primer hybridization will become less stringent and the efficiency of amplification will increase. Moreover, the target DNA specifically amplified in the first step of the touchdown PCR will tend to act as a preferential target for further DNA amplification, even if the annealing temperature falls a few degrees below the optimum T_m of the primer pair(s) being used. Further, if any non-specific PCR products are actually generated in the second step, they will be greatly outnumbered by the specifically amplified target DNA present in the final reaction products.

Optimized touchdown PCRs in combination with hot-start protocols using time release polymerases (see Section 12.4 above) help to ensure the maximum specificity of PCR amplification. If problems with the yield of PCR product are observed using the touchdown PCR protocol, it may be possible to decrease the annealing temperature used in the second step of the PCR protocol in order to improve PCR yield [Hecker and Roux, 1996]. It is also possible to simplify touchdown PCR protocols by using a steeper decrease in annealing temperature per first step cycle (so called *step-down* PCR protocols) whilst maintaining PCR specificity and efficiency. Touchdown PCR has also been successfully applied to amplify DNA of inferior quality from archival (archaeological or post-mortem) samples.

The reverse process or "touch-up" PCR has also been described as a method for increasing PCR specificity. Touch-up PCR protocols normally utilize specially designed "molecular beacon" primers that contain self-complementary sequences or "stem-loop" secondary structure (e.g. Touch-Up and Loop Incorporated Primers or "TULIPS" [Ailenberg and Silverman, 2000]). During touch-up PCR, the annealing temperature is gradually increased per cycle (e.g. from 60°C to 72°C over six thermocycles), whereby the increasing thermal energy gradually facilitates the disassociation of the primer self-complementary "stem-loop" structure, allowing the primer sequences to specifically hybridize to the intended target sequence. A standard PCR at the calculated T_m of the primer (which will be lower than the annealing temperature previously used) is then performed for a further 30–40 cycles to increase the yield of specific amplimers. Designing these special touch-up primers requires specific computer software, as both the T_m of primer loop disassociation and T_m of primer-template hybridization need to be taken into consideration.

12.9 Multiplex PCR

Multiplex PCR protocols utilize multiple PCR primer pairs, all included in the same PCR mix, to amplify several different target regions at the same time. The upper limit on the number of primer pairs that can be included in a single multiplex PCR mix depends on the reaction conditions (annealing temperature, magnesium ion concentration, dNTP concentration, amplimer sizes, template concentration, etc.), and most importantly the absence of interactions between individual primers (cross-hybridization) within the reaction mix. Multiplex PCR primers are usually slightly longer (23–28 nucleotides) than those used in single-target "standard"PCR protocols, and are selected on the basis of similar T_m and similar (at least 40% of the total primer sequence) guanosine/cytosine nucleotide content. Multiplex PCRs require extensive optimization and titration of reaction components, and several different thermostable DNA polymerases may need to be tried [Moretti et al., 1998; Kebelmann-Betzing et al., 1998]. It should be noted that the concentrations of the different primer pairs used do not have to be equimolar for successful optimization, not even within individual PCR primer pairs [Berg et al., 2000]. In some more complex cases, the addition of certain additives, e.g. DMSO, glycerol or formamide, may help in obtaining specific amplification products due to the ability of these chemicals to interfere with DNA secondary structure formation which may influence primer-target hybridization accessibility and efficiency.

Multiplex PCR protocols are frequently used in clinical genetics and in forensic investigations. They facilitate the co-detection of several targets in a single PCR and hence the identification of complex disease genotypes, e.g. Duchenne's Muscular Dystrophy. For forensic purposes, various single nucleotide polymorphisms (SNPs) can be detected using a single test, and microsatellite regions (defined as regions within DNA where short sequences of two, three, or four nucleotides are repeated one immediately after the other) can be screened for variability, thereby saving time with respect to the generation of genetic fingerprint data. A more specialized multiplex PCR protocol for SNP analysis uses different fluorescent labelled ddNTPs, allowing SNPs to be identified using for example, a capillary sequencing machine [Hebert and Brazill, 2004]. Reverse transcription PCR (RT-PCR) protocols can also be multiplexed, allowing differences in expression between two different genes (e.g. a "target" and a "housekeeping control" gene) to be assessed in a single multiplex RT-PCR. Of course, suitable DNA contamination controls have to be included in multiplex RT-PCR protocols similar to those required in standard single primer pair RT-PCR reactions. Recent advances in multiplex RT-PCR protocols also include the development of novel multiplex PCRs using self-quenched fluorescent primers which contain 5′-end hairpin loops, yielding a better "signal to noise" ratio [Nazarenko et al., 2002].

It should be noted that DNA contamination is a much more severe problem in multiplex PCR protocols as opposed to single primer pair PCR protocols, because there are more target sequences to be amplified. Therefore, as with all PCR protocols, anti-contamination control measures should be vigorously implemented.

12.10 PCR Using Degenerate Primers

Degenerate primers are mixtures of PCR primers designed to amplify the same genetic DNA target sequence where small numbers of nucleotide changes between different isolates or individuals is expected. These minor target nucleotide sequence changes may occur between different microbial isolates or species (the "quasi-species" concept), or within the same cell type of different individuals. Degenerate primers are usually a mix of primers where the main "core" nucleotide sequence of the primer is identical to all of the other primers, but where one (or more) of the primer sequences contain changes in one (or more) nucleotides compared to the other primers present within the degenerate primer mix. Under low stringency conditions mismatching and cross hybridization between non-perfectly matching sequences are allowed. This is especially relevant during the first few cycles of the PCR, where lower annealing temperatures, prolonged annealing times and slow temperature ramp rates should be used. However, it may be possible to increase the stringency of PCR thermocycling after the first few (5–10) cycles, without affecting the yield of PCR products, as the majority of amplimers will now contain the original degenerate primer sequences. Degenerate primer mixes may be ordered from commercial suppliers, in which case, standard letter codes are used to indicate which particular nucleotides should be added at a specific primer position. These letter codes are:

R = A or G (puRine) **Y** = C or T (pYrimidine) **K** = G or T (Keto) **M** = A or C (aMino)
S = G or C (Strong-3H bonds) **W** = A or T (Weak-2H bonds) **N** = any base
B = C, G or T **D** = A, G or T **H** = A, C or T **V** = A, C or G

A commonly used PCR protocol involving degenerate primers is the "degenerate oligonucleotide primer" (see Section 5.1.1. as well) or "DOP-PCR" protocol, outlined in Fig. 12.9. The DOP-PCR protocol is able to randomly amplify minute amounts (≤1 μg) of DNA, and has been successfully used in the amplification of purified chromosome preparations and the generation of probes for comparative chromosomal hybridization [Alcock et al., 2003].

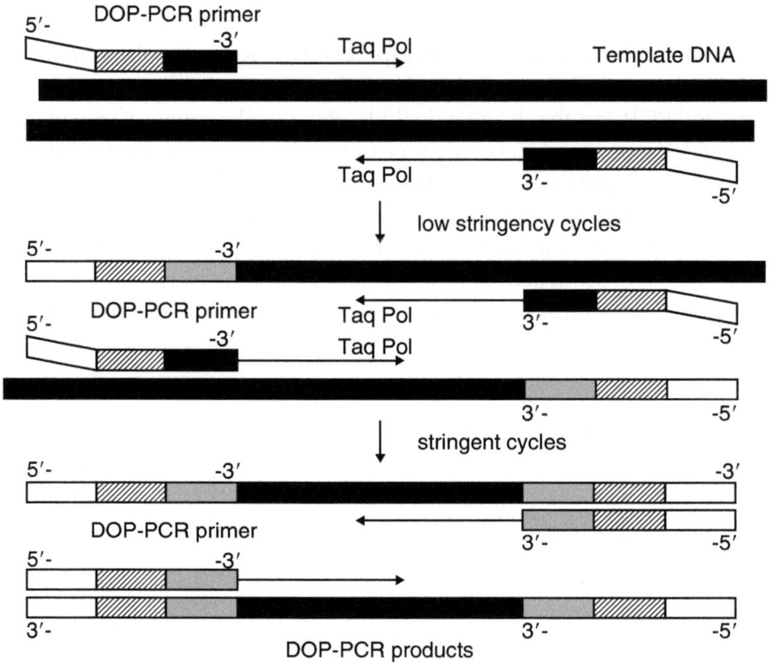

Fig. 12.9 DOP PCR. Degenerate oligomer primer PCR utilizes degenerate primers that contain an exact complementary six-nucleotide region at their 3′-end but contain six mixed nucleotides (where all four nucleotides have all been incorporated in equimolar amounts) in their central regions. At the 5′-end of the primer sits an additional region of nucleotides added to enhance the stability of hybridization interactions. Using DOP PCR primers and protocols generates a large variety of products depending on the complexity of the original template molecule used. Joos. S et al Roche product catalogue for ISH; Reproduced by permission of: 'Roche'

12.11 Repeat and Inter-repeat PCR

Strongly conserved repetitive sequences are found in nearly all genomic DNA molecules, even in organisms with relatively simple genomes, e.g. viruses. In bacteria, repeat DNA sequences or motifs, either in tandem or dispersed as single copies throughout the whole bacterial genome, have been successfully identified, characterized, and used in bacterial typing and epidemiological studies [Van Belkum, 1994]. With respect to the human genome, approximately 30% of all human DNA consists of repeat DNA sequences, which may vary from short (2–30 bp), to intermediate (300 bp), to large repeat units (≥300 bp), whilst mRNA is mostly devoid of these repeats. Highly repetitive sequences in the human genome are found in the telomeric and centromeric regions of chromosomes and short tandem repeat sequences or "STRs" are excellent genotyping targets and not only restricted to the human genome [Benecke, 1997; Singh et al., 2004; Leclair and Scholl, 2005].

Repeat sequences themselves are not usually 100% identical between different isolates or individuals, due to the fact that during DNA replication, DNA dependent DNA polymerases encounter problems attempting to copy repeat DNA sequences and may occasionally incorporate a mismatched nucleotide into the repeat sequence region. As the incorporation of mismatched nucleotides by DNA dependent DNA polymerase is a random process, the detection and characterization of repeat region variability provides an excellent means for identifying specific isolates or individuals. However, *in vivo*, the rate of such random mutations is held in check by several mismatch repair enzymes which act to prevent and correct mismatched nucleotide sequence errors, limiting such mutations to a rate of approximately 10^{-8} to 10^{-9} mutations per cell per generation [Boyer and Farber, 1998]. In fact, extensive repeat region sequence variation could point to a genetic deficiency in the DNA repair mechanism system of the individual presenting with such extensive errors, and microsatellite instability may be used to diagnose several hereditary diseases, including hereditary non-polyposis colorectal cancer (HNPCC) [Kruhoffer et al., 2005].

12.11.1 Repeat PCR

Repeat PCR protocols are designed to determine the length of genetic regions containing tandem repeat sequences [Van Belkum et al., 1995] (Fig. 12.10). Primers are designed that anneal specifically to either the 5′-end or 3′-end region of the repeat sequence to be investigated and are used as conserved PCR "initiating" regions. The length of the amplimer obtained using these protocols will be specific, and in the case of diploid organisms (where the presence of chromosome pairs means that heterozygosity can occur), it may be possible to PCR amplify two distinct repeat region fragments which will be distinguishable upon gel electrophoresis. Loss of one of these heterozygote amplimers (for example after malignant cell expansion during oncogenesis) can be easily detected. The subset of DNA repeat regions known as "short tandem repeats" are particularly suited for "repeat PCR" analysis, as they give rise to distinct and short PCR products, are easily amplified due to their relatively short size, and are easy to separate on standard agarose gels.

12.11.2 Inter-repeat PCR and Random Amplification of Polymorphic DNA (RAPD)

Inter-repeat PCR is based on the fact that repetitive DNA sequences may occur randomly within individual genomes, and that the intervening distances between these sequences in different isolates or individuals may also vary. In this case, PCR amplification from one particular repeat region to another repeat region gives rise

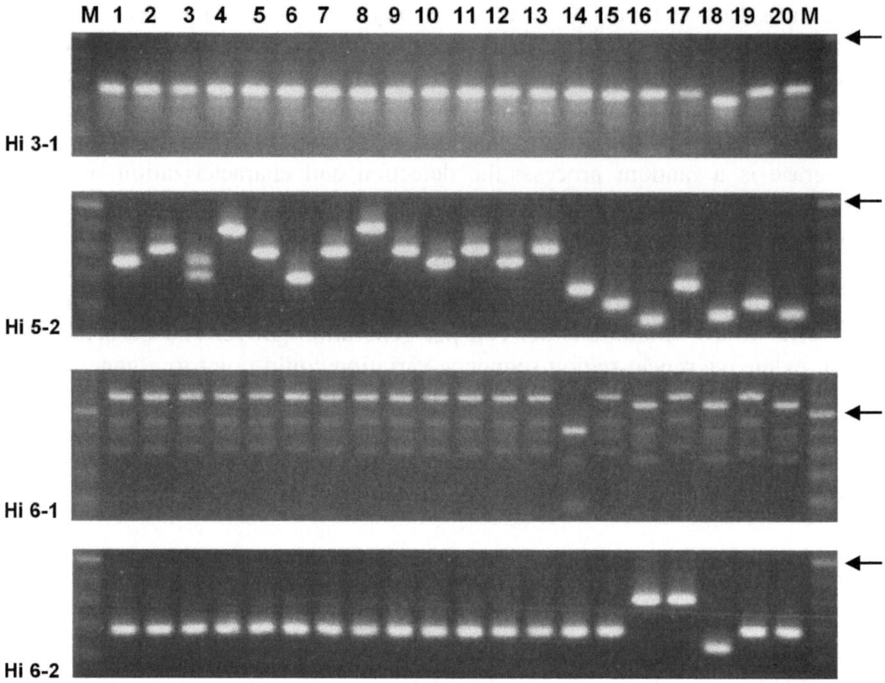

Fig. 12.10 Differences in repeat unit numbers at a particular repetitive DNA locus may be easily established using PCR primers designed to hybridise to neighbouring target sequence motifs [van Belkum et al., 1998]. Shown is an example of four different repeat detecting PCRs for a single set of *Haemophilus influenzae* isolates. Strains 1–13 were isolated during an outbreak of infection and it is obvious that in most (but not all) cases the repeat PCR is identical. Strains 14–20 are genetically unrelated and show a greater diversity. Lane M contains molecular weight markers and the arrow on the right identifies a 100 bp molecule

to variable fragment lengths and therefore variable genetic fingerprints within different isolates or individuals due to inter-repeat region length variation [Van Belkum, 1994] (Fig. 12.11). It should be noted that the presence and position of the repeat region does not have to be known prior to performing the PCR. In order to be successful however, the repeated motifs do need to be present in opposing orientations, i.e. facing in towards each other, so that a single PCR primer (a key feature of inter-repeat PCR protocols) can hybridize to both individual repeat sequences and generate a PCR amplimer. This process is generally referred to as random amplification of polymorphic DNA (RAPD) or arbitrary primed (AP-PCR). Such RAPD and AP-PCR protocols are usually performed with the help of relatively short primers (approximately ten nucleotides in length) which increases the chance of the primer "finding" and hybridizing to an opposing repeat sequence. In order to further increase the probability of primer hybridization, the annealing temperature used in these protocols is kept low (between 25°C and 42°C) so that partially identical stretches of DNA may also act as sites for primer hybridization. RAPD,

AP-PCR and indeed inter-repeat PCR protocols may generate complex mixtures of amplimers (and hence gel electrophoresis banding patterns), which need to be adequately separated for successful analysis. However, in most cases, simple agarose gel electrophoresis is sufficient to generate interpretable band pattern profiles (often referred to as "DNA fingerprints"). Densitometric scanning of the fingerprints may facilitate the automated interpretation of the sometimes dense arrays of DNA fragments. One particular problem encountered using RAPD and AP-PCR protocols is that the reproducibility of the PCR using a particular primer sequence and a specific DNA extract may be variable, and that selection of a suitable primer may be a complicated process involving multiple rounds of trial and error. However, these PCR protocols are convenient, rapid and flexible in the production of DNA fingerprints for genetic identification [Van Belkum, 2002]. PCR-mediated DNA fingerprinting

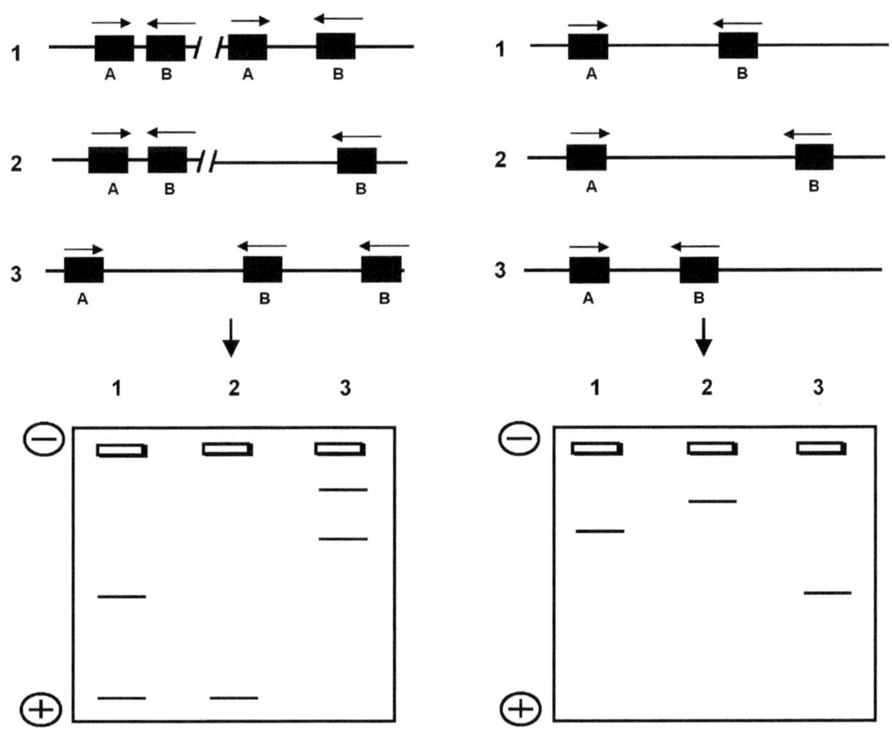

Fig. 12.11 Two variants of DNA fingerprinting mediated by PCR [Van Belkum, 1994]: (a) *Primer binding site heterogeneity*. DNAs 2 and 3 lack a site present in DNA 1. This results in the disappearance of a band in the electropherogram. In this example, multiple PCR primer pairs included in a single reaction mix may enhance the number of polymorphic sites that are detected; (b) *Primer binding site homogeneity*. Results of DNA amplification with primers that anneal to constant binding sites and span a variable segment of DNA. In this example, DNA segments 1 and 3 could contain deletions when compared to DNA segment 2. Alternatively, DNA segment 1 and 2 could harbour insertions, which are absent in DNA segment 3. The upper part of the panel provides the theoretical background for the electropherograms shown

protocols have been mainly used in epidemiological studies, to determine the evolutionary relatedness of different species (both in the eukaryotic and prokaryotic kingdoms), and in the determination of genetic polymorphisms between different individuals (paternity testing, forensic examinations, etc.) and may be readily adapted and applied to high-throughput screening strategies.

12.12 AFLP Fingerprinting

Amplification fragment length polymorphism (AFLP-PCR) analysis is a highly reproducible DNA typing and PCR amplification technique, which utilizes the presence/absence of restriction sites (restriction site polymorphisms) and PCR amplification to distinguish between different species or isolates [Vos et al., 1995]. The starting material for AFLP analysis comprises genomic DNA isolated from an individual organism, cell, tissue etc which is then digested to completion using one or more restriction enzymes (usually a "frequent" cutter and a "rare" cutter, e.g. *EcoRI* and *MseI*), so that a range of several thousands of short restriction fragments are generated of between 100 and 600 nucleotides in length. The best restriction enzymes to be used for a particular organism may be chosen by trial and error, or if the genome sequence is available, then computer software may be used. After complete restriction digestion, specific adapter molecules (short fragments of double-stranded DNA equipped with specific restriction sites and ligation linkers), are ligated to the "sticky" ends generated by the restriction endonuclease enzymes using DNA ligase. If two different restriction enzymes are used, then two different adaptors may be used together and ligated to the DNA fragments. A PCR is then performed using primers specifically designed to bind to the ligated adaptors. If two restriction enzymes have been used, then the PCR primers may be designed in such a way that only those fragments bordered by two dissimilar restriction sites (i.e. bordered by two different ligated adaptors) will be selectively amplified, so that fragments bordered by a single adapter type will not be amplified. After PCR amplification, the amplified fragments are separated by polyacrylamide gel electrophoresis so that the differences in banding patterns between isolates, cells, tissues, etc (due to amplification fragment length polymorphisms) may be detected. Individual fragments may be visualized using fluorescent or radioactively labelled primers or nucleotides, cut out of the gel and sequenced to reveal the genetic location of the restriction site polymorphism.

For many genomes and restriction enzyme combinations, the number of DNA fragments generated is so large that they are not separable, even by polyacrylamide gel electrophoresis. Hence, in many cases, minor modifications may have to be made to the AFLP primer sequences in order to limit the number of fragments PCR amplified. This is usually achieved by adding one or two "selective" nucleotides to the 3′-end of one or more of the PCR primers used. Hence, for a single primer with a single 3′-end extension, the number of fragments amplified will be reduced by one quarter, whilst for two primers each containing a single selective 3′-end nucleotide the number of fragments amplified will be reduced by one sixteenth, under stringent thermocycling conditions (assuming a random and equal distribution of all four nucleotides within the genomic sequence to be tested by AFLP). In essence, AFLP is a high-throughput genotyping procedure, where multiple DNA samples may be analyzed and compared in parallel. Due to the versatile application of diverse combinations of restriction enzymes and PCR primers, the AFLP is, next to whole genome sequencing, the method of choice for high density mapping of genetic polymorphisms.

As well as being useful in determining the genetic relatedness of microbial isolates or individuals, AFLP banding patterns may also be used to examine the genetic basis of phenotypic differences between very similar organisms, tissues, etc. Genes or gene polymorphisms associated with a particular phenotypic characteristic may be putatively identified by: (i) generating AFLP DNA fingerprints between similar organisms, (ii) searching for band differences associated with that particular phenotype, (iii) cutting out and sequencing these bands from the polyacrylamide gel and finally, (iv) searching for sequence matches on Internet databases (e.g. the National Centre for Biotechnology Information, *http://www.ncbi.nlm.nih.gov/BLAST/* web site). AFLP exhibits several advantages or disadvantages over existing genotyping techniques [Torpdahl et al., 2005]

12.13 Base Excision Sequence Scanning (BESS-T-Scan) for Mutation Detection

BESS is a technique for determining T/A or A/T mutations and generating "genomic fingerprints" allowing differential comparisons between different isolates, species etc, to be made [Hawkins and Hofmann, 1999] (Fig. 12.12). The BESS PCR protocol is performed using a typical PCR mix but with one of the primers labelled at the 5′-end, and partial replacement of the dTTP (deoxythymidinetriphosphate) nucleotide by dUTP (deoxyuraciltriphosphate). This strategy overlaps with the one employed to prevent carry-over contamination. After PCR amplification, amplimers contain dUTP nucleotides, which when treated with the enzyme uracil N-glycosylase will be excised out of the amplimer (uracil N-glycosylase removes the organic base component from each incorporated dUTP nucleotide) creating single stranded "nick sites". These nick sites are subsequently available for digestion by the enzyme endonuclease IV, which cuts the DNA strand where the dUTP base has been removed. Separation of the single-stranded DNA fragments thus generated using denaturing polyacrylamide gel electrophoresis, followed by visualization using a label incorporated at the 5′-end of one of the primers, allows single nucleotide sequence differences between different amplimers to be detected. T/A or A/T mutations will result in the presence or absence of individual fragments respectively, and frame shift mutations, nucleotide deletions and differences in the number of DNA repeats can also be detected. The characteristic band patterns generated by this technique are dependant on the length and conformation of the individual single strands of DNA.

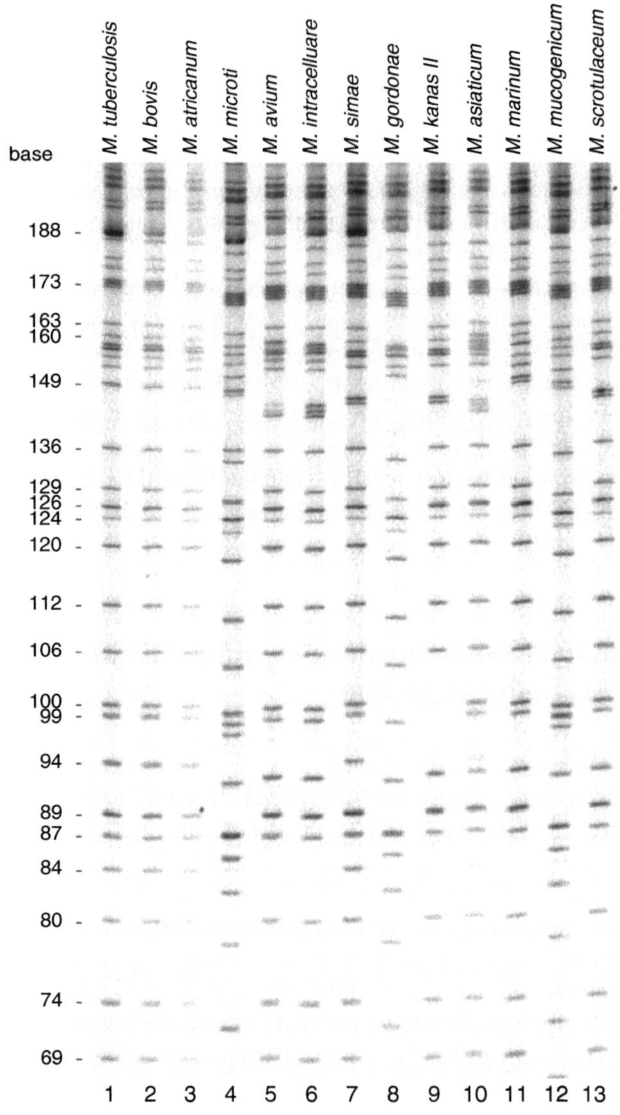

Fig. 12.12 Base excision sequence scanning. A standard PCR mix is prepared except that the nucleotide dTTP is partially replaced by the nucleotide dUTP. Upon PCR amplification, the amplimers contain dUTP, which when treated with the enzyme uracil N-glycosylase creates "nicks" by removing the organic base from each incorporated dUTP nucleotide. These nick sites are subsequently available for digestion by endonuclease IV. Separation of the single-stranded fragments generated using denaturing polyacrylamide gel electrophoresis, and visualisation using for example a label incorporated at the 5′-end of one of the primers, allows T/A and A/T mutations to be detected. Frame shift mutations, nucleotide deletions and differences in the number of DNA repeats, may also be detected. Shown is a BESS T-scan analysis of the 16S rRNA gene of 13 *Mycobacterium* strains. The samples were run on an 8% denaturing polyacrylamide gel (Reproduced from EPICENTRE Forum 5(2): "Single nucleotide typing of *Mycobacterium* species using the BESS T-scan mutation and localisation kit" by GA Hawkins. With permission from Epicentre Technologies Corp.)

12.14 Differential Display RT-PCR (DD-PCR)

Of the 10^5 different genes in a human cell, only 10–15% are actually expressed. Messenger RNA (mRNA) differential display techniques such as DD-PCR have been developed to visualize subtle differences in gene expression, both within different cell types and in the same cell type treated in different ways [Garcia and Castrillo, 2004; Castles et al., 1996]. The key steps in the DD-PCR protocol, begin with the extraction of mRNA transcripts (including DNase I treatment to remove any contaminating genomic DNA), and cDNA synthesis using a primer with a 5′-end comprising 10–12 oligo-dT nucleotides (eukaryotic mRNA molecules contain tracts of poly-adenosine nucleotides or poly-A tails and a 3′-end comprising either an A, C, or G nucleotide and one other nucleotide from A, C, G or T) so that the total length of the primer is approximately 12–14 nucleotides in length. The addition of the two extra nucleotides at the 3′-end, means that only a subset of all the many thousands of mRNA transcripts within the cells being compared are actually reverse transcribed into cDNA (thereby reducing the complexity of the final results to a manageable number of transcripts). After reverse transcription, an RT-PCR is performed using labelled dNTPs (e.g. radioactive ^{35}S-dATP), the 5′-end reverse transcription primer previously used in the reverse transcription reaction, and a 3′-end "arbitrary" primer (a random selection of ten nucleotides containing at least 50–70% GC content; AP-1 in Fig. 12.13). The DD-PCR products are then electrophoresed in a denaturing (results in better separation of bands) polyacrylamide gel and visualized (e.g. by autoradiography). Differences in banding patterns between different cell types or in the same cell type after different treatments may then be observed, and the identity of differentially expressed bands obtained by cutting-out the differentially expressed amplimer from the polyacrylamide gel, followed by cleaning and sequencing the unknown amplimer. If required, to further verify the results obtained using DD-PCR, "Northern blotting" may be performed by extracting the RNA from each of the original specimens tested and probing for the particular differentially expressed mRNA identified by DD-PCR. In this case, a probe is used based on the sequence obtained for the differentially expressed amplimer previously obtained using DD-PCR.

The selective nucleotides at the 3′-end of the reverse transcription primer, as well as the specificity of the arbitrary primer hybridization, help to ensure that the banding patterns obtained by DD-PCR are not too complex to interpret, though it is a question of *trial and error* finding the best combination of reverse transcription primer, arbitrary primer and correct DD-PCR thermocycling protocol to use that will yield an acceptable number of amplimers for investigation. Finding the best 5′-end reverse transcription and arbitrary primer combination may however be a time consuming and tedious process, and only RNA species harbouring an oligo-dA tail can be analyzed (a clear problem for researchers working with prokaryotic organisms). However, even with the "correct" reverse transcription primer, arbitrary primer, and thermocycling conditions, DD-PCR is not without its problems. If heterogeneous mixtures of cell types are used in mRNA isolation (e.g. liver versus kidney tissue) the pool of extracted mRNAs will usually be too diverse to be analyzed by DD-PCR. Therefore, it is advisable to use laboratory cultured (i.e. not

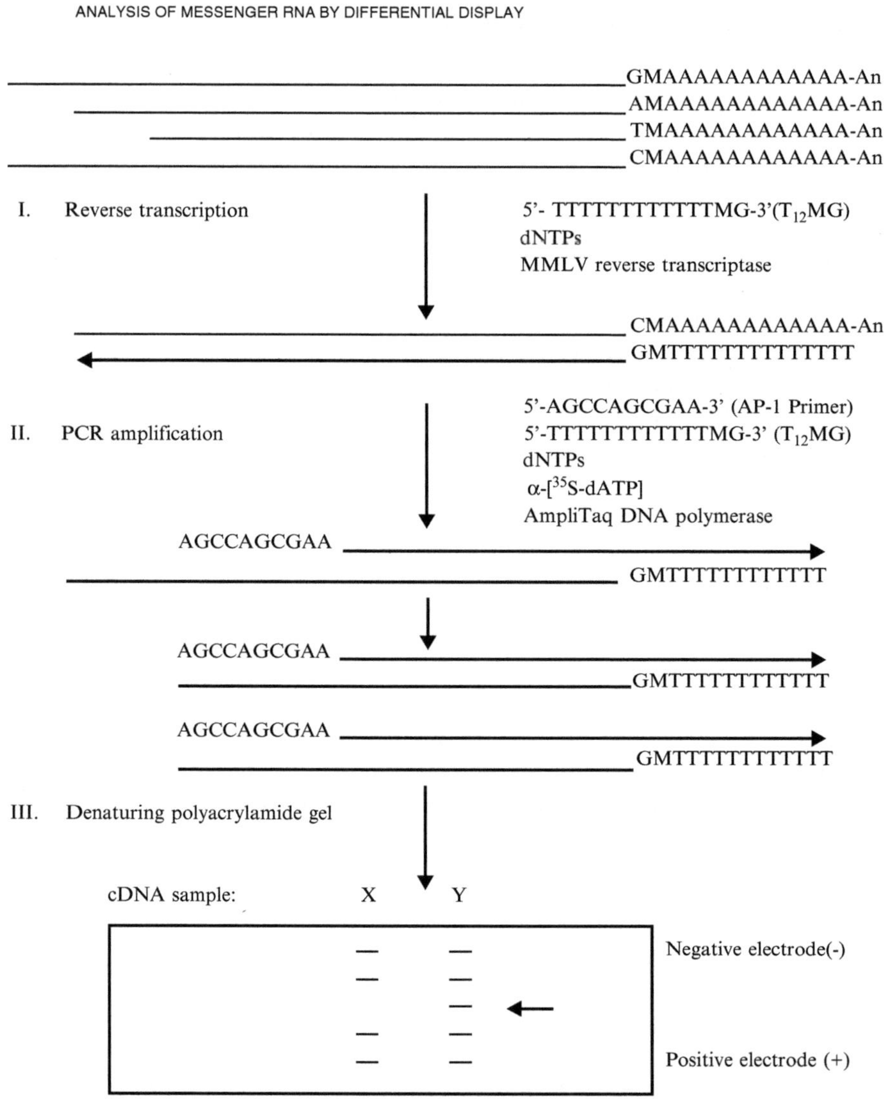

Fig. 12.13 Principle of differential display PCR. An mRNA pool is reverse transcribed into cDNA using a specially designed poly-dT primer which anneals to the 3′-end terminal poly-A tail of eukaryotic mRNA. The reverse transcribed cDNA pool is then PCR amplified using the reverse transcription primer, a randomly chosen (arbitrary) primer and labeled dNTPs. After polyacrylamide gel electrophoresis, differences in mRNA expression profiles may be observed as differences in banding patterns. The origin of these differences may then be determined by sequencing the differentially expressed band(s) (From Liang et al., 1996)

biopsy material) and growth synchronized cells for DD-PCR testing. Alternatively, it is possible to use specific cell types (e.g. tumour cells) which have been excised from paraffin embedded or air-dried coupes using micro-dissection. Finally, the DD-PCR fragments usually only encode the 3′-end non-translated region of the mRNA molecule, rendering the identification of differentially expressed mRNA molecules by DD-PCR problematic without access to complete genomic DNA sequence databases, as translating differentially expressed DD-PCR amplimer sequences and searching in protein expression libraries may not yield the protein actually encoded by the differentially expressed mRNA transcript.

One alternative to DD-PCR is "RNA arbitrary primed-PCR" or "RAP-PCR". This technique is homologous to RAPD PCR, but uses reverse transcribed RNA/cDNA hybrids as PCR template instead of double-stranded genomic DNA. In contrast to DD-PCR, RAP-PCR utilizes a short primer of 10–20 randomly chosen nucleotides in the reverse transcription reaction, which means that problems associated with the poly-dT 3′-end mRNA orientated DD-PCR primer are avoided. In RAP-PCR, the reverse transcription primer may also be used for PCR thermocycling, with thermocycling being performed under low stringency conditions. Labelled RAP-PCR amplimers (labelled for example by incorporating radioactive dNTPs or fluorescently labelled primer in the RAP-PCR mix) may then be separated by denaturing polyacrylamide gel electrophoresis and the banding patterns visualized using for example autoradiography. An appropriate primer combination should be capable of generating at least 50 different RAP-PCR amplimers per reaction in order to adequately assess differences in gene expression.

12.15 The Protein Truncation Test (PTT)

Many inherited diseases (e.g. breast and colon cancer, muscular dystrophy, etc.) result from the accumulation of prematurely terminated proteins, which are often caused by the presence of mutations in genomic DNA that have led to the introduction of stop triplet codons (i.e. UGA, UAA or UAG) into transcribed mRNA molecules [Gite et al., 2003]. These shortened protein products may be detected using the *protein truncation test* or "PTT" [Hogervorst et al., 1995].

The principle of the PTT is shown in Fig. 12.14. The area of a gene in which the translation stop mutation is expected is PCR amplified from either reverse transcribed cDNA or genomic DNA to yield amplimers of 1–2 kb in length. PTT uses a specially designed "sense" primer (i.e. a primer which reads in the correct mRNA reading frame and can be translated directly into mRNA by DNA dependant RNA polymerases), and contains at its 5′-end: (i) a T7 RNA polymerase promoter sequence, (ii) a eukaryotic translation recognition (Kozak) sequence, and iii) an ATG "start" codon. The second PCR primer in the PTT test should be designed to contain a sequence complementary to the terminus of the gene sequence under investigation. Once PCR amplification has been performed, the resultant PTT amplimer is used as a template in a *coupled transcription and translation cell free system* (e.g. TNT Quick coupled Transcription/Translation System, Promega), which generates mRNA and protein via the 5′-end (T7 RNA polymerase recognition site) sequences included in the 5′-end PTT primers. For detection purposes, protein synthesis may be performed in the presence of radioactive amino acids or biotinylated lysine-tRNA and then detected (after SDS-PAGE gel electrophoresis)

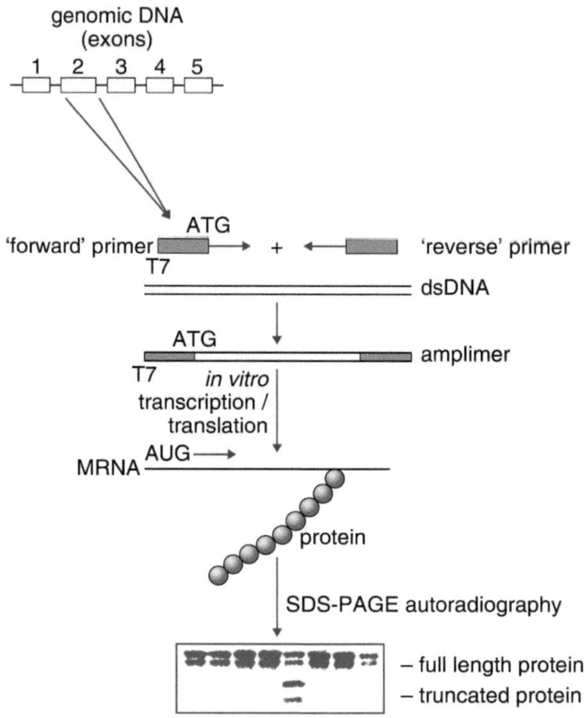

Fig. 12.14 Principle of the protein truncation test. Regions of genes to be analyzed are amplified from genomic, or reverse transcribed cDNA. The upstream primer is designed to include transcription sequences (e.g. for T7 RNA Polymerase) and eukaryotic translation signal sequences (Kozak sequences). Upon incubation of the resultant amplimer in a coupled transcription/translation cell free system (e.g. TNT Quick coupled Transcription/Translation System, Promega) the cDNA is translated back into RNA and then transcribed into protein. Full-length proteins and truncated proteins can then be detected by SDS-PAGE using autoradiography or Western blot techniques

by autoradiography or "Western" blotting using labelled anti-biotin antibodies [Roest et al., 1996]. Any truncated protein products will be visible as shorter polypeptide fragments, when compared to the expected protein fragment size after autoradiography or Western blotting.

Usually, PTT is performed on only a small part of the gene of interest (the part where most of the mutations are expected to occur), not least because eukaryotic genes are usually greater than 50–300 kb in length and the analysis of the entire 50–300 kb coding region is too complex to perform. Also, PTT does not detect genomic mutations that affect mRNA stability, mRNA processing, or mutations that do not introduce "stop" codons into the mRNA (i.e. mis-sense mutations, small insertions and short deletions).

However, PTT is a test that can potentially be adapted to high-throughput automated formats, allowing the processing of large numbers of samples in a single experiment. Also, PTT offers the ability to discriminate between clinically relevant and clinically non-relevant mutations (as the presence of DNA sequence mutations may not actually be clinically relevant, whereas the presence of truncated proteins usually are) with good sensitivity and low false-positive rates.

12.16 Methylation Specific PCR and PCR in the Detection of Mutagens

Methylation of cytosine molecules in higher eukaryotes is a biologically important event, especially in epigenetic processes (i.e. in the developmental and organizational processes of organisms). Such methylation can bring about many effects in an organism including: (i) "gene silencing", (ii) the modulation of mRNA transcription, (iii) inactivation events on the X-sex chromosome, and (iv) occasionally the development of oncogenesis [Lehmann et al., 2003]. The precise mechanisms of methylation and/or protection of GC-rich regions against methylation are not yet fully understood, though it is known that methylation of genomic DNA can be both temporary or permanent in nature and only occurs if a cytosine nucleotide is immediately followed by guanosine (i.e. in the dinucleotide CpG). Also, methylation of cytosine bases in GC-rich sequences of promoter or enhancer regions of the genome (CpG islands) can be used to efficiently block mRNA transcription. Further, it has been shown that patterns of genomic methylation can be transferred from parents to offspring in a process referred to as "imprinting", with the dis-regulation of imprinting having been associated with several growth and behavioural defects, including Beckwith–Wiedemann syndrome, Prader–Willi syndrome, Angelman syndrome and cancer (Wilms tumour) [Reik et al., 2001].

Methylated DNA may be detected using several non-PCR techniques, including single-stranded conformational polymorphism (SSCP) analysis of native DNA to look for the presence of aberrant migration banding patterns upon gel electrophoresis. However, it is also possible to use novel PCR methodologies including methylation-specific PCR [Herman, 1996] or methylation specific AFLP.

In methylation-specific PCR, use is made of the fact that incubation of DNA containing non-methylated cytosine nucleotides for a short period of time at physiological temperature with sodium hydroxide (NaOH), renders the non-methylated cytosine nucleotides sensitive to overnight treatment at 55°C with acidic sodium bisulphite ($Na_2S_2O_3$) and hydroquinone, whereas C5-methylated cytosine nucleotides are not rendered susceptible by this treatment. For non-methylated (susceptible) cytosine nucleotides, such treatment facilitates cytosine deamidation, effectively changing the cytosine into a uracil nucleotide. The presence or absence of cytosine to uracil "metamorphoses" may then be detected using high stringency PCR conditions and a pair of sequence specific PCR primers, one of which contains either (i) a discriminatory adenosine nucleotide at its 3′-end (which will hybridize to a uracil nucleotide if present thereby indicating the original presence of non-methylated cytosine at that 3′-end position) or (ii) a discriminatory guanosine nucleotide at its 3′-end (which will hybridize to an undeaminated C5-methylated cytosine nucleotide if present thereby indicating the original presence of methylated cytosine at that 3′-end position). Methylation specific PCR is complicated by the fact that the PCR primer that is not being used to probe the methylation site must comprise a sequence that is not affected by cytosine/uracil metamorphoses. In effect, this means that this particular primer should be designed from a region of DNA whose sequence contains no cytosine nucleotides (the primer sequence will contain no guanosine nucleotides). If multiple methylation sites are expected to be present within a short DNA sequence region (i.e. approximately 10–30 nucleotides), then guanosine or adenosine nucleotide substitutions may also be introduced into the "core" of the PCR primer as well as the 3′-end. However, using this methodology may require a large increase in the number of PCR primer combinations that have to be used. For example, to search for two methylation sites using the same primer sequence, may require primers containing 5′-nAnA–3′, 5′-nGnG–3′, 5′-nAnG–3′, and 5′-nGnA–3′ modifications (where "n" equals a given number of nucleotides), in order to probe all possible methylation combinations.

In the laboratory, methylation specific PCR may be used to track regions of inactivation on the chromosome or to make methylation charts for entire chromosomes, thereby providing clues as to the expression state of specific chromosomal regions. DNA from paraffin embedded tissues can also act as a template for methylation-sensitive PCR, though sensitivity may be improved by DNA precipitation, possibly supported by addition of carrier DNA or glycogen.

Methylation is a specific, biologically programmed nucleic acid modification mechanism. However, several chemical agents (e.g. antibiotics such as the anthracyclins) may act to non-specifically modify DNA or RNA, often resulting in nucleic acid damage or mutagenesis. The capability of such chemicals to facilitate DNA damage or mutagenesis can be assessed by observing a reduced PCR amplification yield in bacteriophage λ or P53 amplimer model systems [Jenkins and Parry, 2000]. High-throughput PCR testing of chemical mutagens is also an option, as many mutagens require co-factors (bleomycin requires Fe^{2+} ions, mitomycin C requires Cu^{2+} ions, etc.). By coupling the target DNA to 96-well microtitre plates, adding varying concentrations (or mixtures) of the suspected mutagenic agent(s) to each well, and subsequently performing high-throughput (nested) PCR and gel electrophoresis, either a decrease in PCR yield (mutagenic effect) or no decrease in yield (no mutagenic effect) may be observed. Novel generations of antibiotics may be (de-)selected on the basis of their mutagenicity using such PCR methods [Hotta, 1995].

12.17 Breakpoint PCR

An increasing number of diseases, and especially oncogenic malignancies, are associated with chromosomal translocations (reciprocal exchange of chromosomal regions). These translocations may be associated with the development and initiation of oncogenesis and are scientifically and diagnostically interesting (e.g. the *Philadelphia chromosome* reciprocal translocation between chromosomes 9 and 22, t(9;222); chronic myeloid leukaemia, etc.). The breakpoint PCR is a simple PCR test designed to specifically amplify translocated regions of DNA. In essence, one primer of the breakpoint PCR pair is designed to anneal to a DNA sequence on the "recipient" chromosome and the other to a DNA sequence on the translocated chromosomal fragment. This means that an amplification product is obtained only when the (correct) translocation is present in the target DNA. A primer pair designed to amplify another (unaffected) region of DNA on the recipient chromosome may be used as an amplification control and included in the same breakpoint PCR mix resulting in a "multiplex breakpoint PCR" protocol. Since the regions where chromosomal breakage and translocations occur (the breakpoints) are not always exactly identical, breakpoint PCR primers may need to be spaced in an appropriate fashion so that the entire catalogue of already identified translocation sites associated with a particular disease are covered. However, this could mean that a "long and accurate" breakpoint PCR thermocycling methodology may need to be performed in order to generate long (5–30 kb) PCR amplimers. Breakpoint PCRs work well for mixtures of translocated cells in a background of healthy cells, since only the translocated cells will generate amplification products. If chromosomal translocations are accompanied by extensive chromosomal deletions, then the breakpoint PCR may yield a "false negative" result, in which case the use of comparative genomic hybridization (CGH), or array-CGH, may be a more appropriate diagnostic technique [Verhagen et al., 2000]. CGH may also be a better option if gene duplications during translocation are also usually observed.

12.18 Site Directed Mutagenesis by PCR

The selective introduction of nucleotide changes (mutations) within a known DNA sequence is a process called site-directed mutagenesis or SDM. The goal of such SDM may include (i) generating new restriction digestion sites, (ii) generating cloning sites, (iii) to change the sequence of a transcribed mRNA molecule so that a novel protein is translated, (iv) to introduce stop codons into DNA, (v) to help make fusion genes or vi) to study the effects of mutation events on the efficiency of promoters. SDM using PCR may be utilized to generate mutations by both nucleotide deletion and/or nucleotide substitution mechanisms.

Short nucleotide deletions may be introduced into DNA using a "degenerate" PCR primer, which at its 3′-end is complementary to the DNA sequence to be mutated, but lacks the nucleotides to be deleted near its 5′-end. After PCR amplification, the products will lack the deleted nucleotides. Due to the degenerate nature of the primer sequence (i.e. it will not perfectly hybridize to the intended target DNA sequence), the T_m of thermocycling must be lowered during the first few cycles of PCR in order to allow primer hybridization.

Degenerate primers may also be used to introduce nucleotide substitutions into specific DNA sequences, however the use of degenerate primers in a standard SDM deletion protocol (as mentioned above), would limit nucleotide substitutions to the outer limits (5′-ends) of the DNA to be amplified (dependent on the length of the degenerate primer to be used). Therefore, to facilitate nucleotide substitutions near the centre of a particular gene sequence, two separate PCRs need to be performed, with the two corresponding amplimers overlapping in the region where the mutation is to be introduced (see Fig. 12.15). Post-amplification, the two amplimers are purified away from excess dNTPs, enzyme, etc., and added to a single new PCR mix without PCR primers. The mix is then heated to denature the two amplimers and the temperature will be allowed to decrease so that cross-hybridization of the amplimers with each other may occur. Resultant hybrid molecules with single stranded ends extending beyond the common hybridization region will be filled in by the thermostable DNA polymerase present in the reaction mix, generating an elongated double-stranded template. The addition of 5′-end and 3′-end primers designed to hybridize to the termini of the new elongated DNA molecule, followed by PCR thermocycling, leads to amplification of the new elongated hybrid DNA construct, which may then for example, be ligated into a cloning vector for eventual protein expression studies or used to transform recipient cells, etc.

12.19 PCR Amplimers for Cloning and Expression

PCR is a powerful method for synthesizing large amounts of specific DNA fragments or complete genes (with or without their corresponding promoter regions). Further, the design of PCR primers with added 5′-end restriction enzyme recognition sequences followed by restriction enzyme digestion and ligation has allowed the cloning of such amplimers into vectors that contain identical restriction enzyme recognition sequences. However, PCR applications are constantly changing and improving with regard to speed, throughput and cost. For example, the introduction of “TA-cloning kits” (TA Cloning kit, Invitrogen) have been the greatest advance in PCR amplimer cloning and expression in recent years. Such kits include both special enzyme mixes and linear vectors containing 3′-end thymidine nucleotide “overhangs”, and make use of a property of the majority of thermostable enzymes (including Taq but not the proofreading enzyme Pfu) to add an extra adenosine (or other) nucleotide to the 3′-end of PCR amplification products. Ligation of PCR amplimers to a special TA cloning plasmid than allows rapid insertion of the PCR product into a cloning/sequencing vector, without the need for PCR primers containing restriction enzyme recognition sites and multiple digestion and ligation experiments. Such advances in PCR cloning applications means that the construction of cumbersome random genomic or random cDNA libraries, along with labour intensive library screening, restriction mapping and sub-cloning, may now be avoided. The only requirement however, is that at least part of the sequence of the region to be amplified is known (in order to design suitable PCR primers). Once cloned into a relevant vector, PCR amplimers may be expressed in either prokaryotic or eukaryotic cells

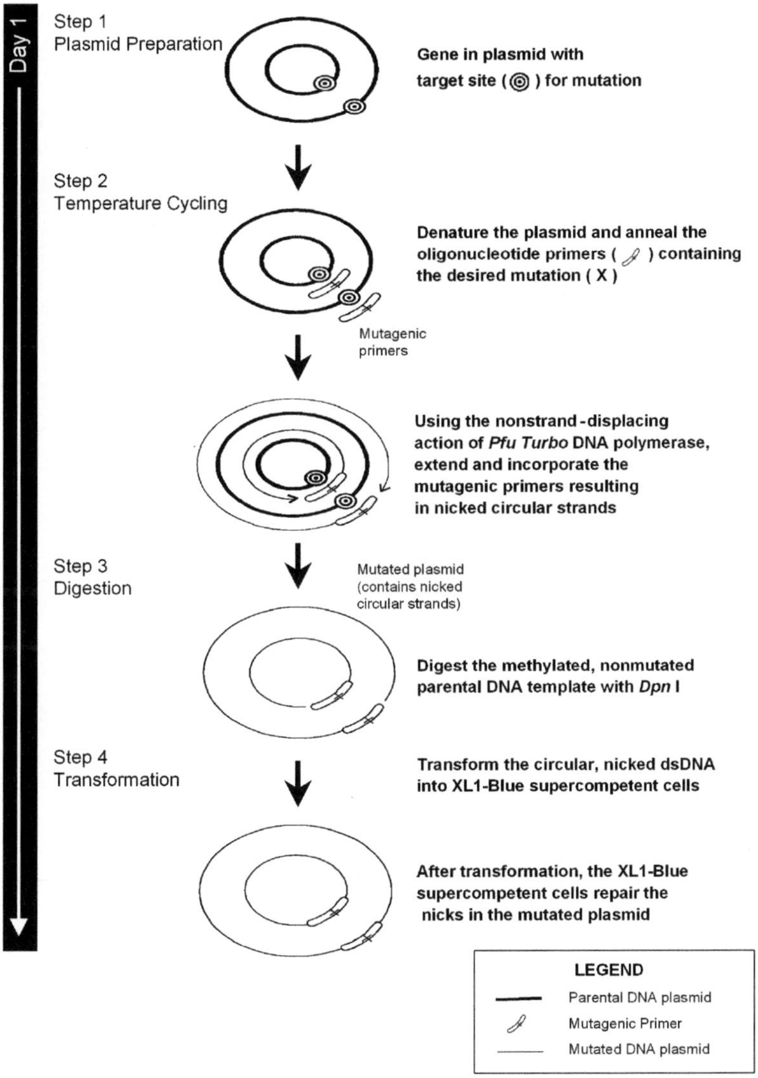

Fig. 12.15 Site directed mutagenesis by PCR. The target sequence is copied into two overlapping fragments containing the required nucleotide substitution(s) which are then co-amplified in a second PCR. In this way, nucleotide substitutions (mutagenesis) may be introduced into the target DNA sequence (www.stratagene.com/manuals)

(including bacterial, yeast, insect and mammalian cells). This depends on the expression system and promoter used. For this purpose, many commercially available expression vectors have been developed, e.g. the CheckMate Mammalian Two-Hybrid System and the the pALTER-MAX system for use in bacterial cells (Promega), and the BaculoDirect Expression System (Invitrogen) (Fig. 12.16). Alternatively, a

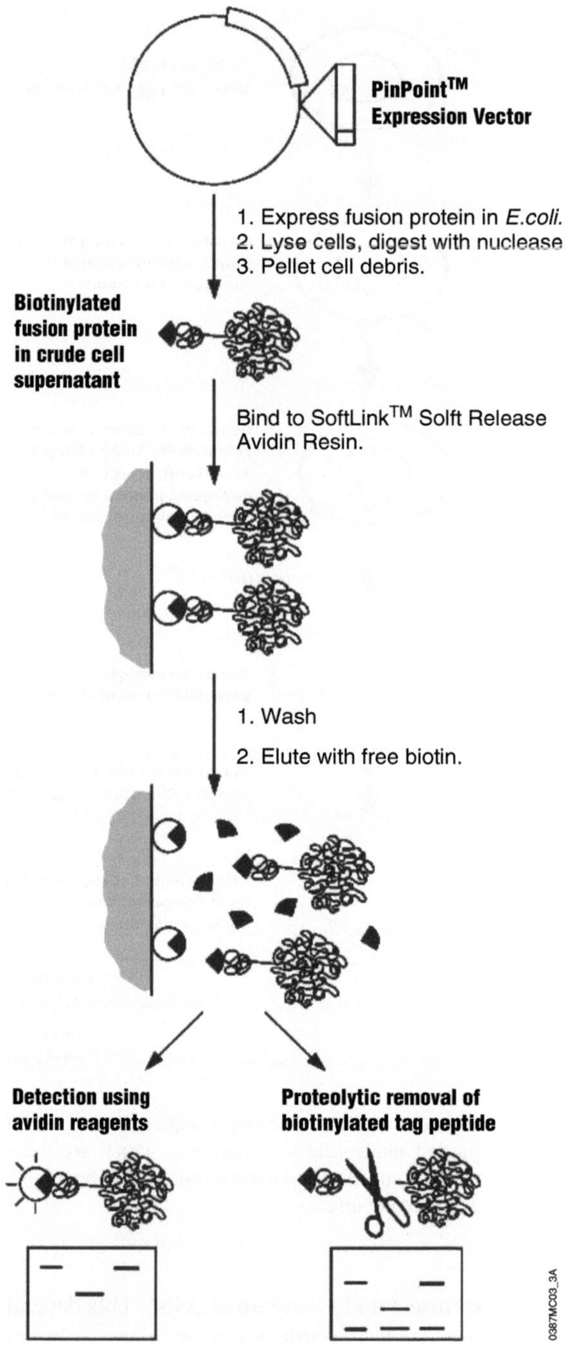

Fig. 12.16 The cloning of an amplified gene by PCR into a commercially available vector allows expression of the encoded protein. In this example, the vector has been designed to generate a fusion protein allowing easier isolation and manipulation of expressed protein (Promega Protocols)

gene may be PCR amplified and cloned into a vector which puts it under the control of a set of regulatory elements allowing for cell free lysate coupled transcription and translation, indeed some commercially available coupled transcription and translation systems are specifically designed for protein expression from the appropriate PCR products, e.g. TNT T7 quick for PCR system (Promega).

Large quantities of protein may also be generated from PCR products cloned into specific vectors by incorporating for example, a thioredoxin tag (to increase expressed protein solubility) in the vector, e.g. the His-Patch ThioFusion expression System (Invitrogen), whilst polyhistidine (a sequence that encodes 6–8 histidine residues) encoding vectors allow the purification of PCR amplimer generated proteins using nickel chelate resins in the form of columns or plates (Sigma). Alternatively, in the Pinpoint Vector and Expression Systems (Promega), the PCR product is cloned into a vector and synthesized as a fusion protein coupled to a 22 kDa polypeptide that is naturally biotinylated in *E. coli*. This means that the final fusion protein can be immediately coupled to streptavidin-coated beads for further purification or immediately used in downstream processes such as coating ELISA plates for sero-prevalence studies. For detection purposes, PCR amplimer cloning vectors are available which allow PCR amplimers to be expressed as fusion proteins containing such easily detectable "tags" as β-galactosidase, chloramphenicol acetyltransferase (CAT), enhanced green fluorescent protein (EGFP), luciferase and glutathione-s-reductase (GST).

12.20 SAGE

SAGE or "Serial Analysis of Gene Expression" (Fig. 12.17) is a method for mapping the expression of genes by constructing multiple repeats (of approximately 100 bp in length) of small (approximately 10 bp) cDNA "Tags" [Velculescu et al., 1995; Velculescu et al., 2000]. In the SAGE format, mRNA (representing all of the genes that were being expressed at the time of mRNA isolation) is extracted from cells and attached to a solid support. Next, cDNA synthesis is performed using oligo-dT or random hexamer primers, yielding double-stranded cDNA copies of the original mRNA transcripts available. These double-stranded cDNAs are then digested using the restriction enzyme *Nla*III generating a GTAC overhang. Two specific DNA adapters containing a recognition sequence for a second restriction enzyme (usually *BsmF*I), are then ligated to one half of the *Nla*III digested cDNA (one adapter per half), and further digested using *BsmF*I. As *BsmF*I cuts 10–14 bp downstream of the recognition sequence, small Tag/adapter products are liberated from the solid support. Next, the 5′-ends of the *BsmF*I digested products are "filled in" and products from each of the two adapters are pooled and ligated to form a "ditag". PCR amplification is then performed using the two adapter sequencers as primers and the PCR products again digested using *Nla*III. The "sticky ended" PCR amplimers are then ligated into long concatemers comprising multiple copies of the same ditag. These concatemers are then cloned

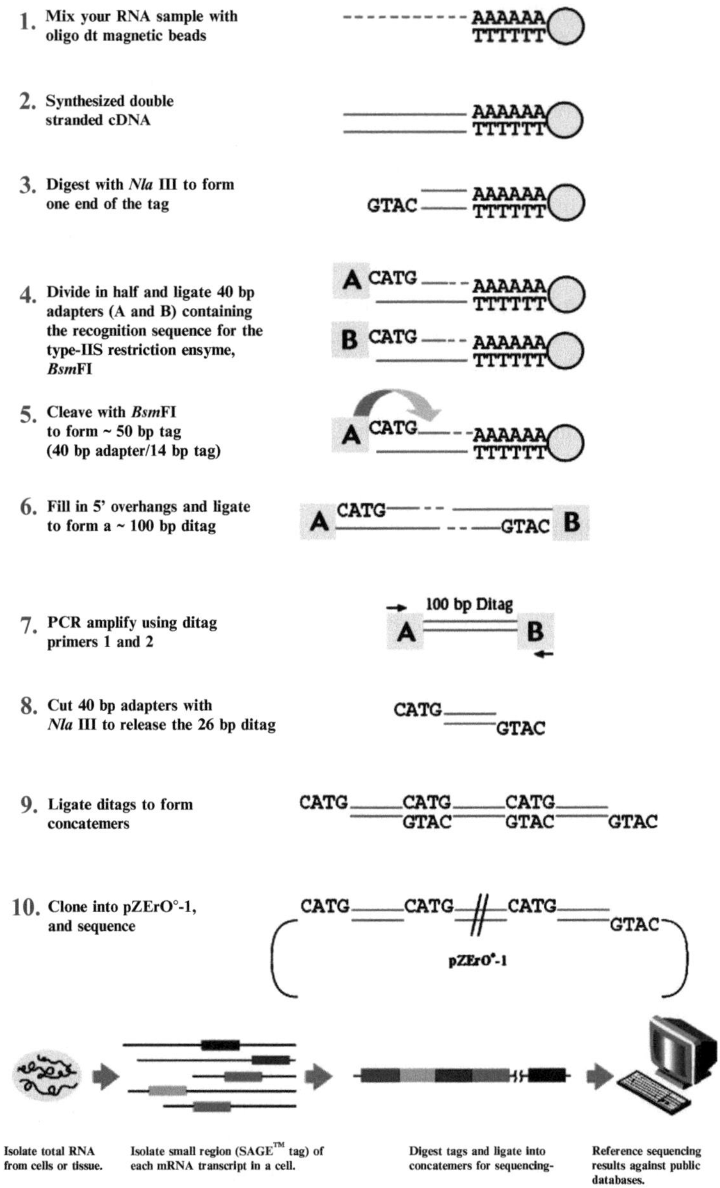

Fig. 12.17 Schematic presentation of the various steps of the Serial Analysis of Gene Expression (SAGE) protocol (see text in this chapter for more precise written explanation and www.invitrogen.com for more detail on the I-SAGE kit for Genome Wide Expression Analysis) (Copied from Invitrogen) (*see Color Plates*)

into a plasmid vector, sequenced, and the 10 bp tag compared to currently available databases (e.g. SAGEMAP at www.ncbi.nlm.nih.gov/projects/SAGE/) [Lash et al., 2000; Lal et al., 1999]. Such tags usually contain enough sequence information to reliably identify the source of the original complete mRNA transcript, and the level of gene expression is indicated by the frequency of occurrence of a particular 10 bp tag. Commercial kits are currently available to perform this technique, e.g. I-SAGE Longkit (Invitrogen).

12.21 PCR Inhibition by DNA Specific Antibiotics and Mutagens

There exists a large multitude of agents capable of selectively damaging double stranded DNA. The degree of damage can be reflected in the decrease in the amount of the PCR product formed. The damaging agents can be found in food products and product formed during incineration of wood or charcoal. In addition, many chemotherapeutic agents do have DNA-mutagenic characteristics. This includes various antibiotics as well. Quality control of novel medications does include assessment of the compounds' mutagenic potential. DNA damage can be screened at high throughput using 384-wells ELISA plates. The wells contain a fixed amount of a biotinylated PCR product and these are subjected to a variety of (combinations of) chemicals to be explored. A subsequent quantitative nested PCR indicates the degree of DNA damage incurred by the chemicals. The same assay, of course, may also be used to identify compounds with anti-mutagenic effects. When testing antibiotics the effect of metal co-factors needs to be considered. Mitomycin, for instance, requires divalent copper ions for optimal activity.

References

Ailenberg M, Silverman M. 2000. Controlled hot start and improved specificity in carrying out PCR utilizing touch up and loop incorporated primers (TULIPS). Biotechniques 29:1018–1024.

Alcock HE, Stephenson TJ, Royds JA, Hammond DW. 2003. Analysis of colorectal tumour progression by micro-dissection and comparative genomic hybridization. Genes Chromosome Cancer 37:369–380.

Benecke M. 1997. DNA typing in forensic medicine and in criminal investigations: a current survey. Naturwissenschaften 84:181–188.

Berg H, Ziegler K, Piotukh K, Baier K, Lockau W, Volkmer-Engert R. 2000. Biosynthesis of the cyanobacterial reserve polymer multi-L-arginyl-poly-L-aspartic acid (cyanophycin): mechanism of the cyanophycin synthetase reaction studied with synthetic primers. Eur J Biochem 267:5561–5570.

Boyer JC, Farber RA. 1998. Mutation rate of a microsatellite sequence in normal human fibroblasts. Cancer Res 58:3946–3949.

Castles CG, Allred DC, Krieg SL, Benedix MG, Fuqua SA. 1996. RNA from air dried frozen sections for RT-PCR and differential display. Biotechniques 21:425–428.

Chevalier J, Yi J, Michel D, Tang XM. 1997. Biotin and digoxigenin as labels for light and electronmicroscopy in situ hybridization probes: where do we stand? J Histochem Cytochem 45:451–471.

Chou Q, Russell M. Birch DE, Raymond J, Bloch W. 1992. Prevention of pre-PCR mis-priming and primer dimerization improves low-copy number amplifications. Nucleic Acids Res 20:1717–1723.

Garcia J, Castrillo JL. 2004. Differential display RT-PCR analysis of human choriocarcinoma cell lines and normal term trophoblast cells: identification of new genes expressed in placenta. Placenta 25:684–693.

Gite S, Lim M, Carlson R, Olejnik J, Zehnbauer B, Rothschild K. 2003. A high throughput nonisotopic protein truncation test. Nat Biotechnol 21:194–197.

Gyllenstein U. 1989. Direct sequencing of in vitro amplified DNA. In: Erlich H (ed) PCR technology. pp. 45–61.

Hawkins GA, Hoffman LM. 1999. Rapid DNA mutation identification and fingerprinting using base excission sequence scanning. Electrophoresis 20:1171–1176.

Hebert NE, Brazill SA. 2004. Microchip capillary gel electrophoresis with electrochemical detection for the analysis of known SNPs. Lab Chip 3:241–247.

Hecker KH, Roux KH. 1996. High and low annealing temperatures increase both specificity and yield in touchdown and stepdown PCR. Biotechniques 20:478–485.

Herman JG, Graf JR, Myohanen S, Nelkin BD, Baylin SB. 1996. Methylation specific PCR: a novel PCR assay for methylation status of CpG islands. Proc Natl Acad Sci USA 93:9821–9826.

Higuchi RG, Ochman H. 1989. Production of single stranded DNA templates by exonuclease digestion following the PCR. Nucleic Acids Res 17:5865.

Hogervorst FB, Cornelis RS, Bout M, Van Vliet M, Oosterwijk JC, Olmer R, Bakker B, Klijn JG, Vasen HF, Meijers-Heijboer H. 1995. Rapid detection of BRCA1 mutations by the protein truncation test. Nat Genet 10:208–212.

Hotta K, Zhu CB, Phomsuwansiri P, Ishikawa J, Mizuno S, Hatsu M, Imai S. 1995. PCR inhibition assay for DNA-targeted antibiotics. J Antibiot 48:1267–1272.

Jenkins GJ, Parry JM. 2000. PCR-based methodology for the determination of the photoprotection afforded by sunscreen application. Biotechniques 29:1318–1326.

Joos S, Schutz B, Bentz M, Lichter P. Roche product manual "ISH to whole chromosomes". In: Procedures for In Situ Hybridization to Chromosomes, Cells, and Tissue Sections; Detection of Chromosomal Imbalances using DOP–PCR and Comparative Genomic Hybridization (CGH). p95–103. Also available at https://www.roche-applied-science.com/fst/publications.jsp? page=/PROD_INF/MANUALS/InSitu/InSi_toc.htm

Jouquand S, Cheron A, Galibert F. 1999. Microsatellite analysis using a two step procedure for fluorescence labeling of PCR products. Biotechniques 26:902–905.

Kaboev OK, Luchkina LA, Tretíakov AN, Bahrmand AR. 2000. PCR hot start using primers with the structure of molecular beacons (hairpin-like structures). Nucleic Acids Res 28:94.

Kainz P. Schmiedlechner A, Strack HB. 2000. Specificity-enhanced hot-start PCR: addition of double-stranded DNA fragments adapted to the annealing temperature. Biotechniques 28(2):278–282.

Kebelmann-Betzing C, Seeger K, Dragon S, Schmitt G, Moricke A, Schild TA, Henze G, Beyermann B. 1998. Advantages of new Taq DNA polymerase in multiplex PCR and time release PCR. Biotechniques 24:154–158.

Kellogg DE, Rybalkin I, Chen S, Mukhamedova N, Vlasik T, Siebert PD, Chenchik A. 1994. TaqStart antibody: hot start PCR facilitated by a neutralizing monoclonal antibody directed against Taq DNA polymerase. Biotechniques 16:1134–1137.

Kruhoffer M, Jensen JL, Laiho P, Dyrskjot L, Salovaara R, Arango D, Birkenkamp-Demtroder K, Sorensen FB, Christensen LL, Buhl L, Mecklin J, Jarvinen H, Thykjaer T, Wikman FP, Bech-Knudsen F, Juhola M, Nupponen NN, Laurberg S, Andersen CL, Aaltonen LA, Orntoft TF. 2005. Gene expression signatures for colorectal cancer microsatellite status and HNPCC. Br J Cancer 92:2240–2248.

Kutyavin IV, Afonina IA, Mills A, Gorn VV, Lukhtanova EA, Belousov ES, Singer MJ, Walburger DK, Lokhov SG, Gal AA, Dempcy R, Reed MW, Meyer RB, Hedgpeth J. 2000. 3′-minor groove binder DNA probes increase sequence specificity at PCR extension temperatures. Nucleic Acids Res 28:655–661.

Lal A, Lash AE, Altschul SF, Velculescu V, Zhang L, McLendron RE, Marra MA, Prange C, Morin PJ, Polyak K, Papadopoulos M, Vogelstein B, Kinzler KW, Strausberg RL, Riggins GJ. 1999. A public database for gene expression in human cancers. Cancer Res 59:5403–5407.

Lash AE, Tolstoshev CM, Wagner L, Schuler GD, Strausberg LR, Riggins GJ, Altschul SF. 2000. SAGEmap: a public gene expression resource. Genome Res 10:1051–1060.

Leclair B, Scholl T. 2005. Application of automation and information systems to forensic genetic specimen processing. Expert Rev Mol Diagn 5:241–250.

Lehmann K, Seemann P, Stricker S, Sammar M, Meyer B, Suring K, Majewski F, Tinschert S, Grzesik KH, Muller D, Knaus P, Nurnberg P, Mundlos S. 2003. Mutations in bone morphogenetic protei receptor 1B cause brachydactyly type A2. Proc Natl Acad Sci USA 100:12277–12282.

Liang P. 1996. Analysis of messenger RNA by differential display. In: Krieg PA (ed) A laboratory guide to RNA. Wiley, New York, pp. 224–235.

Lin Y, Jayasena SD. 1997. Inhibition of multiple thermostable DNA polymerases by a heterodimeric aptamer. J Mol Biol 27:100–111.

Liu WH, Kaur M, Wang G, Zhu P, Zhang Y, Makrigiorgos GM. 2004. Inverse PCR based RFLP scanning identifies low level mutation signatures in colon cells and tumours. Cancer Res 64:2544–2551.

Maxam AM, Gilbert W. 1977. A new method for sequencing DNA. Proc Natl Acad Sci USA 74:560–564.

Mizuguchi H, Nakatsuji M, Fujiwara S, Takagi M, Imanaka T. 1999. Characterization and application to hot start PCR of neutralizing monoclonal antibodies against KOD DNA polymerase. J Biochem 126:762–768.

Moretti T, Koons B, Budowle B. 1998. Enhancement of PCR amplification yield and specificity using Amplitaq Gold DNA polymerase. Biotechniques 25:716–722.

Mullis KB. 1991. The PCR in an anemic mode: how to avoid cold oligodeoxyribonuclear fusion. PCR Methods Appl 1:1–4.

Nazarenko I, Lowe B, Darfler M, Ikonomi P, Schuster D, Rashtchian A. 2002. Multiplex quantitative PCR using self quenched primers labeled with a single fluorophore. Nucleic Acids Res 30:37.

Petitjean J, Vabret A, Dina J, Gouarin S, Freymuth F. 2005. Development and evaluation of a real time RT-PCR assay on the LightCycler for the rapid detection of enterovirus in cerebrospinal fluid specimens. J Clin Virol 5:1271–1282.

Reik W, Dean W, Walter J. 2001. Epigenetic reprogramming in mammalian development. Science 293:1089–1093.

Roest PA, Bout M, van der Tuijn AC, Ginjaar IB, Bakker E, Hogervorst FB, van Ommen GJ, den Dunnen JT. 1996. Splicing mutations in DMD/BMD detected by RT-PCR/PTT: detection of a 19AA insertion in the cysteine rich domain of dystrophin compatible with BMD J Med Genet 33(11):9350939.

Sanger F, Nicklen S, Coulson AR. 1992. DNA sequencing with chain-terminating inhibitors. Biotechnology 24:104–108.

Singh A, Gaur A, Shailaja K, Satyare BB, Singh L. 2004. A novel microsatellite (STR) marker for forensic identification of big cats in India. Forensic Sci Int 141:143–147.

Sonoki T, Willis TG, Oscier DG, Karran EL, Siebert R, Dyer MJ. 2004. Rapid amplification of immunoglobulin heavy chain switch (IGHS) translocation breakpoints using long-distance PCR. Leukemia 18:2026–2031.

The year of sequencing. 2008. Nature Methods 5(1):11–21.

Torpdahl M, Skov MN, Sandvang D, Baggesen DL. 2005. Genotypic characterization of *Salmonella* by multilocus sequence typing, pulsed field gel electrophoresis and amplified fragment length polymorphism. J Microbiol Methods 63:173–184.

Van Belkum A. 1994. DNA fingerprinting of medically important microorganisms by use of PCR. Clin Microbiol Rev 7:174–184.

Van Belkum A. 2002. Molecular typing of microorganisms: at the centre of diagnostics, genomics and pathogenesis of infectious diseases? J Med Microbiol 51:7–10.

Van Belkum A, Scherer S, Van Alphen L, Verbrugh H. 1998. Short sequence DNA repeats in prokaryotic genomes. Microbiol Mol Biol Rev 62(2):275–293.

Velculescu VE, Vogelstein B, Kinzler KW. 2000. Analysing uncharted transcriptomes with SAGE. Trends Genet 16:423–425.

Velculescu VE, Zhang L, Vogelstein B, Kinzler KW. 1995. Serial analysis of gene expression. Science 270:484–487.

Verhagen PC, Zhu XL, Rohr LR, Cannon-Albright LA, Tavtigian SV, Skolnick MH, Brothman AR. 2000. Microdissection, DOP PCR and comparative genomic hybridization of paraffin-embedded familial prostate cancers. Cancer Genet Cytogenet 122:43–48.

Vos P, Hogers R, Bleeker M, Reijans M, Van de Lee T, Hornes M, Frijters A, Pot J, Peleman J, Kuiper M, Van Montagu M. 1995. AFLP: a new technique for DNA fingerprinting. Nucleic Acids Res 23:4407–4414.

Wu WM, Tsai H J, Pang JH, Wang HS, Hong HS, Lee YS. 2005 Touchdown thermocycling program enables a robust single nucleotide polymorphism typing method based on allele-specific real-time polymerase chain reaction. Anal Biochem 339(2):290–296.

Chapter 13

In Situ PCR Amplification (ISA) – Major Considerations, Sample Processing and Applications

The vast majority of PCR protocols have been developed to amplify soluble DNA or RNA, which has been extracted from its *in situ* environment, i.e. from clinical or environmental samples. However, the process of removing target nucleic acid from its *in situ* environment necessarily means that questions such as: is the nucleic acid specific for a particular diseased tissue type; is the nucleic acid derived from extra-cellular or internalised bacteria or parasites; is viral DNA integrated into the host chromosome or present in episomal form? etc, cannot easily be answered. *In situ* hybridisation (ISH) and *in situ* amplification (ISA) techniques help provide the answers to these questions, as the nucleic acid is not extracted during testing. In addition, *in situ* protocols are the assays of choice when investigating the local expression of particular genes. Many pathology laboratories maintain extensive historical collections of well-preserved, most frequently paraffin embedded, tissue specimens. In addition, collections of native, deep-frozen samples are usually available as well. These archives have most probably previously been investigated using pathological staining techniques and/or ISH tests. However, non-PCR-based ISH protocols tend to exhibit a limited sensitivity, even though several non-nucleic acid-based immunohistochemical signal amplification methodologies have been developed (e.g. use of tyramine derivatives [Speel et al., 1999]). This is why the development of *in situ* amplification (*in situ* PCR) techniques has been so important. These methods allow for the investigation of both tissue sections and cell suspensions, such that nucleic acid targets may be detected in an exquisitely specific manner in their native, tissue- or cell-embedded environment. However, this technology itself also faces a range of technical obstacles, including (1) the fact that DNA targets must be made accessible to PCR enzymes and reagents, (2) PCR inhibitors should be removed or diluted without extensive tissue damage, (3) PCR amplimers must remain at the site of their synthesis, (4) non-specific amplification must be prevented, (5) target DNA degradation may be a problem even before PCR thermocycling begins, and (6) the PCR thermocycling conditions should not destroy the physical status of the sample tissue or its intracellular compartmentalization. Perhaps most important is the initial processing and preparation of (tissue) samples even before PCR thermocycling begins, steps being necessary to control any (pre-) autolytic processes that may occur during post-mortem tissue processing. In this respect, the extent of fixative penetration, duration of fixation, type of fixative used, nuclease and protease inhibition in the specimens, the presence of DNA degrading metal ions, acidification of the fixative, extensive cross-linking of proteins and nucleic acids etc, should all be optimised prior to PCR processing [Golenberg et al., 1996]. In addition, immersion in paraffin, the use of aggressive detergents, the presence of picric acid or mercury containing fixatives, hypo-osmosis, electrostatics, and heat treatment can all frustrate the application of ISA. These considerations are especially important with respect to the preservation of RNAs, and mRNAs in particular, and it is always necessary to control for both false positive and false negative results in ISA protocols. It should also be remembered that the effect of fixation is not always beneficial, as the use of fresh specimens is advisable in certain situations [Jackson et al., 1990].

This chapter aims to summarise the most important considerations for the implementation of ISA (*in situ* PCR) protocols in the clinical laboratory. Advances in both tissue processing and ISA protocols have helped to establish a technology that generates reliable results, provided that tissue sections are handled correctly, before, during and after ISA processing.

13.1 Tissue Processing – Nucleic Acid Fixation/Extraction

Tissue processing comprises the complete series of actions required to transform fresh (tissue) specimens into permanent, storable preparations that are well prepared for subsequent microscopic and ISA analysis. The most important variables, affecting the outcome of the fixation process on the whole, are the temperature of the fixatives (Fig. 13.1), their chemical composition, the fixation method (perfusion or immersion), the nature of the tissue or cell preparation, and the application of dehydration steps. If the material to be fixed is ultimately destined to be embedded in paraffin, then it will have to endure prolonged contact with fixatives, dehydrating compounds, organic solvents, and liquid paraffin (at 56°C). Many of these steps have the capacity to damage nucleic acids, sometimes to such an extent that successful ISH and ISA amplification is not possible at all. When processing tissue samples, it is necessary to consider whether ISH, ISA or both techniques will be used, largely because there is as yet no standard single protocol by which all (pathological) materials may be successfully processed.

13.1.1 Fixation

Efficient "fixation" of nucleic acids in tissue specimens helps guarantee their relatively intact retention during tissue processing and is essential to prevent their large-scale degradation through hydrolysis by cellular nucleases. Several different methods for fixing nucleic acids are currently available, including:

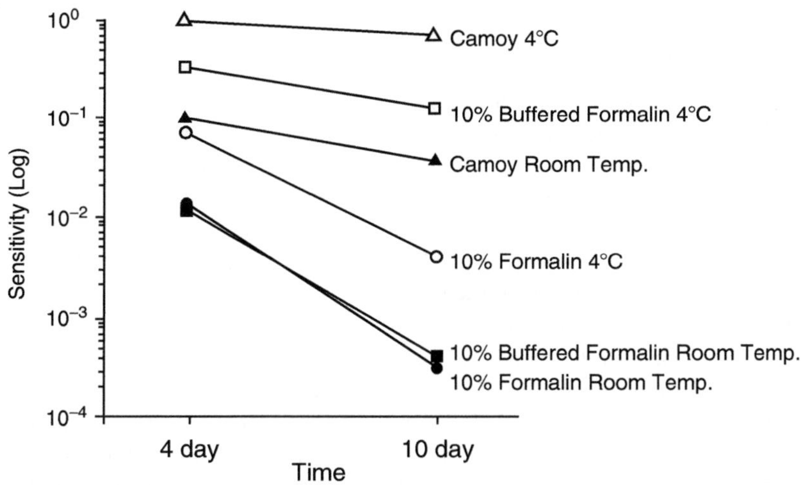

Fig. 13.1 Effect of fixation at 4°C on the sensitivity of PCR. Normal spleen was fixed in three kinds of fixatives at 4°C for 4 or 10 days. The sensitivity of PCR was calculated and the values at 4°C were plotted (open symbols). For comparison the values obtained at an incubation temperature of 25°C were also plotted (closed symbols) (From Honma et al., 1993).

Freezing – the most widely used and also most straight forward physical form of fixation, with various studies having shown that fixation at low temperature improves the quality of nucleic acid templates, possibly due to the combined effects of nuclease inhibition and the prevention of (pre)autolytic catabolism (Fig. 13.1). Best results are obtained by using very small tissue specimens and by immediately storing the frozen samples at –70°C, a temperature well below the re-crystallisation point of most proteins. "Snap" freezing (freezing within only a few seconds) in the presence of cryo-preservatives such as isopentane allows for the excellent preservation of both labile mRNA and genomic DNA within their intracellular environments. Material stored in –70°C archives, may be sectioned using a cryomicrotome at temperatures of –15°C to –25°C. Finally, frozen or lyophilised materials yield excellent quantities of nucleic acids in standard DNA or RNA extraction protocols (for use in standard PCR protocols if required).

Chemical fixation – involves the modification of cell- and tissue-associated compounds by precipitation or polymerisation using reactive chemicals. This normally prevents tissue degradation due to catabolic reactions that may occur after the cessation of energy production, or the lack of a blood supply.

Coagulating fixatives – include methanol, ethanol or acetic acid/ethanol mixtures, which act by removing the water from cells and by precipitating proteins, though nucleic acids will co-precipitate in the presence of excess protein. Non-dissociated acetic acid and other acids are especially useful in precipitating chromosomal proteins without chemically cross-linking the proteins to the chromosomal

INTERMEZZO

Chapter 13 is partially based on results obtained during our own model studies on HPV16 containing cell lines (CasKi ATCC CRL 1550, SiHa ATCC HTB25 and HeLa 229 ATCC CCL-2.1, containing 300, 1 and 0 copies of HPV16 DNA per cell, respectively; the HeLa line contains 25 copies of HPV18 per cell). In addition, an epitheloid (PTEC), a lung carcinoma (A495) and fibroblast (Helgron) cell line was used. Cultured cells were grown until confluence, harvested by scraping and washed in Hank's balanced salt solution. The cell pellet was then immersed in ice-cold fixative of choice (n = 8). Cells were fixed at 4°C, dehydrated using two xylene steps and embedded in Paraplast tissue embedding material at 56°C. Cells to be used in standard PCRs were brought to 70% ethanol after 2, 4, 20 or 168 hours of incubation in the fixative of choice. The HPV16 specific PCR generated a 395 basepairs long product. For *in situ* PCR, sections of 8 µm thickness were mounted on silane coated IS-PCR slides using the Perkin Elmer InSitu PCR System 1000 assembly tool and Amplicover discs and clips. Usually *in situ* PCR was performed using a wet start procedure. This means that the samples were loaded on the slides in 100 µl of the PCR mixture without the primers and Taq polymerase. This was replaced by the complete amplification mixture at the start of PCR. Slides were subjected to various ISA conditions, including: (1) different pretreatment strategies, section thickness, low melting-point agarose, incubation in HCl, use of detergents, DTT, lecithin and/or triethanolamine – acetic anhydride combinations, (2) different PCR protocols – addition of oligo dT, herring sperm carrier DNA, bovine tRNA, bovine serum albumin, hotstart PCR, annealing temperature (T_m), number of thermocycles, and (3) different post-PCR washes. Cycling was performed on a Gene Amp *In Situ* PCR System 1000 thermocycler. Synthesis of non-specifically amplified DNA was assessed by immunohistochemical staining of incorporated DIG-labelled nucleotides in the negative control samples that were subjected to IS-PCR in the absence of primers.

General protocol: Deparaffinized sections of the cell pellet were first preequilibrated in a buffer containing 20 mM Tris.HCl, 10 mM EDTA for 10 minutes, followed by the addition of an appropriate amount of proteinase K dissolved in the same buffer (the proteinase K concentration being predefined by the fixation procedure used). Digestion was stopped by the addition of 0.2% glycin in phosphate buffered saline. HPV16 specific PCR was performed (*In Situ* PCR Core Kit, Boehringer Mannheim, Germany) using 17 cycles of PCR (soak file: 3 minutes 70°C, and cycling: 1 minute 94°C, 1.5 minutes 55°C) in the presence of 4.5 mM magnesium chloride, and using DIG-dUTP as a label and HPV specific primers to generate amplicons of 96 (Q1/2) and 395 (Q1/Q3) bp (kindly supplied by Dr. W.G.V. Quint, DDL, Delft, The Netherlands), as well as 119 bp (HPV 16 primer 1 and primer 2 and OL16) [O'Leary, 1994]. PCR amplimers were immediately fixed by

(continued)

coagulation in ice cold 100% ethanol and washed in 80% ethanol at room temperature. DIG was detected using Fab-anti-DIG antibodies and alkaline phosphatase as a reporter molecule. The enzyme was detected via NBT/BCIP catalysis under standard conditions. To asses non- specific DNA synthesis and/or diffusion of amplicons, the PCR mixture overlaying the sections was submitted to agarose gel electrophoresis and ethidium bromide staining. This procedure was used as a quality control indicator for each ISA assay.

nucleic acid. Ethanol in amounts higher than 70% precipitates pure DNA at −20°C or −70°C (the lower the temperature, the better the precipitation), which is further enhanced by addition of salts such as sodium acetate. This process is reversible, which facilitates the extraction of relatively intact nucleic acids from rehydrated paraffin sections. Fixation in ethanol or acetone at room temperature is widely used for histological purposes. Mixtures of coagulating fixatives (ethanol/acetic acid, ethanol/acetic acid/chlorophorm (Carnoy's fixative), methonol/chlorophorm/acetic acid (Methacarn)) beautifully fix chromosomes in mitotic nuclei, and chromatin in interphase nuclei, by precipitation of the chromosomal proteins without chemical cross-linking. mRNA tends to survive fixation using these coagulating fixatives in good condition (Fig. 13.2) [Urielli et al., 1992; Shibutani et al., 2000], and there also exists commercially available coagulating fixatives of unknown composition, that essentially show the same qualities [Ben-Ezra et al., 1991; Giannella et al., 1997]. Finally, proteins subjected to coagulating fixatives remain susceptible to proteolytic digestion once extracted from the cellular environment. This property is advantageous with respect to nucleic acid extraction protocols for non-*in situ* PCR methodologies. Coagulating fixatives are the fixatives of choice when performing *in situ* RT-PCR [Koopmans et al., 1993; Goldsworthy et al., 1999; Shibutani et al., 2000].

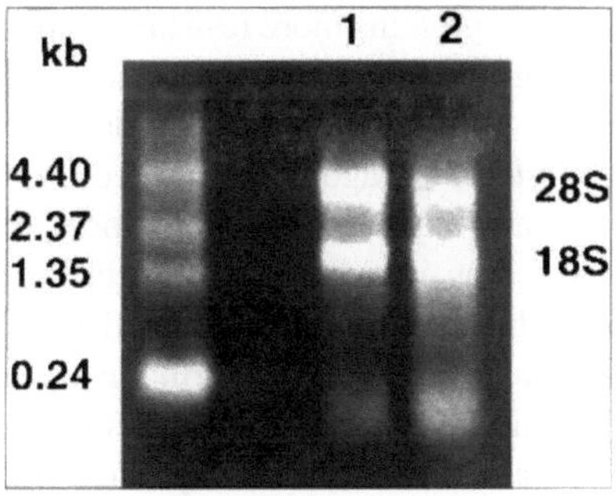

Fig. 13.2 Denaturing agarose electrophoresis of total RNA isolated from paraffin-embedded methacarn fixed tissue (lane 2) compared with a similar extract from a frozen section (lane 1). On the left molecular size markers are indicated in kilobases (Reproduced from Shibutani et al., 2000. With permission from Nature Publishing Group)

Cross-linking fixatives – include aldehyde-type fixatives (paraformaldehyde, glutaraldehyde) and oxidative-type fixatives (osmium tetraoxide, mercury dichloride) [Hopwood, 1975]. The first compounds react through redox reactions generating a multitude of chemical by-products [Kiernan, 1999]. Paraformaldehyde at 4% in 0.05–0.1 M phosphate buffer pH 7.4 and used at 4°C is excellent for fixing small tissue samples, with neither short nor prolonged fixation times interfering with normal PCR efficiency. Due to their intrinsic structural association with nuclear/cellular proteins, most nucleic acid molecules may be physically trapped within protein structures via the use of these cross-linking fixatives. This form of physical entrapment works best for genomic DNA due to its nuclear localisation and intricate association with nuclear histone proteins. However, extensive entrapment will result in nucleic acids becoming more insoluble and less accessible to *in situ* amplification reagents. Hence, it is often necessary to achieve a balance between entrapment and solubility, especially with respect to mRNA molecules. Fixation by perfusion is the most effective method when using cross-linking fixatives, though perfusion cannot be applied directly to clinical specimens. Unlike proteins, nucleic acids generally display little reactivity towards even the strongest cross-linking fixatives such as gluteraldehyde or osmium tetraoxide. This is most probably due to the coating proteins that are initially modified and as such protect nucleic acids from further reactions. Formalin is a crosslinking fixative that induces methylene bridges between histones [Mason et al., 1985], histones and DNA or DNA–DNA complexes [Feldman, 1973], a process that increases the apparent melting temperature of the embedded DNA [Brutlag, 1980]. Formalin is effective at concentrations as low as 0.2%, with the resultant degree of crosslinking, in for example purified histones, able to completely block any attempts at mild proteolytic digestion (when applying 20 μg per ml of pronase [Brutlag et al., 1969]). However, in order to be chemically modified by formalin, the double stranded helix of DNA molecules must first be denatured, which means that DNA is only fixed by formalin at temperatures above the melting temperature of DNA [McGhee and Hippel, 1977]. Initially, AT-rich, or unstable regions are modified, including hydroxymethylation, a process that may also occur in limited amounts at room temperature [Karlsen et al., 1994; Nuovo et al., 1994]. Several publications have clearly demonstrated the relationship between duration of formalin fixation and loss of large amplifiable DNA fragments, and that both DNA and RNA quality diminish when incubations longer than 12 hours (overnight) are used [Greer et al., 1991a, b; Ben-Ezra et al., 1991; Karlsen et al., 1994; Masuda, 1999], However, not all tissue types are equally vulnerable. Long-term fixation in formalin renders proteins more resistant to the action of proteinases and also leads to irreversible changes in the nucleic acids themselves. This renders ISA and conventional PCR ineffective and requires special adaptations to these experimental protocols. The fact that long-term storage in unbuffered formalin was widely used, was a major frustration in the early days of ISA development, though special adaptations to the experimental protocol may in specific circumstances help overcome this problem. For example, it is possible to identify genomic HPV16 DNA and its transcription products even after one week of buffered 4% formalin fixation (Fig. 13.3). Interestingly, viral and bacterial nucleic acids appear to be less susceptible to the deleterious effects that may occur from formalin fixation, probably due to these nucleic acids being less intimately associated with proteins.

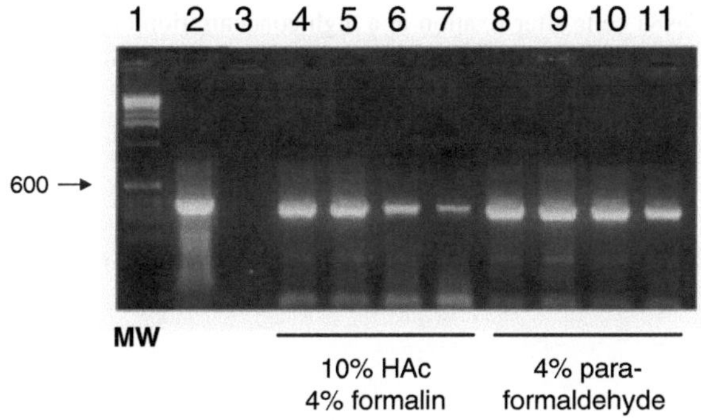

Fig. 13.3 HPV16 PCR on a DNA extract from sections containing CasKi cells after fixation in 10% acetic acid and 4% commercial formalin (lanes 4–7), or 4% fresh paraformaldehyde buffered in 0.05 M phosphate at pH 7.4 (lanes 8–11). From left to right – the tissue used for the 4 PCR amplification reactions were fixed for 2 hours, 4 hours, 20 hours and 7 days, respectively. Lane 2 contains amplicons derived from a positive plasmid control template, lane 3 a water negative control. Lane 1, marked MW at the bottom, shows a molecular weight marker with the 600 base-pair molecule highlighted by an arrow

Metal-containing fixatives (mercury, copper, chromium and osmium salts included) yield varying results when *in situ* amplification assays are performed. Although initially used for their excellent morphology-conserving properties and maintenance of immunogenicity, these fixatives now have limited popularity. The DNA damaging effects of these fixatives may be partially reversed by desublimation, as may the application of Lugol's iodine solutions in combination with sodium thiosulphate, and a short post-fixation step of less than 2 hours with buffered formalin [Sukpanichnant et al., 1993]. However, other authors failed to reproduce these optimisation steps [Tbakhi et al., 1999]. The same authors report that the level of inhibition is dependant on the metal used, with zinc ions for example, being far less detrimental than mercury ions.

13.1.2 Type of Nucleic Acid

Essential differences exist in the effect that particular fixatives have on either RNA or DNA nucleic acids. Eukaryotic genomic DNA may readily be fixed within histone complexes, and ribosomal RNA (rRNA) is usually well protected in the ribosomes in which they are embedded. However, the fixation of mRNA may be more problematical, with the time allowed for fixation being critical [Koopmans et al., 1993]. The use of low amounts of crosslinking fixatives such as glutardialdehyde has been shown to improve the retention of mRNA and DNA templates *in situ* [Moench et al., 1985; Urieli et al., 1992] (Fig. 13.4), and mercury containing fixatives are also recommended for use with RNA-specific tests [Foss et al., 1994; Shibutani et al., 2000]. It should also be noted that the ratio of amplifiable RNA and DNA

Fig. 13.4 Human CasKi cells after fixation in a high concentration of glutaraldehyde (2.5%). Southern blot using a radioactive phosphorus HPV16 probe after PCR amplification of HPV16 DNA. Fixation times were 2, 4, 20 and 7 days (lanes 1–4, respectively). This type of test is semi-quantitative at best

present in a processed tissue sample may differ widely dependant on the type of fixative used, and that the type of fixative may influence whether genomic DNA is "unremovable" from tissue sections required for RT-PCR. Such RT-PCR testing using fixed pathological specimens should be performed with extreme care and its outcome should be subjected to critical analysis using appropriate internal positive controls, e.g. housekeeping gene mRNA targets (Section 8.2).

Both viral and bacterial nucleic acids tend to be associated with proteins to a lesser extent than their eukaryotic counterparts, and the *in situ* detection of microorganisms in tissue sections is an important option in the detection of infection. For example, *Mycobacterium leprae*, causal agent of leprosy, can be easily detected in paraffin embedded tissue sections, whether the material has been fixed in coagulating or precipitating fixatives [Fiallo et al., 1992]. For viruses, the use of acetone or formalin containing fixatives appeared to be equally successful [Koopmans et al., 1993]. It has to be remembered that contamination of tissue sections with extraneous microorganisms, derived from the laboratory environment itself, may complicate the interpretation of *in situ* PCR results.

An adequate means for assessing the integrity of nucleic acids isolated from paraffin embedded tissue sections is to perform simple agarose gel electrophoresis of a portion of the extract [Jackson et al., 1990]. Denaturing agarose electrophoresis is the most useful option for assessing the quality of RNA extracts, with the visible presence of the small and large subunit ribosomal RNA molecules indicating a good quality extract (Fig. 13.2) [Masuda, 1999; Shibutani et al., 2000]. However, the presence of intact ribosomal RNA does not always correlate to the presence of intact mRNA molecules, which tend to be less stable.

13.1.3 Detrimental Effects of Various Fixatives on Nucleic Acids

Several studies have compared the effect of various tissue fixatives on the amplification of nucleic acids [Bramwell et al., 1998; Ben-Ezra et al., 1991; Giannella et al., 1997], such that it is now known that fixatives deleteriously affect nucleic acids via a variety of different mechanisms. For example, low pH, and the presence of picric acid and mercury ions, all result in the hydrolysis of nucleic acids molecules, which may lead to molecule fragmentation or the formation of adducts, all in a time and concentration dependant manner [Dubeau et al., 1986; Nuovo and Silverstein, 1988; Beland et al., 1984; Bramwell and Burns, 1988; Greer et al., 1991a, b, 1994]. Formalin is a denaturing agent that chemically modifies (by the addition of methylol groups) exocyclic aminogroups (A, C and G) and iminogroups (T and U) of both DNA and RNA [McGhee and Hippel, 1975, 1977]. The reaction velocity of denaturation is dependant on fixation temperature and formalin concentration, but is deemed to be quite low in general, with total inhibition of amplification occurring when the fixation reaction is allowed to proceed for more than 1 week. This effect is due to the slow condensation reaction, in contrast to the high-speed adduction products generated by glutaraldehyde, a reaction that reaches completion in 2 hours [Pearse, 1980]. The preferential addition of methylol groups to adenine and cytosine bases by formalin [Masuda, 1999], has led to the suggestion that isolation of mRNA from formalin fixed material and subsequent cDNA synthesis via the poly-A tail (using oligo-dT primers) may not be optimal. Formalin does not lead to excessive DNA degradation and it normally leaves fragments larger than 30 kilobasepairs intact [Karlsen et al., 1994], though this is only the case when the formalin is buffered. In non-buffered formalin solutions, the average length of degraded fragments will be less than 1,000 basepairs in length, if still present at all, with the quantity of high molecular weight DNA being reduced to 57% of the initial amount within 12 hours and only 35% after 36 hours [Dubeau et al., 1986; Inoue et al., 1996]. Formic acid forms in all non-buffered formalin solutions, causing not only chemical-based but also low-pH mediated damage, though extended fixation in formalin solutions may be attempted for (RT)PCR assays where the target DNA or RNA is less than 100 base pairs in length [Kleter et al., 1998; Koopmans et al., 1993]. However, many collections of pathological specimens have been processed using formalin-based protocols and it will be impossible to use all of these collections for molecular analyses, due to nucleic acid degradation [Tournier et al., 1987; Tokuda et al., 1990; Honma et al., 1993].

Until approximately 20 years ago, picric acid containing fixatives (e.g. Bouin) were quite popular as they yield well-conserved morphology and excellent staining in for example "trichrome" staining protocols. However, picric acid hydrolyses nucleic acids to approximately 100 basepairs long fragments [Nuovo and Silverstein, 1988; Greer et al., 1991a, b; Tbakhi et al., 1999]. Only a very specific extraction methodology using lithium carbonate and amplification of very short target sequences tends to facilitate PCR using picric acid containing fixatives [Longy et al., 1997]. Consequently, Bouin treated specimens are not suitable for *in situ* PCR applications, except when tissues have been fixed for 6 hours or less [Camilleri-Briet et al., 2000].

Chemical modification of nucleic acids also induces steric (conformational) hindrance, resulting in reduced primer annealing, both in reverse transcription and in PCR amplification reactions. Further, DNA polymerase may actually "fall off" of its target DNA during replication, or may "cross over" to another strand altogether, producing bizarre PCR products resulting in false positive or negative results [Greer et al., 1994]. The influence of this steric hindrance may be reduced by using an extended pre-denaturation step at 94°C prior to commencement of PCR thermocycling.

As an important note, nucleic acid fragmentation seems to be a universal characteristic of all currently known fixatives, though in some cases, fragmentation can be advantageous (similar to limited restriction enzyme digestion) in improving PCR yield by improving accessibility of PCR reaction ingredients to the target nucleic acid [Kitazawa et al., 1989].

13.1.4 Effects of Tissue Processing Steps (Decalcification, Dehydration, Intermedium Application, Embedding) and Storage of Paraffin Blocks

The gradual loss of nucleic acid molecules is largely unavoidable during both tissue processing and long-term storage of fixed specimens. These processes result in a cumulative reduction of the amount of target nucleic acid present, which may be especially troublesome with respect to retrospective studies using archived specimens stored for five years or more. Of course, the successful analysis of such archived material would help provide an insight into the historical prevalence and pathogenicity of existing diseases, novel diseases, medical syndromes, and infectious diseases [Hewett et al., 1994]. Therefore, knowledge regarding the effect of various tissue-processing steps during specimen archiving is important in attempting to limit the amount of nucleic acid damage that occurs.

Decalcification, may be required in order to remove tough, calcified parts from tissues (e.g. bone fractions in cartilage) so that such tissues may be successfully embedded in paraffin. Decalcification using the chelator EDTA does not harm nucleic acids, though the use of faster acting and more aggressive agents such as formic acid, hydrochloric acid and nitrous acid result in large scale DNA degradation, possibly inducing false negative amplification results. Nucleic acid extraction from paraffin embedded tissues that have been decalcified using EDTA is relatively easy, especially if specific DNA isolation kits are used (e.g. Qiagen QIAamp tissue kit) [Wickham et al., 2000].

Dehydration and intermedium compounds are used in the process of substituting water for paraffin in pathological specimens, and as such have a profound effect on both lipid and protein integrity. Lipids are hydrophobic molecules that are most efficiently extracted using toluene and xylene, or less efficiently with ethanol or acetone. These agents permeabilize the cell membranes, rendering their internal compartments available for hydrophilic agents. If dehydration is performed above 10°C, conformational changes in cellular proteins may also occur, with even higher temperatures, e.g. 56°C (the temperature at which paraffin liquefies), promoting the complete denaturation of proteins. The use of such dehydration and intermedium reagents does not lead to the destruction of DNA, although extreme desiccation will lead to a certain degree of non-specific fragmentation, and removal of the protective water "shell" surrounding the DNA may lead to conformational changes, especially if elevated temperatures (56°C) are subsequently used. In practice, ethanol is the most frequently used dehydrating reagent, being used in essentially all paraffin-embedding protocols. Reagents such as the dehydrating acetone enhance protein coagulation if used simultaneously with other coagulating fixatives, which in ISH methodologies may lead to the loss of a hybridization focus and hence hybridization localization detail. However, yields of extracted DNA (to be used for example in non-IS-PCR amplification methodologies) are significantly improved using this treatment. The presence of any "coagulation" effects, which may occur during dehydration treatment, may be minimised by incorporating a preceding cross-linking formalin fixation step for a minimum period of 24 hours prior to dehydration.

Embedding fixed tissues in paraffin also results in the loss of nucleic acids and PCR efficiency, especially if protocols using an extended period are used. Any fragmentation of nucleic acids during this phase is probably due to the effect of shearing forces that can occur during the embedding process. The application of low-melting-point paraffin sometimes enables the detection of nucleic acid molecules that would otherwise be lost using normal melt-point paraffin [Pavelic et al., 1996]. It is known that long-term storage of paraffin blocks leads to progressive loss of nucleic acid, though this effect is not necessarily detrimental to *in situ* PCR, as freshly cut sections of stored paraffin embedded tissues contain amplifiable DNA even after years of storage. The greatest success occurs when short targets of maximum 250–300 bp are amplified [Tokuda et al., 1990; Greer et al., 1994; Hewett et al., 1994]. However, if storage exceeds 15 years, then a notable reduction in PCR efficiency becomes apparent, though some investigators report successful PCR amplification on samples stored for up to 40 years [Chuaqui et al., 1999, Howe et al., 1997; Gall et al., 1993]. The detrimental effect of storage is more obvious for RNA templates [Tournier et al., 1987; Finke et al., 1993], as even a relatively short storage period of 3 years, results in a significant drop in the RT-PCR sensitivity. This effect seriously hinders the applicability of RT-PCR to many archival materials. In any case, it is important that the RNA extraction protocol be as efficient as possible, [Masuda, 1999].

13.1.5 Effects of Histological and Histochemical Staining

Microdissection is a technology whereby samples for PCR amplification may be collected from histological or cytological specimens and fixed onto microscopy slides. Cells of interest can be collected manually (using a scalpel) or by laser capture techniques [Shibutani et al., 2000; Sukpanichnant et al., 1993; Rubin, 2001]. This methodology has been extremely successful for the molecular analysis of Papanicolaou stained cell smears derived from cervical swab specimens, especially when organic stains, including hematoxylin, have been removed by extraction [Greer et al., 1994]. Hematoxylin inhibits PCR more significantly than methyl green, crystal violet or nuclear fast red [Burton et al., 1998]. Microdissection of materials that have not been de-paraffinised improves the ease of manipulation.

Successful PCR on immunophenotype-specific cell groups has been described [Alcock et al., 1999]. This shows that immunophenotyping does not generally deleteriously affect DNA quality.

13.2 Differences in Approach for ISH, ISA and Standard PCR

Usually, molecular techniques are performed with microtome-cut tissue sections of 4–15 μm as starting materials. For standard PCR amplifications, one or more of these sections are collected in individual Eppendorf tubes, the DNA or RNA extracted, and PCR amplification performed. For ISH and ISA applications, the tissue sections are immobilised on microscope slides and testing performed with the cellular and tissue matrix intact, allowing signal to accumulate at the actual cellular site where the target DNA/RNA is situated. ISA, ISH and PCR protocols require different strategies with respect to the fixation of nucleic acid in tissue specimens; the retention and fixation of target molecules at their original cellular location being an absolute requirement for ISH and ISA protocols. Moreover, the fixation process should be able to withstand the aggressive steps required for ISH/ISA amplification and detection, yet should not generate cross-linking that is too extensive and which may limit the accessibility of ISH/ISA reaction ingredients. In contrast, standard PCR testing mainly requires that the fixation process does not cause extensive nucleic acid hydrolysis or fragmentation (N.B. too extensive cross-linking of the nucleic acid molecules will also inhibit PCR efficiency). Our own studies have shown that ethanol/acetic acid fixation generates excellent retention of nucleic acids for PCR, irrespective of the duration of fixation, and is equally as reliable as formalin and paraformaldehyde fixation. Gluteraldehyde fixation generates similar results, though both approaches were found to be less successful using ISH on the same material. Until now there has only been a single fixative described that supports RNA ISH, DNA ISH, RT-PCR and standard PCR protocols, namely: immersion in acetic acid/formalin under ‘physiological salt’ conditions. The fixation time used may vary between 2 hours and 7 days. This protocol does not allow for ISA applications.

13.2.1 Different Types of Tissue Preparations

A variety of differently processed tissue specimens are currently utilised in ISH, ISA and PCR analysis, including frozen and fixed specimens. Frozen tissue sections may be used without processing for direct nucleic acid extraction and PCR studies, but will need to be fixed (coagulating fixatives are routinely used) before ISH methodologies can be attempted [Brooks et al., 1993; Chiu et al., 1992; Patel et al., 1994]. If microtomy is used, then DNA target molecules will most likely be directly accessible to probes without any further pre-treatment, since the plasma membrane will have been most likely sliced open during processing. When using ISH protocols to detect RNA molecules, specimens will have to be first fixed using strongly cross-linking fixatives, e.g. formalin or glutardialdehyde [Moench et al., 1985; Urieli-Shoval et al., 1992; Lloyd et al., 1991]. The use of such strongly cross-linking fixatives may require probe adaptation as high degrees of cross-linking necessitate smaller probe molecules in order to efficiently penetrate and permeate through the tissue section. Formalin fixed frozen sections rather than paraffin embedded sections have been recommended for use in ISA methodologies [Nuovo, 1994]. This negates the requirement for paraffin infiltration at elevated temperatures that can lead to breakage of DNA and subsequent problems related to DNA repair activity. Relatively few groups have reported successful ISA protocols for formalin fixed paraffin embedded archival tissues [Chiu et al., 1992; Embretson, 1994; Long, 1993; Nuovo et al., 1991b; Nuovo et al., 1992a; Nuovo, 1993]. Papanicolau stained cytological smears are excellent sources of nucleic acid for PCR, though in such circumstances, the cells will first need to be scraped from the glass matrix prior to nucleic acid extraction. Cytological preparations in the form of cytospins or smears on glass slides may be destained or microdissected and successfully used for ISA studies (Cackwell 2000). Tests have also been developed for cell suspensions *per se*. [Bagasra, 1993; Chiu et al., 1992; Haase et al., 1990; Komminoth, 1992; Long, 1993; Patterson et al., 1993; Seidman et al., 1994].

13.2.2 DNA and RT-PCR on Paraffin-Embedded Tissue Sections

Paraffin is a hydrophobic compound which does not mix with normal extraction buffers, and which hinders efficient nucleic acid extraction. An essential first step in the isolation protocol is therefore the use of organic solvents to dissolve the paraffin, followed by a series of extractions and centrifugation procedures prior to the addition of substitution buffer. Such procedures, besides being labour intensive, are also prone to contamination, since reaction vessels need to be continually opened and closed. Alternatively, paraffin may be melted by immersing tissue sections in extraction buffer, which is then boiled or subjected to microwave-induced heating (45 seconds at 500W). After a short cooling period the solidified paraffin will float to the top of the extraction buffer, where it can be easily removed. At the same time, proteolytic enzymes may be added [Banerjee et al., 1995]. However, boiling temperatures can damage DNA, so heating at lower temperatures (37–56°C) with overnight incubation in buffers containing proteinase K, SDS, EDTA or Triton X-100 may be an alternative.

Prior to standard PCR amplification, fresh tissues are usually digested with proteolytic enzymes in the presence of detergent and an adequate buffer containing additional stabilising compounds. In contrast, the use of fixed tissue sections in standard PCR amplification protocols usually requires a more aggressive approach to proteolytic digestion than is customary using fresh tissue preparations (though the direct amplification of DNA from formalin fixed, paraffin wax embedded tissue sections has been reported [Cawkwell and Quirke, 2000]). This is the result of protein cross-linking during fixation, which protects contaminating proteins from proteolysis. In this case, additional heating steps will also be required in order to efficiently denature any DNA/protein complexes and inactivate the proteinases that have been added [Frank et al., 1996]. Prolonged protease treatment has been reported to increase the overall yield of DNA for PCR, but decrease the integrity and the average molecular length [Bramwell and Burns, 1988; Jackson et al., 1990; Honma et al., 1993; Sepp et al., 1996]. The use of elevated salt concentrations may improve phase separation between DNA and protein [Howe et al., 1997], and phenol extraction may also be helpful (phenol extraction also removes some PCR inhibitors [An and Fleming, 1991]). However, DNA present within regions of high heterochromatin concentration are barely extractable [Dubeau et al., 1986], a problem compounded by increased storage times. The use of fixed tissues for PCR means that detergents are not required in the DNA extraction step, as the intermedium treatment and dehydration steps used will have already effectively removed any cellular lipids. However, detergents do affect surface tension and may help the nucleic acid extraction process. In order to obtain reliable PCR results from fixed tissue specimens an extra series of PCR cycles or the use of a nested PCR protocol is usually necessary unless the intended PCR amplimers are extremely short [Karlsen et al., 1994].

With respect to RT-PCR, it has been demonstrated that 3–6 extractions may significantly increase the yield of RNA for standard PCR [Camilleri et al., 2000]. However, changes induced by tissue fixation (especially protein cross-linking) increases the probability of DNA contamination, which means that there is an increased probability of false positive results. Therefore, appropriate controls, or the use of relevant primers (e.g. primers which target two separate exons) are mandatory [Weireich et al., 1997]. Various commercial isolation kits are available for extracting nucleic acids from archived tissue for PCR analysis.

Finally, remnants of the original tissue fixation process may lead to inactivation of thermostable DNA polymerases. Aldehyde-based fixatives for example block the active site of the polymerase enzyme and fixation may result in the generation of new PCR inhibitory compounds. In order to prevent this, either efficient or adequate washing steps need to be implemented immediately after the fixation process or during nucleic acid extraction.

13.2.3 Improvement of PCR Efficiency Using Fixed Tissue Sections

Several methods have been reported to improve the efficiency of PCR amplification of nucleic acids extracted from fixed tissue sections. For example, it has been shown that PCR results obtained from the microdissection of Bouin-fixed tissues are improved if HPLC purified primers are used [Weirich et al., 1997]. Further, it

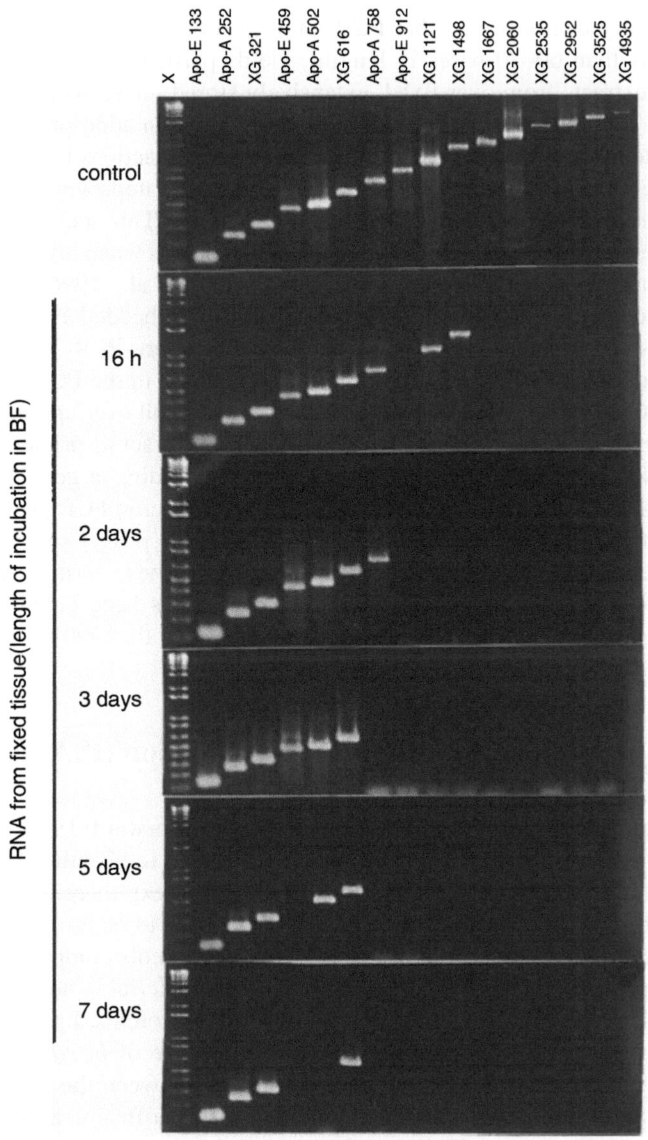

Fig. 13.5 Examples of RT PCR results using RNAs retrieved in various manners from liver tissue incubated in BNF for various lengths of time. Sixteen primer pairs, which amplify 133–4,035 base-pairs from three mRNAs (apolipoprotein ApoA; apolipoprotein E, ApoE; alpha-2 macroglobulin, MG), were tested. The target mRNA and expected product size are given above each lane. Each panel represents the indicated length of incubation in BF before dehydatation and storing at 4°C for 1 month. The RNA from fresh frozen liver tissue was used as a control. M, 1 kilobase DNA ladder (Reproduced from Masuda et al., 1999. With permission from Oxford University Press)

has been clearly demonstrated that PCR amplification of short stretches of DNA (down to less than 60–70 bases in length), yields particularly good results, even when DNA extracts from over-fixed, extensively stored materials are used [Kleter et al., 1998; Kleter et al., 1999; Greer et al., 1991a, b]. In addition, methods have been presented that allow for reversion of the chemical reactions taking place upon fixation (Fig. 13.5) [Masuda, 1999]. Another option for improving PCR yields is the application of reconstructive PCR methodologies. This technique has been especially developed to increase the yield of PCR products when highly fragmented DNA preparations are to be amplified [Golenberg et al., 1996]. The method combines well with fragmented DNA from paraffin-embedded fixed tissue samples. A key step in this technique is the application of a pre-PCR 20-cycle thermocycling protocol where primers are not initially included in the PCR reaction mix. This means that (once melted into single strands) any small overlapping DNA fragments within the nucleic acid extraction will themselves act as primers for double stranded DNA amplification by the Taq polymerase, yielding larger DNA products which may then be used in a standard PCR (i.e. with specific PCR primers added). Larger fragments can also be synthesised using multiple primer sets that generate DNA segments with overlapping cohesive termini. These termini then act as primers for further PCR amplification. This process has been coupled with ISH detection to detect lentiviral DNA within cells [Haase et al., 1990].

13.3 An Introduction to *In Situ* Amplification (ISA)

The principal description of ISA is simple and straightforward: ISA comprises all methods based on DNA or RNA amplification that aim to visualise limited numbers of nucleic acid molecules within their cellular context, thereby defining their tissue distribution. ISA methods are usually considered to be more sensitive than ISH approaches. The main problem faced by ISA protocols compared to conventional PCR protocols, is that the nucleic acid target material is not purified away from potential contaminants and PCR inhibitors, e.g. proteins, lipids and mucous substances, prior to ISA processing. Therefore, the use of *in situ* amplification techniques requires an efficient balance to be made between the conservation of tissue/cellular morphology and the maintenance of sufficient accessibility for PCR primers and enzymes. However, increasing the tissue accessibility for PCR primers and enzymes results in a concomitant increase in the possibility that target nucleic acid will diffuse away from its original site. Further, as detailed in previous sections, the processing of tissues and cells in order to fix nucleic acids at their respective tissue locations will also influence the success rate of *in situ* amplification assays.

General ISA methodologies involve the use of cells or tissue specimens that have been fixed to for example glass slides, and which are subjected to limited proteolysis followed by specialised IS-PCR protocols using dedicated IS-PCR thermocyclers. Any amplimers synthesised will remain, at least in part, at the site of their synthesis due to electrostatic attraction, allowing their localization to be determined using post-PCR detection techniques. In fact, amplimer retention occurs through electrostatic interaction between positively charged proteins and amplimers such that the highest affinity of retention is displayed by cytoplasmic proteins and to a lesser extent chromosomal proteins. Therefore, it is essential to keep proteins in their correct conformational state during tissue processing, as native, fully hydrated proteins seem to have the highest affinity for PCR amplimers. Various experiments have indicated that ISA is far less susceptible to nucleic acid contamination than classical PCR techniques, largely due to the fact that unlike amplimers derived from the intended target nucleic acid, non-specific amplimers derived from contaminating nucleic acid will be dispersed throughout the ISA reaction medium and not remain localized to a specific cellular location.

There are two basic variants of the ISA technique, the most essential difference being the period when amplimer labelling occurs. During *in situ* PCR (IS-PCR) a label is incorporated directly during PCR thermocycling *per se*, which may be achieved by incorporating labelled PCR primers or labelled nucleotides in the PCR reaction mix. This label may then be directly visualised after PCR thermocycling using direct fluorescence or immunohistochemical procedures [Stein et al., 1997]. The specificity of the labelling reaction is mainly determined during the annealing step and requires that stringent and optimized PCR conditions (e.g. hot-start procedure, high annealing temperatures) be used. As with all PCR protocols, negative controls are essential in IS-PCR, especially if labelled nucleotides are being used. Basically, labelled nucleotides may be incorporated into non-specific sites by Taq polymerase resulting in false positive results and false localization of the intended target sequence. In the second ISA variant, PCR *in situ* hybridisation or PCR-ISH, *in situ* PCR thermocycling is performed first and the immobilised amplimers are subsequently detected using specifically labelled DNA hybridization probes. Popular PCR-ISH labels include digoxygenin and biotin, though the use of methodologies utilizing immunogold has also been published [Zehbe et al., 1992; Stamps et al., 2003]. Additional positive and negative hybridisation control reactions should of course be incorporated into every PCR-ISH run.

IS-PCR and PCR-ISH protocols are able to specifically detect for example a single viral genome copy per cell, with many clinical applications having been described. These applications include the detection of: the human immunodeficiency virus HIV [Bagasra, 1993; Bagasra et al., 1996; Patterson et al., 1993; Seidman et al., 1994], the human papillomavirus HPV [Walker et al., 1996; Nuovo et al., 1991a, b; Nuovo et al., 1992a, b; Sallstrom et al., 1993; Zehbe et al., 1992], cytomegalovirus CMV [Zawilinska et al., 2006; Ray et al., 1995, 1996], the hepatitis B and C viruses [Shieh et al., 1999; Benkoel et al., 2004], mouse mammary tumour provirus MMTV [Chiu et al., 1992], maedi-visna virus [Carrozza et al., 2003], lentiviruses in general [Haase et al., 1990] and measles virus [Nuovo et al.,

1992c; Ray et al., 1996]. Protocols for the ISA-mediated detection of non-human bovine immunodeficiency virus have been also described [Zhang et al., 1997]. IS-PCR/PCR-ISH techniques have also been utilised to detect chromosomal sequence motifs [Long, 1993] and mRNA molecules [Kaminska et al., 2005; Brooks et al., 1998; Patel et al., 1994].

13.4 Considerations in the Development of ISA Protocols

Many questions need to be answered during the development of *in situ* amplification assays. For example, does one choose IS-PCR or PCR-ISH? Also, which target sequence, type of nucleic acid, specimen type, and fixative should be chosen? What amplimer size is considered most likely to be useful? These are just some of the considerations that need to be taken into account in the development of ISA protocols. Information useful in helping to provide an answer to these questions is presented below.

13.4.1 IS-PCR or PCR-ISH

As mentioned above, the fundamental difference between IS-PCR and PCR-ISH is the stage at which detection labels are incorporated into the PCR amplimer. There is some evidence to suggest that PCR-ISH methodologies are more specific than IS-PCR methodologies due to the additional specificity obtained using a separately labelled hybridization probe. However, the stringency conditions required to allow specific hybridization of the PCR-ISH probe may affect cellular morphology and hence localization of the PCR amplimer. Several authors have shown that IS-PCR is a good option when applied to cell suspensions and formalin-fixed frozen sections, mainly due to the reduced number of experimental steps required and the fact that suppression of the Taq polymerase repair mechanism (which non-specifically incorporates labelled nucleotides into non-target nicked DNA) may be feasible using these specimen types [Nuovo et al., 1994; Long, 1993]. For paraffin embedded tissue sections Nuovo [1994] indicated that the introduction of a dideoxynucleotide in the reaction mix could suppress Taq polymerase-mediated non-specific extension repair mechanisms. The enzymatic incorporation of dideoxynucleotides blocks the 3′ terminus of small DNA fragments, subsequently reducing repair opportunities (see also [O'Leary et al., 1996]).

In conclusion, the major advantage of IS-PCR over PCR-ISH is that these tests, once operational, are less time- and money-consuming. In addition, the use of hazardous chemicals (e.g. formamide) is reduced for IS-PCR protocols. However, PCR-ISH protocols are generally considered to be more specific and more broadly applicable.

13.4.2 Diffusion of Nucleic Acids

Cell and tissue processing drastically affect the accessibility of target nucleic acid for enzymes, primers and other important PCR components. It also influences the success of retaining subsequent PCR amplimers at their original location. Cells from cytospins or in suspension may be used with or without their cellular membranes intact, as methanol or acetone may be used to dissolve lipid membranes. In contrast, cross-linking fixatives render the membrane essentially impenetrable. Tissue sections, without exception, contain at least some cells from which a major part of the cell membrane has been cut away, rendering their nucleic acids quite accessible to PCR reagents. The major disadvantage of this accessibility is that target nucleic acid may be relatively easily washed out of the cellular compartment. Studies using cellular supernatants have shown evidence for the presence of target nucleic acids, which have leaked out of cells, in these solutions [Nuovo et al., 1994]. This process also occurs even if the proteolytic pre-digestion step is minimiized. When using sliced sections, PCR products diffuse in all directions depending on the temperature, concentration gradients and, of course, the density of the intra- and inter-cellular networks. This leakage of PCR products tends to result in extra-cellular, randomly localised PCR products. In fact, low levels of specific amplimers may be found in the supernatant of false negative ISA reactions (i.e. when no signal is observed in a "true positive" specimen). Finally, it is interesting to note that the infection status of a cell may be an important factor in allowing ISA reaction mix ingredients to diffuse into cellular compartments, as cells infected with HIV exhibit different behaviour to similar cells infected with chromosomally integrated human papillomaviruses.

13.4.3 The Correct Fixative

Fixation is one of the most important variables in ISA, being required to provide cellular stability (which will allow the cellular integrity to withstand the harsh ISA or ISH reaction conditions), as well as aiming to block the activity of any nuclease enzymes present within the specimen. Ideally, the least aggressive fixation procedure should always be used. Cross-linking aldehyde fixatives provide the physical rigidity to cope with ISA conditions, but tend to restrict primer/Taq/probe accessibility to the target nucleic acid. The use of such fixatives also presents consequences with respect to histone-histone, histone-DNA and DNA-DNA cross-linking [Moench et al., 1985; Mason et al., 1985; Feldman et al., 1973]. Accessibility to ISA reaction mix products may be investigated by rigorously extracting nucleic acids from the fixed cells, and then performing a standard DNA PCR. If the standard DNA PCR yields a positive signal in the absence of an ISA signal, then ISA reaction mix accessibility is most likely a serious problem [Sagawa, 1997]. Single cells that have been briefly immersed in coagulating fixatives may not be sufficiently stabilised to effectively retain ISA products [Nuovo et al., 1994]. Some of the problems associated with coagulating fixatives can be overcome by adding an overnight fixation step in diluted formalin. Pepsin treatment, to break down the proteinaceous cellular membrane, has also been described as an additional approach. Fixation at 4°C in 4% paraformaldehyde or 10% formalin-acetic acid (FAA) for a period between 2 and 24 hours apparently yields excellent results for ISH (once post-fixation proteinase K pre-treatment had been optimised), whilst the FAA treatment does not apparently work for the ISA technique.

Permeafix and Strecks Tissue Fixative are non-crosslinking commercially available fixatives that are mainly used for cytological specimens [Uhlmann, 1998; Patterson et al., 1993]. Permeafix does not require additional proteolysis steps, whilst Strecks treatment requires only limited digestion prior to for example mRNA detection by ISH. Only Strecks fixative appears to be compatible with ISA for the detection of HIV mRNA and DNA [Patterson et al., 1998]. However, it has proven difficult to reproduce these findings as it appears that Strecks-fixed DNA may not fully withstand a hot-start procedure without being significantly degraded.

13.4.4 Damage Caused by Paraffin Embedding

ISA studies using freeze-sectioned material have concluded that double stranded DNA is damaged when subjected to temperatures above 56°C. Further, protein denaturation and melting of AT-rich DNA domains may occur at these temperatures, with the possibility of breakage of the double stranded DNA. Renaturation of melted regions may occur upon lowering the temperature below 56°C, but any nicks introduced into the DNA will remain. The use of low melting point paraffin, which liquifies between 43°C and 45°C, solves most if not all of these problems (Fig. 13.6). Technical problems are less frequent if paraffin embedded tissue sections are desiccated for 1 hour at 33°C and stored in a refrigerator at 4°C until required.

13.4.5 Detachment of Cells and Tissue Sections

Proteolysis, heat denaturation (e.g. during IS-PCR and ISH thermocycling) and enduring contact with moving organic/water-based solvents, may detach cells and tissues from protein-glycerin coated glass microscopy slides. Proteolysis reactions will dissolve protein-glycerin "glue", whereas heating will denature and hence interfere with the interaction between protein-glycerin and proteins present within cells or tissue sections. Of note, the adherence of cells and tissues is strongly enhanced if glass slides are pre-treated with 0.1% poly-L-lysin or organosilane. These compounds can be equipped with aldehyde groups by incubation with gluteraldehyde (so-called APES coating, [Bagasra et al., 1998]). These aldehyde groups are able to couple to amino groups of proteins present in cells and tissues

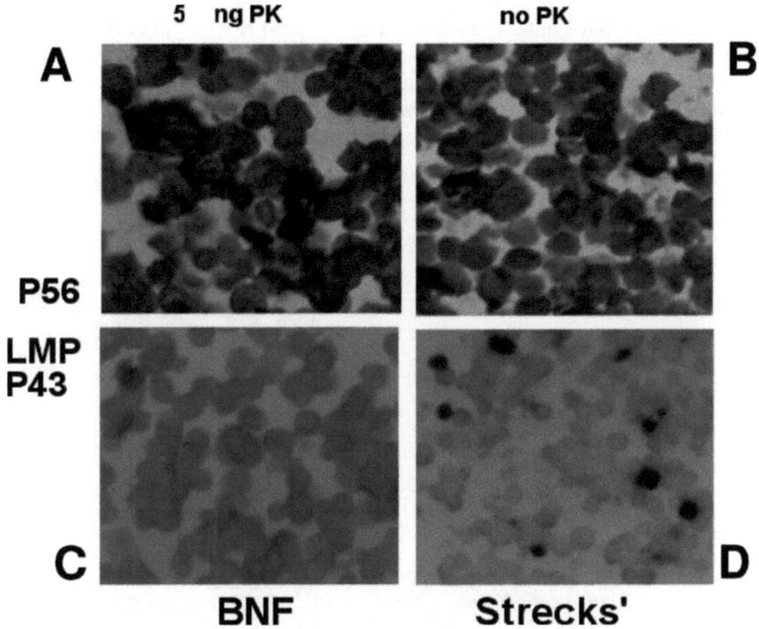

Fig. 13.6 Differences in non-specific DNA synthesis in sections of SiHa cell pellets infiltrated with paraffin at different temperatures. Panels A and B show the effect of embedding cells with a commercially available paraffin formulation of Paraplast (infiltration at 56°C), whilst Panels C and D show the effect of embedding cells in low melting point paraffin (Merck) (infiltration at 43°C). Cells were fixed in buffered neutral formalin containing 5 ng/ml phosphate buffered neutral formalin (BNF) (panels A and C) or Streck fixative (panels B and D). Non-specific DNA synthesis (after performing a standardised IS-PCR protocol without any PCR primers) is very apparent in the nuclei of cells that have been heated to 56°C (Paraplast). Much less non-specific DNA synthesis is observed when the temperature during paraffin impregnation was kept at 43°C (low melting point paraffin) (*see Color Plates*)

resulting in almost immediate covalent binding. Useful alternatives to traditional glass microscope slides include the commercially available Superfrost$^+$ (Menzel-Glaser, Braunschweig, Germany) and ProbeOn Plus (FisherBiotech, Fisher Sceintific International, USA) slides.

13.4.6 Specimen Proteolysis

Specimen fixation is only the first in a series of steps that need to be implemented in order to process cell and tissue specimens for ISA and ISH protocols. Proteolysis is an enzymatic procedure designed to enhance the accessibility of embedded and (partly) hidden nucleic acid target molecules and to render the tissue section or cell permeable to probes, primers and enzymes. Excessive cross-linking during fixation may prevent the protease enzyme accessing proteins present within the cell or

Fig. 13.7 The effect of proteinase K overdigestion on non-specific PCR DNA synthesis as detected in the IS-PCR mix with standard agarose electrophoresis. When overdigested (4, 8 and 12 ng/ml), large quantities of non-specific DNA synthesis are observed. MW = 100 bp molecular weight marker, with the 600 bp fragment indicated

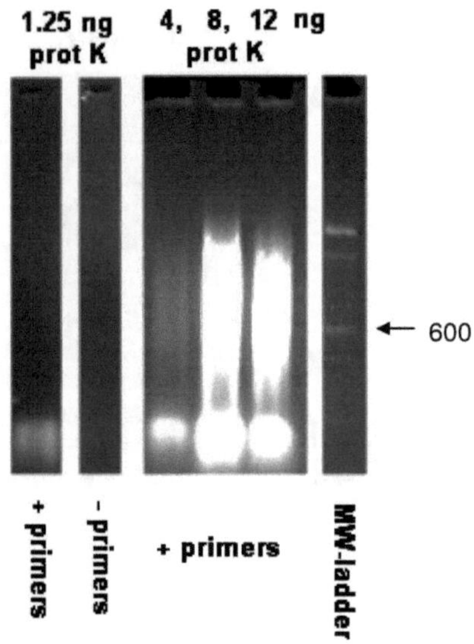

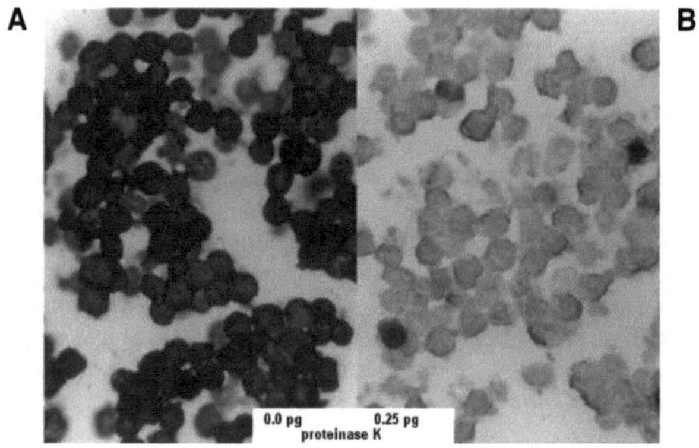

Fig. 13.8 Diffusion artiefacts in SiHa cells processed in low melting point paraffin after fixation in Streck fixative. If cells are processed according to the protocol of Patterson (1993), using 0.25 pg/ml proteinase K, proteins become digested to such extend that they loose amplicon binding capacity. This leads to diffusion of any amplicons (Panel B). Without proteinase K pretreatment, massive accumulation of amplicons takes place (Panel A)

tissue, whilst on the other hand too much proteolytic digestion may destabilize the cellular structure. Specimens where only short fixation periods have been used may not require a proteolytic digestion step [Komminoth, 1992; Komminoth et al., 1994]. Various types of proteases have been used in *in situ* protocols, including proteinase K, pronase B, digitonin, chymotrypsin, trypsin or pepsin. [Brahic and Haase, 1978; Walboomers et al., 1988; Unger et al., 1991], though it should be remembered that nuclease free proteinases should always be used. The protocol provided by these authors involved 15 hours of fixation in formalin combined with a 90-minute pepsin digestion for solid tissues. Commercial protease prepations are usually supplied in 0.01 N HCl which may damage nucleic acids, and if found to be the case, buffered enzyme preparations may be useful in generating successful ISA results [Mee, 1997; Nuovo, 1999; O'Leary et al., 1994]. After incubation and digestion, the action of the proteolytic enzyme may be blocked via the addition of 2% glycine or histidine. These chemicals circumvent the need for heat-based protease denaturation steps, helping to prevent tissue/cellular damage. From the authors' own experience, very low concentrations of carefully titrated proteinase K (1–3.5 ng/ml depending of the experiment) have been found to be useful in allowing penetration of ISA reaction ingredients whilst preventing the diffusion of amplicons and the synthesis of non-specific DNA products (Fig. 13.7). Sensitivity towards proteolytic digestion strongly depends on the fixative used. Therefore, extensive titration is necessary. Nanogram traces of proteinase K after fixation in Strecks' [Patterson, 1993] may lead to the loss of amplicon-binding molecules (e.g. proteins) in the nucleus (Fig. 13.8).

13.4.7 Acetylation and Other Forms of Tissue Section Pre-treatment

Partial hydrolysis of proteins by proteases results in an increased number of exposed, chemically promiscuous carboxy- and amino-groups. Protonated amino-groups can couple to nucleic acid phosphates, an effect which may be considered as having both negative (shielding the DNA making it less accessible to the DNA polymerase) and positive (amplimers may be retained at the site of synthesis because of these chemical interactions) advantages. Acetylation is a technique that can be used to inactivate both the amino- and the carboxy-groups of proteins, thereby neutralizing any possible chemical interference during the ISA process [Kiernan et al., 1990]. It has been reported that the acetylation of entire slides (e.g. the amino-groups on the lysine-coated APES slides) may help in preventing non-specific probe hybridization in ISH protocols, [Hayashi et al., 1978; Brahic et al., 1984], though this phenomenon is not easily reproducible. This approach was not useful in our model system utilizing SiHa cells.

Lecithin (0.01–1% in PCR buffer) may be useful in blocking non-specific DNA binding events during ISH and ISA [Teo and Shaunak, 1995a, b], though this effect should be tested in combination with various concentrations of magnesium. From our experience, lecithin induces high yields of non-specific DNA synthesis, especially at higher magnesium concentrations, e.g. 4.5 mM (Fig. 13.9).

13.4.8 Pre-treatment of Preparations for IS-PCR

Triton X-100 alongside other solvents such as methanol and acetone, dissolve membrane lipids facilitating cellular permeabilisation and increasing accessibility for primers, dNTPs and polymerases. The inclusion of Triton X-100 during paraffin embedding usually guarantees complete lipid extraction. Mild hydrolysis in

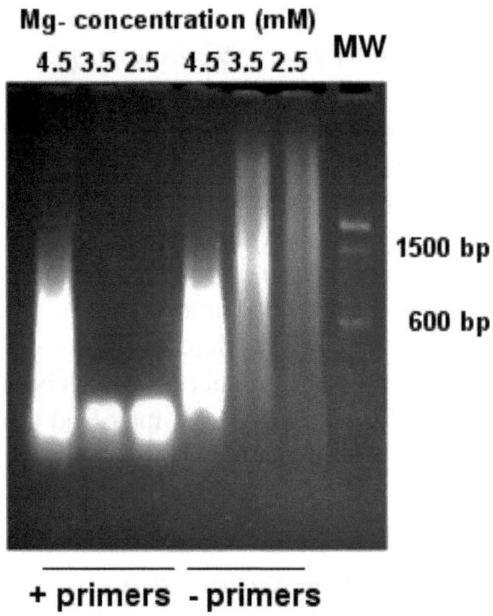

Fig. 13.9 Non-specific DNA synthesis after pre-PCR coating of tissue section slides with lecithin (1 mg/ml for 2 hours at 37°C in PCR buffer). Varying amounts of magnesium chloride in the IS PCR mix, in the presence or absence of PCR primers were used. DNA fragments were obviously present in the IS-PCR mix with standard agarose electrophoresis. Mwt = 100 bp molecular weight marker

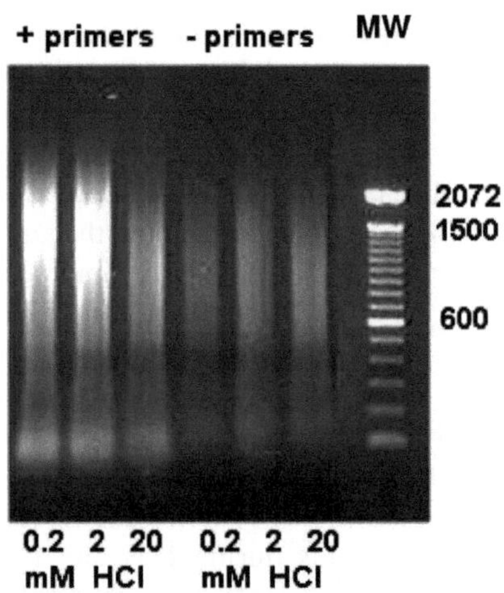

Fig. 13.10 Non-specific DNA synthesis after pre-PCR treatment sections with 0.2, 2 and 20 mM HCl. Washing was performed for 90 seconds with PBS and 0.01% Triton X100 subsequently. IS PCR was performed in the presence and absence of primers. DNA fragments were present in the IS-PCR mix with standard agarose electrophoresis. Mwt = 100 bp molecular weight marker

hydrochloric acid also improves the diffusion of nucleic acid probes and primers into cellular compartments. However, care should be taken when using hydrochloric acid pre-treatments, as HCl has the ability to depurinate, and hence destroy nucleic acids. For example, in ISH protocols, it has been demonstrated that HCl concentrations over 0.2 mM start to become detrimental to the assay. Some investigators advocate the pre-treatment of sections with HCl, triton X100, and microwaving [An et al., 1999; O'Leary et al., 2000]. From the authors' own experience, HCl treatment brings about excessive damage to nucleic acids (Fig. 13.10).

A post-fixation step using 2–4% paraformaldehyde for 5–10 minutes is applied in many ISH protocols (including our own protocols) in order to fix any areas "loosened" from the glass slide and to preserve the morphological detail. However, diffusion of PCR reagents may be a consequence of using this procedure.

13.4.9 Testing for Loss of Amplimers Due to Leakage from Their Site of Production

Amplimers that have defused away from their local site of synthesis may yield false results with respect to their original cellular location. Further, amplimers which have diffused away from their original focus may also act as templates for PCR, resulting in a decrease in the concentration of necessary PCR reagents at the intended target site. This process will act to reduce the sensitivity of the ISA methodology. Of course, such amplimer leakage is easy to detect if the target sequence is known to be present in either the nucleus or the cytoplasm of the cell (if present at all). In any case, a simple PCR reaction using a fraction of the supernatant after ISA processing will reveal whether amplimer leakage from the cell has occurred. Amplimer leakage *per se* would be expected to be most troublesome when small amplimers are generated. However, ISA protocols generating immobilized amplimers of between 68 (!) and 700 nucleotides in length have been published [Patel et al., 1994; vande Vijver et al., 1995; Nuovo et al., 1993; O'Leary et al., 1994; Nuovo et al., 1991 a, b; Long, 1993]. From our own studies, amplicons of 65, 96, 199 and 395 basepairs have been generated with product sizes of 96 and 119 basepairs being optimal.

The use of amplimer diffusion-limiting compounds and procedures has also been described, including the application of a thin covering layer of agarose, limiting the number of PCR cycles to approximately 17 cycles (see also boxed "General protocol" section near the beginning of this chapter), and the addition of certain labels to the PCR primers [Mee, 1997]. The amount of amplimer that can be lost to leakage depends of course on the amount of amplimer that is actually synthesised.

13.4.10 Miscellaneous IS-PCR Considerations

Inhibitory compounds (for example remaining fixative), present on the surface of glass microscopy slides or in the cells or tissues of specimens, may act to effectively inhibit IS-PCR reactions. Also, reactive sites within tissues or present at the surface of glass slides may deplete essential ISA reaction components (magnesium ions, for example, can be bound by any negatively charged ion present within the specimen preparation, whilst polymerase enzymes may adhere to glass slides, cellular surfaces, etc.), meaning that magnesium ion, Taq polymerase and primer concentrations may have to be increased in ISA protocols by a factor of 5–10 compared to conventional PCR. There is some evidence

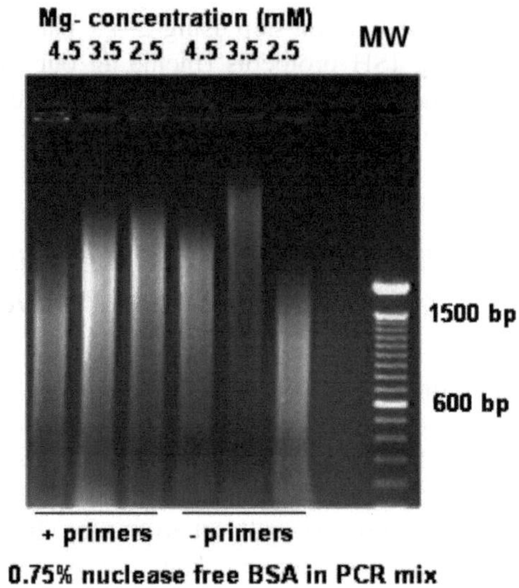

Fig. 13.11 Non-specific DNA synthesis after addition of nuclease free BSA into the PCR mixture. Various concentrations of magnesium ions were included with and without the random PCR primers OL94. DNA fragments were present in the IS-PCR mix as revealed by standard agarose electrophoresis. MW = 100 bp molecular weight marker

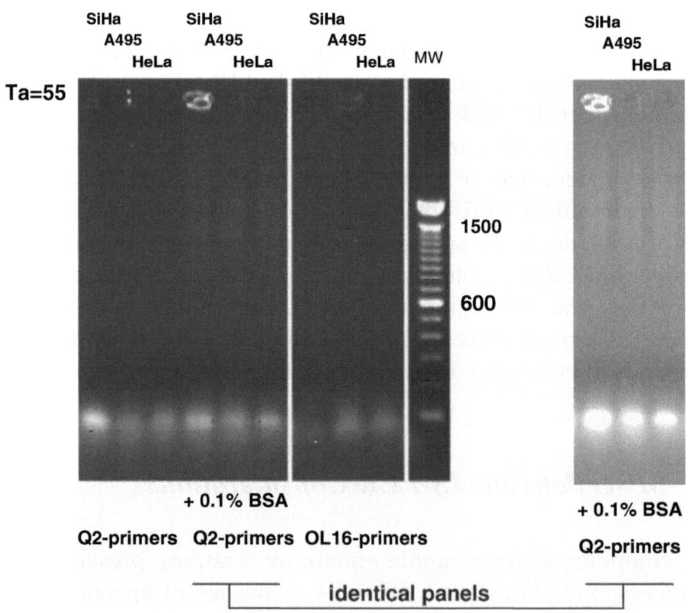

Fig. 13.12 Blocking of non-specific DNA synthesis by the addition of bovine liver tRNA type IX, using primer sets Q1/2 and HPV 16 primer 1 and primer 2 OL94 at Ta = 55°C (see Section 13.1, INTERMEZZO), for the synthesis of a 96–119 bp amplimer, respectively. SiHa HPV 16 positive cells and A549 and HeLa negative controls were used. The addition of nuclease free BSA to the PCR mix may generate non-specific DNA product synthesis as observed after standard agarose electrophoresis (NB. the panel right shows lane 4–6 with increased contrast in order to highlight the resultant faint smear of PCR fragments. Mwt = 100 bp molecular weight marker)

to suggest that nuclease-free bovine serum albumin (at 0.01–2%) can function as an efficient blocking agent, inhibiting polymerase binding to glass and cellular surfaces. However, it also increases the likelihood of non-specific DNA synthesis (Fig. 13.11). Other blocking reagents, such as herring sperm DNA, single stranded oligo-dT or bovine liver tRNA, help prevent the electrochemical attraction of primers, probes and amplimers to positively charged ions present within the specimens. Indeed, bovine liver tRNA is particularly effective in preventing the diffusion of amplicons from their site of synthesis in the nucleus and also in rendering the cytoplasm less likely to bind newly synthetised DNA or renaturing GC rich stretches (Figs. 13.12 and 13.14). In contrast, dithiothreitol (DTT), Tween 20 and NP40 do not appear to result in significant increases in the quality of ISA results.

Most authors restrict the number of thermocycles performed in IS-PCR protocols in order to help preserve tissue morphological detail, and reduce the chance of non-specific DNA synthesis to occur. Further, the hybridisation step in PCR-ISH protocols is itself sensitive enough to detect low copy numbers of amplimer. Other authors however prefer the use of larger numbers of PCR cycles, due to the fact that exponential amplification often only occurs in the later stages of PCR thermocycling (the first few thermocycles of IS-PCR often being highly inefficient due to non-specific binding and inhibition of PCR reaction mix components). The most common ISA protocols advise between 15 and 20 cycles of amplification. However, the optimum number of cycles will vary per ISA application and should be determined by optimisation. Finally, a post-PCR fixation step may be useful, e.g. the application of ice-cold absolute ethanol [Mougin et al., 1997; Bettinger, 1999].

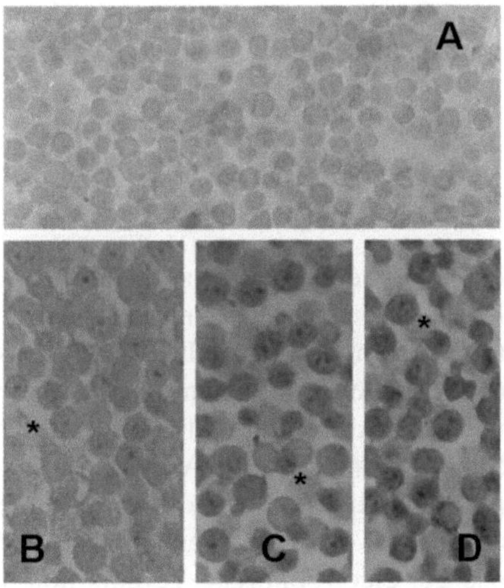

Fig. 13.13 IS-PCR amplification without primers controls for non-specific DNA synthesis. The panels show the results of direct *in situ* PCR (without primers) on BNF fixed SiHa cells embedded in low melting point paraffin and predigested with 3.5 ng proteinase K in T20E1 buffer at pH 7.4. Panel A. Optimized fixation results in an absence of non-specific DNA synthesis. Panels B–D. Experiments performed at optimal digestion but without the carrier tRNA. Panel B. Nucleolar artefacts may be seen. Panel C. With greater cellular damage, signals in both nucleoli and entire nucleus may be seen. Panel D. Short boiling in PBS in a microwave oven leads to even greater non-sepcific DNA synthesis. NB. Cells without nuclei remain negative for PCR amplification products (asterisks)

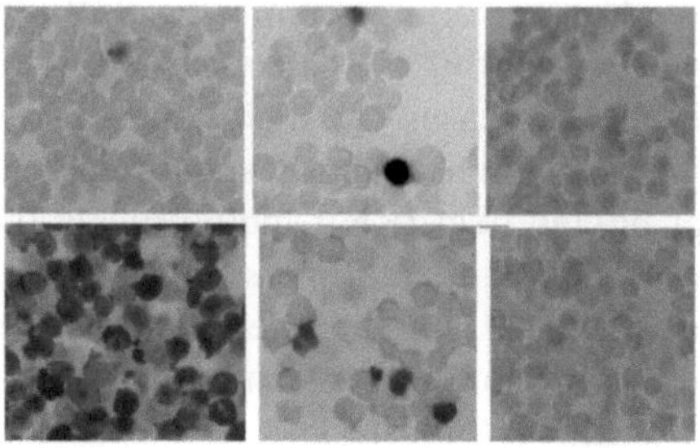

Fig. 13.14 IS-PCR using 3.5 ng proteinase K as a pre-treatment, a wet hot-start protocol (Section 13.6.13) and tRNA added to the PCR mixture (17 cycles, annealing temperature at 55°C and 8 μm sections). The IS-PCR procedure is fully optimised and 50 pmoles of Q1/2 HPV16 specific primers (see Section 13.1, INTERMEZZO) were used during amplification. The first row shows negative control IS-PCR amplifications (without PCR primers) as an indication of non-specific DNA synthesis in SiHa, Hela and A549, HPV negative cells. The second row shows IS-PCR amplifications (using HPV16 specific primers) in SiHa (HPV positive) cells and HeLa/A495 (HPV negative) cell controls

13.4.11 Mispriming

In IS-PCR protocols, it is recommended to select the highest possible annealing temperature in order to prevent mispriming reactions from taking place. Further, the inclusion of a hot-start step prior to PCR thermocycling *per se*, as well as specialised hot-start polymerases, e.g. AmpliTaq Gold is recommended (Section 12.4). The disadvantage of using a hot-start step is that the high temperature incubation may inflict damage upon the target nucleic acid to be amplified. Mispriming and non-specific DNA synthesis (see Section 13.4.12.) is less of a problem for PCR-ISH protocols, due to the fact that the hybridisation step should distinguish between non-specific and specific PCR products. The only caveat is that the non-specific amplification must not seriously deplete the pool of PCR reactants required for specific amplification.

13.4.12 Primer Independent Non-specific DNA Synthesis

Primer independent non-specific DNA synthesis remains a major problem in IS-PCR assays, occurring via a variety of mechanisms. In the SiHa model system, such non-specific DNA synthesis generally progresses from the nucleolus, to the nucleus, and eventually to the cytoplasm (Fig. 13.13), being most severe when DNA products diffuse into the reaction mixture (Figs. 13.7, 13.9–13.12). Such non-specific DNA synthesis occurs in the absence of PCR primers and products diffuse into the overlaying IS-PCR reaction mix. Subsequently simple agarosegel electrophoresis reveals non-specific reaction products. Simple agarose gel electrophoresis reveals non-specific amplification products. A natural

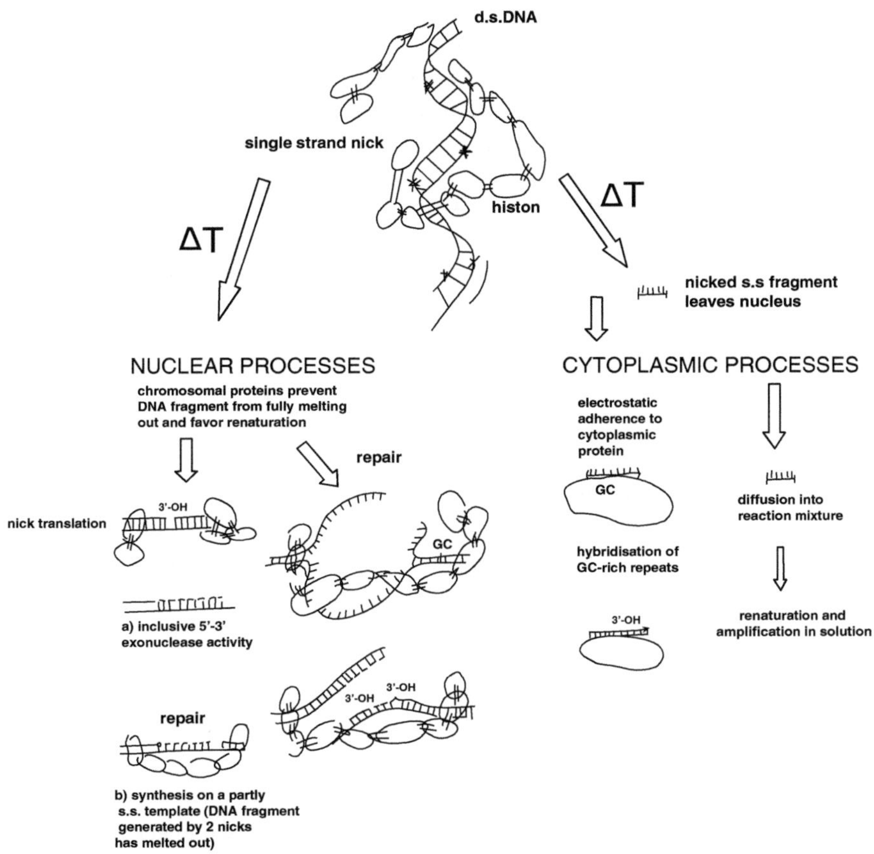

Fig. 13.15 Treatment conditions that may play a role in generating IS-PCR artefacts. Cross-linkage of chromosomal proteins may stabilize local DNA molecules. Formalin fixation and high temperatures (e.g. liquid paraffin at 56°C) may lead to both single-stranded and double-stranded DNA damage. Proteolysis helps digest cellular proteins rendering the cell accessible for PCR reagents and Taq polymerase, however a proportion of the protein should remain intact in order to maintain the integrity of the cell during IS-PCR thermocycling. During the heat–denaturation step of PCR thermocycling, single stranded DNA fragments containing breaks (or "nicks") may become attached to proteins so that their 3′-OH end acts as a binding site for DNA repair enzymes. This repair mechanism may occur at the end of every PCR thermocycle leading to an accumulation of "primer-independent signal", especially when many single-stranded nicks are present. Other affects of single-stranded nicks include: (1) nick-translation (i.e. the removal and repair of nucleotides) by the Taq polymerase as part of its 5′-3′exonuclease activity, (2) in the absence of stabilizing proteins, denatured single-stranded DNA fragments may break due to mechanical stress forces leaving 3′-end extremities exposed that may facilitate new DNA synthesis by Taq polymerase, (3) two single-stranded breaks next to each other may be "filled-in" by Taq polymerase. DNA fragments may also diffuse out of their cellular compartments, leading to PCR amplification and signal detection away from their original site. Non-covalent adhesion of DNA to nuclear or cellular proteins may lead to the formation of a "solid phase" whereby GC-rich fragments may renature during PCR thermocycling. s.s = single stranded

cause for non-specific amplification is programmed cell death (apoptosis), a genetically encoded mechanism which specifically degrades DNA and which begins in the regions lying between the nucleosomes. Apoptosis can be specifically tracked by the TUNEL (terminal deoxynucleotidyl transferase-mediated dUTP nick end-labeling) method [Negoescu et al., 1996], and is frequent in lymphoid tissues, in epithelia, in organs, and also in some cases of viral infection. Apoptosis may result in false positive IS-PCR results generating products in negative control reactions [Sallstrom et al., 1993]. Necrotic cells may also be a source of false positive IS-PCR results [Muro-Cacho, 1997; Teo and Shaunak, 1995a, b; Solomon, 1995; Sallstrom et al., 1993; Komminoth 1992]. Cell necrosis occurs for example in tissue culture, at the stage where cell cultures reach confluence and when nutrients become a limiting factor. The severity of this effect differs from cell line to cell line, but is most obvious in HeLa cells (Fig. 13.14).

Histones, either single or in combination, promote the renaturation of complementary pieces of single stranded DNA [Cox and Lehman, 1981], the renatured strands potentially acting as priming sites for DNA amplification. Nucleases present in the nucleus are no longer under cellular control once the cell has died, and rapid fixation may be necessary in order to prevent target nucleic acid degradation due to nuclease enzymes. Lowering the temperature during fixation not only decreases nuclease activity [Tokuda et al., 1990], but also prevents extensive crosslinking. Finally, it is known that excess proteolysis will accentuate

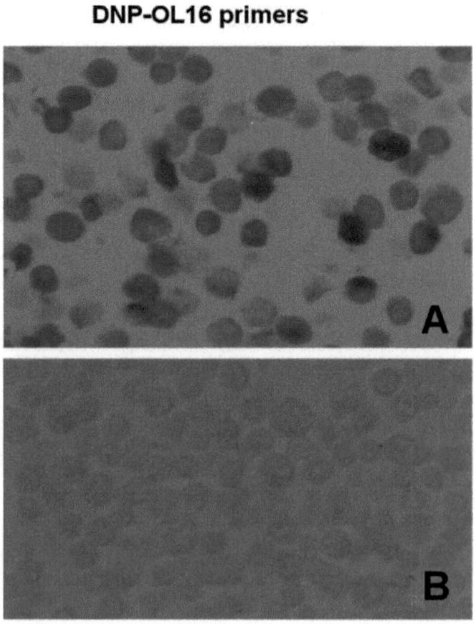

Fig. 13.16 Optimised IS-PCR using 50 pmol of DNP-labelled OL94 primers that have been detected using sheep anti-DNP/AP antibody with subsequent visualisation using a NBT/BCIP/PVA mix. Twenty-five cycles were used for IS-PCR amplification (A: SiHa cells; B: A549 negative control cells)

non-specific DNA synthesis during ISA [Bernard et al., 1994]. From our own studies, mechanical stress during treatment with melted paraffin at 56°C (which leads to partial denaturation of (AT rich) DNA strands), tends to be responsible for the majority of non-specific DNA synthesis problems, as well as electrovalent attraction between "overdigested proteins" and partly renatured DNA fragments (Fig. 13.15).

For *in situ* RT-PCR protocols, problems associated with DNA-mediated non-specific amplification products are less of an issue, as IS RT-PCR processing *per se* requires the removal of the genomic DNA present in cells (achieved using DNase treatment) [Stamps et al., 2003; Nuovo, 2001; Nuovo, 1996a, b].

Pre-PCR incubation of tissue sections or cells with DNA polymerase and dideoxynucleotides may help to suppress non-specific amplification and signals. Alternatively, specifically labelled IS-PCR primers will indicate the presence of specific amplification products and are advantageous over the use of labelled nucleotides that may be incorporated into any nicks found in non-target DNA generating a non-localised IS-PCR signal (Fig. 13.16).

13.4.13 Evaporation of Reactants During IS-PCR/ Wet Hot Start Procedure

During ISA thermocycling, samples are subjected to various high temperature incubation steps. These steps facilitate the evaporation of key PCR reactants, which may affect the reaction dynamics and equilibrium of the PCR. Evaporation therefore needs to be prevented and various systems have been developed to achieve this aim. The use of IS-PCR reaction chambers that can be sealed is the most popular option, involving the use of rubber rings and specific cover slips. Some companies now offer specific sealing kits and tapes, e.g. HybriSlip Hybridization Covers (Invitrogen). Loading the slides into *in situ* PCR machines at 70°C leads to progressive DNA damage due to dehydration. This process is preventable by loading the samples in a re-hydrated state (Fig. 13.17). Re-hydration of the specimen in PCR reaction mix initiates a sort of "wet hot-start" procedure that works well to prevent this phenomenon, especially if tRNA is included in the IS-PCR mix.

13.4.14 Cell Thickness and ISA

The range of cell sizes likely to be encountered for ISA analysis range from 7 μM (lymphocytes) to 20–25 μM. It has been demonstrated that efficient ISA amplification occurs when the section thickness lies in the order of 8 μM.

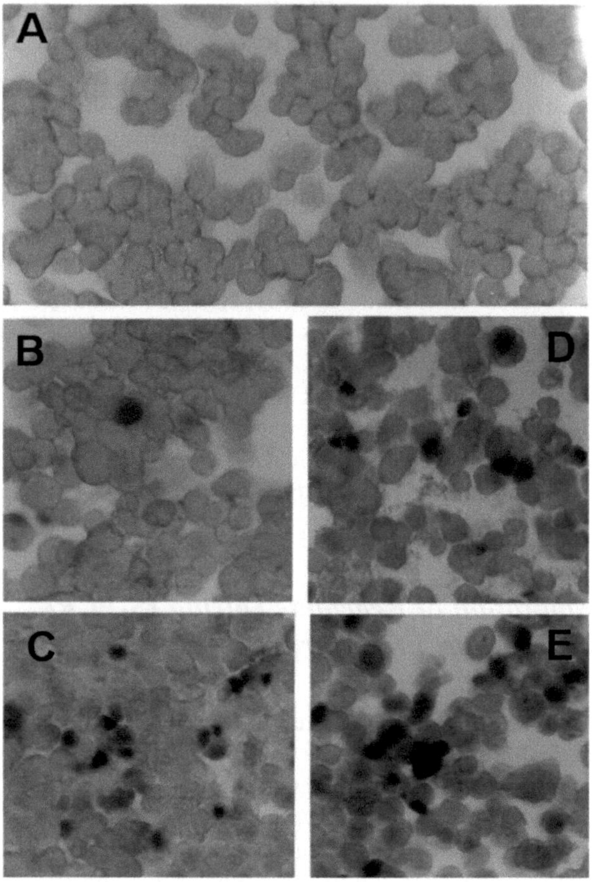

Fig. 13.17 Direct *in situ* PCR in Streck fixative fixed SiHa cells embedded in low melting point paraffin without proteinase K pre-digestion. Non-specific DNA synthesis controls (wet hot-start IS-PCR reactions without PCR primers) are shown in panels A–C, which also show the effect of tissue drying before IS-PCR thermocycling (Panels A-C represent drying times of 0, 2 and 4 minutes, respectively). Increased drying times lead to an increase in damage to the nuclear DNA and therefore an increase in non-specific DNA synthesis. Other treatments may result in even greater nuclear damage and hence larger quantities of non-specific DNA synthesis, e.g. dry hot-start (panel D, see Section 13.6.13) or microwave treatment (panel E).

13.4.15 Choosing a Hybridisation Control for Testing Amplimer Specificity

Many authors claim that ISA protocols must be validated via a hybridisation step, in order to check for amplimer specificity (PCR-ISH). Indeed, post-PCR hybridisation is strongly recommended for single cell suspensions where non-specific

DNA synthesis is common [Sagawa, 1997]. The optimum hybridisation reaction mix to be used post-PCR thermocycling has to be empirically determined. The most popular labels for PCR-ISH hybridisation probes remain digoxygenin, FITC, biotin and radioactive labels (still regularly used because of their exquisite sensitivity).

13.4.16 Choice of the PCR Processor

In the early days of *in situ* PCR, the capacity (and supply) of PCR thermocycling machines suited for ISA applications was limited, with a maximum throughput of 2–4 slides per ISA thermocycling run. Presently, the Gene Amp *In Situ* PCR System 1000 by Perkin Elmer accommodates ten slides with three reaction wells present on each glass slide. Alternatively, the MJ Research PTC-200 machine has space for 32 slides and the PTC 100 room for 16 slides. Some of these machines work via hot air, whilst others utilise metal Peltier elements to effect temperature changes. Finally, the technology to prevent evaporation of reactants from heated slides and specimens varies dependant on the type of *in situ* thermocycling machine being used.

13.4.17 Choice of the Final Detection Method

The label used in ISA protocols essentially defines the detection method. A popular method used is conjugated alkaline phosphatase, though the quality of the NBT/BCIP substrates does need to be of a high standard if significant background staining is to be avoided. Further, NBT, which is a hydrogen acceptor, may interact with sulphhydryl groups that may become exposed during the ISA procedure, generating a lack of signal focus. Alkaline phosphatase staining can be improved by preventing the diffusion of the stained products using 10% polyvinylalcohol [Kiernan et al., 1989; De Block, 1993].

13.5 ISA Optimisation

The main goal of optimisation is to prevent false-negative results whilst maintaining assay sensitivity, and includes an assessment of the efficiency of the extraction method as well as the PCR conditions utilised. The process of ISA optimisation is essentially identical to that required for standard PCR protocol optimisation (as described in Chapters 5–8). The process includes optimisation of chemical (magnesium, dNTP, primer and polymerase concentrations, the type of buffer and pH) and physical (inclusion of an initial hot-start step, the number of PCR cycles, annealing

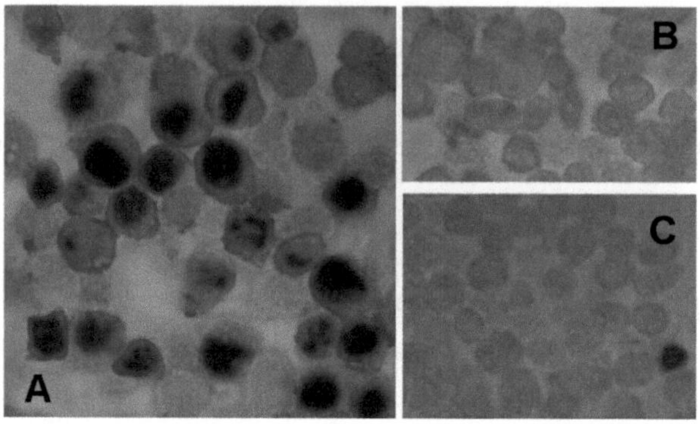

Fig. 13.18 Optimised IS-PCR using 3.5 ng of proteinase K as a pre-treatment in a wet hot- start PCR containing tRNA. The program used 17 cycles of PCR at an annealing temperature of 55°C. Panel A. IS-PCR on SiHa (HPV positive) cells using OL16 primers and showing the nuclear localisation of the amplicons (NB. Some cells do not contain the target and remain unstained). Panel B. A549 (HPV negative) control cells in the presence of OL16 primers. Panel C. SiHa cells that have undergone IS-PCR thermocycling in the absence of PCR primers showing that sporadic cells contain nuclear artefacts of non-specifically synthesized DNA (*see Color Plates*)

temperature, elongation temperature, etc.) characteristics. Of course, ISA optimisation is further complicated by the need to consider cell/tissue pre-treatment steps, type of fixative, time of fixation and post-PCR treatment steps (including the hybridisation conditions to be used in PCR-ISH protocols). An assessment of the effect of PCR thermocycling conditions on the cellular morphology may be made by performing PCR without Taq polymerase, primers and dNTPs. This can be combined with pre-PCR treatment of tissue sections or cells with varying amounts of protease and the addition of tRNA (Fig. 13.18).

A simple method for assessing the sensitivity of the ISA assay being optimised is to obtain identical, or at least similar, cell types which have been shown to be (using another established assay protocol) either positive or negative for the ISA assay being optimised. By mixing these positive and negative cells in different ratios, it should be possible to calculate the sensitivity of the ISA assay compared to currently available testing procedures (N.B. ISA sensitivity = number of true positive ISA results divided by the sum of the true positive ISA results and the false negative ISA results).

Finally, it may be advisable to perform initial ISA optimisation experiments using embedded scraped cell layers rather than tissue sections so that the PCR protocol is first established before moving on to consider the effects of fixation etc. Also, once the initial optimisation process is complete, it is wise to repeatedly test a small but representative panel of clinical specimens (containing specimens which have yielded false positive and false negative results in other assays), in order to determine the threshold values, reproducibility etc of the new ISA assay (Chapter 11).

13.6 ISA Controls

Essential process controls for IS-PCR/PCR-ISH include positive/negative controls comprising clinical material previously proven to be positive/negative using both the IS-PCR/PCR-ISH assay and a completely different assay format. It is also recommended that a universally present target sequence (housekeeping gene), e.g. the β-globin or α-actin gene should also be targeted (for example using the same clinical material on a separate glass slide but included in the same batch of PCR specimens), in order to control for false negative IS-PCR/PCR-ISH results [Koopmans et al., 1993]. The inclusion of a negative control sample used in combination with a PCR mix without primers is also necessary in order to control for DNA repair activity and the generation of non-specific PCR products. A further negative control using a PCR mix containing a non-target primer pair provides an assessment of PCR specificity.

Reverse transcription ISA (RT-ISA) protocols raise some specific problems with respect to optimisation including variables in DNase treatment and controlling for correct reverse transcription. The types of positive and negative controls required in RT-ISA protocols include those used in IS-PCR/PCR-ISH protocols, with the addition of a standard RT-PCR on RNA extracted from the specimen to corroborate any negative RT-ISA results. Any positive control reactions utilising housekeeping gene mRNA (as a measure of possible false negative RT-ISA results) should be designed using housekeeping genes whose mRNA expression occurs at a similar level to that of the gene under investigation [Radonic et al., 2004].

References

Alcock HE, Stephenson TJ, Royds JA, Hammond DW. 1999. Related Articles, Links A simple method for PCR based analyses of immunohistochemically stained, microdissected, formalin fixed, paraffin wax embedded material. Mol Pathol 52(3):160–163.

An SF, Fleming KA. 1991. Removal of inhibitor(s) of the polymerase chain reaction from formalin fixed, paraffin wax embedded tissues. J Clin Pathol 44(11):924–927.

An SF, Groves M, Giometto B, Becket AA. 1999. Detection and localisation of HIV-1 DNA and RNA in fixed adult AIDS brains by PCR in situ hybridisation techniques. Acta Neurol 98:481–487.

Bagasra O. 1993. Polymerase chain reaction in situ: intracellular amplification and detection of HIV-1 proviral DNA and other specific genes. J Immunol Methods 158:131–145.

Bagasra O, Lavi E, Bobroski L, Khalili K, Pestaner JP, Tawadros R, Pomerantz RJ. 1996. Cellular reservoirs of HIV-1 in the central nervous system of infected individuals: identification by the combination of in situ polymerase chain reaction and immunohistochemistry. AIDS 10(6):573–585.

Bagasra O, Bobroski LE, Amjad M, Pomerantz RJ, Hansen J. 1998. Amplification of in situ PCR techniques to human tissues. In Herrington CS, O'Leary JJ (eds) PCR 3 A practical approach. PCR in situ hubridization. Chapter 7. IRL, Oxford, pp. 117–137, ISBN 0–19–963632-X.

Banerjee SK, Makdisi WF, Weston AP, Mitchell SM, Campbell DR. 1995. Microwave-based DNA extraction from paraffin-embedded tissue for PCR amplification. Biotechniques 18(5):768–770, 772–773.

Beland FA, Fullerton NF, Heflich RH. 1984. Rapid isolation, hydrolysis and chromatography of formaldehyde-modified DNA. J Chromatogr 308:121–131.
Ben-Ezra J, Johnson DA, Rossi J, Cook N, Wu A. 1991. Effect of fixation on the amplification of nucleic acids from paraffin-embedded material by the polymerase chain reaction. J Histochem Cytochem 39(3):351–354.
Benkoel L, Biagini P, Dodero F, De Lamballerie X, De Micco P, Chamlian A. 2004. Immunohistochemical detection of C-100 hepatitis C virus antigen in formaldehyde-fixed paraffin-embedded liver tissue. Correlation with serum, tissue and in situ RT-PCR results. Eur J Histochem 48(2):185–190.
Bernard C, Mougin C, Bettinger D, Didier JM, Lab M. 1994. Detection of human papillomavirus by in situ PCR in paraffin-embedded cervical biopsies. Mol Cell Probes 8:337–343.
Bettinger D. 1999. Human papillomavirus detection by non-isotopic in situ hybridisation, in situ hybridisation with signal amplification and in situ PCR. Eur J Histochem 43:185–198.
Brahic M, Haase AT. 1978. Detection of viral sequences of low reiteration frequency by in situ hybridization. Proc Natl Acad Sci USA 75(12):6125–6129.
Brahic M, Haase AT, Cash E. 1984. Simultaneous in situ detection of viral RNA and antigens. Proc Natl Acad Sci USA 81(17):5445–5448.
Bramwell NH, Burns BF. 1988. The effects of fixative type and fixation time on the quantity and quality of extractable DNA for hybridization studies on lymphoid tissue. Exp Hematol 16(8):730–732.
Brooks EM, Sheflin LG, Spaulding SW. 1998. Secondary structure in the 3′ UTR of EGF and the choice of reverse transcriptase affect the detection of message diversity by RT-PCR. In: Siebert P (ed) The PCR technique: RT-PCR. . Eaton, New York, pp. 107–119.
Brutlag D, Schlehuber C, Bonner J. 1969. Properties of formaldehyde-treated nucleohistone. Biochemistry 8(8):3214–3218.
Brutlag DL. 1980. Molecular arrangement and evolution of heterochromatic DNA. Annu Rev Genet 14:121–144.
Burton MP, Schneider BG, Brown R, Escamilla Ponce N, Gulley ML. 1998. Comparison of histologic stains for use in PCR analysis of microdissected paraffin embedded tissues. Biotechniques 24:86–92.
Camilleri-Briet S, Devez F, Tissier F, Ducruit V, Le Tourneau A, Diebolf J, Audouin J, Molina, T. 2000. Quality control and sensitivity of Polymerase Chain Reaction techniques for the assessment of immunoglobulin heavy chain gene rearrangements form fixed- and paraffin-embedded samples. Ann Diagn Pathol 4: 71–76.
Carrozza ML, Mazzei M, Bandecchi P, Arispici M, Tolari F. 2003. In situ PCR-associated immunohistochemistry identifies cell types harbouring the Maedi-Visna virus genome in tissue sections of sheep infected naturally. J Virol Methods 107(2):121–127.
Cawkwell L, Quirke P. 2000. Direct multiplex amplification of DNA from formalin fixed, paraffin wax embedded tissue sections. J Clin Pathol 53:51–52.
Chiu KP, Cohen S, Morris D, Jordan G. 1992. Intracellular amplification of proviral DNA in tissue sections using the PCR. J Histochem Cytochem 40:333–341.
Chuaqui R, Cole K, Cuello M, Silva M, Quintana ME, Emmert-Buck MR. 1999. Analysis of mRNA quality in freshly prepared and archival Papanicolaou samples. Acta Cytol 43(5):831–836.
Cox MM, Lehman IR. 1981. Renaturation of DNA: a novel reaction of histones. Nucleic Acids Res 9(2):389–400.
De Block M. 1993. RNA-RNA in situ hybridisation using DIG-labelled probes: the use of high molecular weight polyvinyl alcohol in the alkaline phosphatase indoxyl-nitroblue tetrazolium reaction. Anal Biochem 215:86–89.
Dubeau L, Chandler LA, Gralow JR, Nichols PW, Jones PA. 1986. Southern blot analysis of DNA extracted from formalin-fixed pathology specimens. Cancer Res 46(6):2964–2969.
Embretson J. 1994. PCR amplification of viral DNA and viral host cell mRNA in situ. In: Mullis KB (ed) The polymerase chain reaction. Eaton, New York, pp. 55–64.

Feldman MY. 1973. Reactions of nucleic acids and nucleoproteins with formaldehyde. Prog Nucleic Acid Res Mol Biol.13:1–49.

Fiallo P, Williams DL, Chan GP, Gillis TP. 1992. Effects of fixation on polymerase chain reaction detection of *Mycobacterium leprae*. J Clin Microbiol 30(12):3095–3098.

Finke J, Fritzen R, Ternes P, Lange W, Dolken G. 1993. An improved strategy and a useful housekeeping gene for RNA analysis from formalin-fixed, paraffin-embedded tissues by PCR. Biotechniques 14(3):448–453.

Foss RD, Guha-Thakurta N, Conran RM, Gutman P. 1994. Effects of fixative and fixation time on the extraction and polymerase chain reaction amplification of RNA from paraffin-embedded tissue. Comparison of two housekeeping gene mRNA controls. Diagn Mol Pathol 3(3):148–155.

Frank TS, Svoboda-Newman SM, Hsi ED. 1996. Comparison of methods for extracting DNA from formalin-fixed paraffin sections for nonisotopic PCR. Diagn Mol Pathol 5(3):220–224.

Gall K, Pavelic J, Jadro-Santel D, Poljak M, Pavelic K. 1993. DNA amplification by polymerase chain reaction from brain tissues embedded in paraffin. Int J Exp Pathol 74(4):333–337.

Giannella C, Zito FA, Colonna F, Paradiso A, Marzullo F, Alaibac M, Schittulli F. 1997. Comparison of formalin, ethanol, and Histochoice fixation on the PCR amplification from paraffin-embedded breast cancer tissue. Eur J Clin Chem Clin Biochem 35(8):633–635.

Goldsworthy SM, Stockton PS, Trempus CS, Foley JF, Maronpot RR. 1999. Effects of fixation on RNA extraction and amplification from laser capture microdissected tissue. Mol Carcinog 25(2):86–91.

Golenberg EM, Bickel A, Weihs P. 1996. Effect of highly fragmented DNA on PCR. Nucleic Acids Res 24(24):5026–5033.

Greer CE, Lund JK, Manos MM. 1991a. PCR amplification from paraffin-embedded tissues: recommendations on fixatives for long-term storage and prospective studies. PCR Methods Appl 1(1):46–50.

Greer CE, Peterson SL, Kiviat NB, Manos MM. 1991b. PCR amplification from paraffin-embedded tissues: effects of fixatives and fixation times. Am J Clin Pathol 95:117–124.

Greer CE, Wheeler CM, Manos MM. 1994. Sample preparation and PCR amplification from paraffin-embedded tissues. PCR Methods Appl 3(6):S113–S122.

Haase AT, Retzel EF, Staskus KA. 1990. Amplification and detection of lentiviral DNA inside cells. Proc Natl Acad Sci USA 87:4971–4975.

Hayashi S, Gillam IC, Delaney AD, Temer GM. 1978. Acetylation of chromosomes squashes of *Drosophila melanogaster* decreases the background in autoradiographs from hybridisation with ^{125}I-labelled RNA. J Histochem Cytochem 26:677–679.

Hewett PJ, Firgaira F, Morley A. 1994. The influence of age of template DNA derived from archival tissue on the outcome of the polymerase chain reaction. Aust N Z J Surg 64(8):558–559.

Honma M, Ohara Y, Murayama H, Sako K, Iwasaki Y. 1993. Effects of fixation and varying target length on the sensitivity of polymerase chain reaction for detection of human T-cell leukemia virus type I proviral DNA in formalin-fixed tissue sections. J Clin Microbiol 31(7):1799–1803.

Hopwood D. 1975. The reactions of glutaraldehyde with nucleic acids. Histochem J 7(3):267–276.

Howe JR, Klimstra DS, Cordon-Cardo C. 1997. DNA extraction from paraffin-embedded tissues using a salting-out procedure: a reliable method for PCR amplification of archival material. Histol Histopathol 12(3):595–601.

Inoue T, Nabeshima K, Kataoka H, Koono M. 1996. Feasibility of archival non-buffered formalin-fixed and paraffin-embedded tissues for PCR amplification: an analysis of resected gastric carcinoma. Pathol Int 46(12):997–1004.

Jackson DP, Lewis FA, Taylor GR, Boylston AW, Quirke P. 1990. Tissue extraction of DNA and RNA and analysis by the polymerase chain reaction. J Clin Pathol Jun; 43(6):499–504.

Kaminska D, Tyran B, Mazanowska O, Letachowicz W, Kochman A, Rabczynski J, Szyber P, Patrzalek D, Chudoba P, Klinger M. 2005. Intragraft mRNA expression of cytokines and growth factors in human kidney allograft biopsies by in situ RT-PCR analysis Transplant Proc 37(2):767–769.

Karlsen F, Kalantari M, Chitemerere M, Johansson B, Hagmar B. 1994. Modifications of human and viral deoxyribonucleic acid by formaldehyde fixation. Lab Invest 71(4):604–611.

Kiernan JA. 1999. Histological and histochemical methods: theory and practice. Third Edition. Butterworth-Heinemann/Pergamon Press, Oxford.

Kitazawa S, Takenaka A, Abe N, Maeda S, Horio M, Sugiyama T. 1989. In situ DNA-RNA hybridization using *in vivo* bromodeoxyuridine-labeled DNA probe. Histochemistry 92(3):195–199.

Kleter B, van Doorn LJ, ter Schegget J, Schrauwen L, van Krimpen K, Burger M, ter Harmsel B, Quint W. 1998. Novel short-fragment PCR assay for highly sensitive broad-spectrum detection of anogenital human papillomaviruses. Am J Pathol 153(6):1731–1739.

Kleter B, van Doorn LJ, Schrauwen L, Molijn A, Sastrowijoto S, ter Schegget J, Lindeman J, ter Harmsel B, Burger M, Quint W. 1999. Development and clinical evaluation of a highly sensitive PCR-reverse hybridization line probe assay for detection and identification of anogenital human papillomavirus. J Clin Microbiol 37(8):2508–2517

Komminoth P. 1992. In situ PCR detection of viral DNA, single copy genes and gene rearrangements in cell suspensions and cytospins. Diagn Mol Pathol 1:85–97.

Komminoth P, Adams V, Long AA, Roth J, Saremaslani P, Flury R, Schmid M, Heitz PU. 1994. Vergleich von histologie, immunohistochemie, RT-PCR, in situ hyridisierung und in situ PCR zum nachweis von hepatitis C virus in paraffin-eingebettem lebergewebe. Verh Deustche Ges Path 78:220–225.

Koopmans M, Monroe SS, Coffield LM, Zaki SR. 1993. Optimization of extraction and PCR amplification of RNA extracts from paraffin-embedded tissue in different fixatives. J Virol Methods 43(2):189–204.

Lloyd RV, Jin L, Chandler WF. 1991. In situ hybridization in the study of pituitary tissues. Pathol Res Pract 187(5):552–555.

Long AA. 1993. Comparison of indirect and direct in situ PCR in cell preparations and tissue sections. Detection of viral DNA, gene rearrangements and chromosomal translocations. Histochem 99(2):151–162.

Longy M, Duboue B, Soubeyran P, Moynet D. 1997. Method for the purification of tissue DNA suitable for PCR after fixation with Bouin's fluid. Uses and limitations in microsatellite typing. Diagn Mol Pathol 6(3):167–173.

Mason JC, Venables PJ, Smith PR, Maini RN. 1985. Characterisation of non-histone nuclear proteins cross reactive with purified rheumatoid factors. Ann Rheum Dis 44(5):287–293.

Masuda N. 1999. Analysis of chemical modification of RNA from formalin fixed samples and optimization of molecular biology application for such samples. Nucleic Acids Res 27:4436–4443.

McGhee JD, von Hippel PH. 1975. Formaldehyde as a probe of DNA structure. 1. Reaction with exocyclic amino groups of DNA bases. Biochemistry 14(6):1281–1296.

McGhee JD, von Hippel PH. 1977. Formaldehyde as a probe of DNA structure. 3. Equilibrium denaturation of DNA and synthetic polynucleotides. Biochemistry 16(15):3267–3276.

Mee AP. 1997. Quantification of vitamin D receptor mRNA in tissue sections demonstrates the relative limitations of in situ RT-PCR. J Pathol 182:22–28.

Moench TR, Gendelman HE, Clements JE, Narayan O, Griffin DE. 1985. Efficiency of in situ hybridization as a function of probe size and fixation technique. J Virol Methods 11(2):119–130.

Mougin C, Didier JM, Bettinger D, Madoz L, Coumes-Marquet S, Lab M. 1997. In situ PCR to cells and wax sections: an alternative to in situ hybridization for the detection of low copies of HPV and CMV DNA. In: Gosden JR (ed) Methods in molecular biology, vol 71. PRINS and In Situ PCR Protocols. Humana, Totowa, NJ.

Muro-Cacho A. 1997. In situ PCR: overview of procedures and protocols. Fron Biosci 2:15–29.

Negoescu A, Lorimier P, Labat-Moleur F, Drouet C, Robert C, Guillermet C, Brambilla C, Brambilla E. 1996. In situ apoptotic cell labeling by the TUNEL method: improvement and evaluation on cell preparations. J Histochem Cytochem Sep; 44(9):959–968.

Nuovo GJ, Silverstein SJ. 1988. Comparison of formalin, buffered formalin, and Bouin's fixation on the detection of human papillomavirus deoxyribonucleic acid from genital lesions. Lab Invest 59(5):720–724.

Nuovo GJ. 1991. An improved technique for the in situ detection of DNA after PCR. Am J Pathol 139:1239–1244.

Nuovo GJ, Darfler MM, Impraim CC, Bromley SE. 1991a. Occurrence of multiple types of human papillomavirus in genital tract lesions. Analysis by in situ hybridization and the polymerase chain reaction. Am J Pathol 138(1):53–58.

Nuovo GJ, MacConnell P, Forde A, Delvenne P. 1991b. Detection of human papillomavirus DNA in formalin-fixed tissues by in situ hybridization after amplification by polymerase chain reaction. Am J Pathol 139(4):847–854.

Nuovo GJ, Becker J, Margiotta M, MacConnell P, Comite S, Hochman H. 1992a. Histological distribution of polymerase chain reaction-amplified human papillomavirus 6 and 11 DNA in penile lesions. Am J Surg Pathol 16(3):269–275.

Nuovo GJ, Gallery F, MacConnell P. 1992b. Detection of amplified HPV 6 and 11 DNA in vulvar lesions by hot start PCR in situ hybridization. Mod Pathol 5(4):444–448.

Nuovo MA, Nuovo GJ, MacConnell P, Forde A, Steiner GC. 1992c. In situ analysis of Paget's disease of bone for measles-specific PCR-amplified cDNA. Diagn Mol Path 1(4):256–265.

Nuovo GJ. 1993. Importance of different variables for enhancing in situ detection of PCR amplified DNA. PCR Methods Appl, Cold Spring Harbor Laboratory Press 2:305–312.

Nuovo GJ. 1994. PCR in situ hybridization: protocols and applications. Second Edition. Raven, New York, ISBN 0-7817-0183.

Nuovo GJ, MacConnell P, Gallery F. 1994. Analysis of nonspecific DNA synthesis during in situ PCR and solution-phase PCR. PCR Methods Appl 4(2):89–96.

Nuovo GJ. 1996a. PCR in situ hybridization: protocols and applications. Third Edition. Lippincott Raven, New York.

Nuovo GJ. 1996b. The foundation of successful RT in situ PCR. Front Biosci 1:c14–c15.

Nuovo GJ. 1999. In situ amplification using universal energy transfer-labelled primers. J Histochem Cytochem 47:273–279.

Nuovo GJ. 2001. Co-labelling using in situ PCR: a review. J Histochem Cytochem 49:1329–1339.

O'Leary JJ. Browne G, Johnson MI, Landers RJ, Crowley M, Healy I, Street JT, Pollock AM, Lewis FA, Andrew A, et al. 1994. PCR in situ hybridisation detection of HPV16 in fixed CaSki and fixed SiHa cell lines. J Clin Pathol 47(10):933–938.

O'Leary JJ, Chetty R, Graham AK, McGee J. 1996. In situ PCR: pathologist's dream or nightmare. J Pathol 178:11–20.

O'Leary JJ, Kennedy M, Luttich K, Uhlmann V, Silva I, Russel J, Sheils O, Ring M, Sweeney M, Kenny C, Bermingham N, Martin C, O'Donovan M, Howells D, Picton S and Lucas SB (2000). Localisation of HHV-8 in AIDS related lymphadenopathy. J Clin Pathol: Mol Pathol 53:43–47.

Patel VG, Shum-Siu A, Heniford BW, Wieman TJ, Hendler FJ. 1994. Detection of epidermal growth factor receptor mRNA in tissue sections from biopsy specimens using in situ polymerase chain reaction. Am J Pathol 144(1):7–14.

Patterson BK. 1993. Detection of HIV-1 DNA and messenger RNA in individual cells by PCR-driven in situ hybridisation and flow cytometry. Science 260:976–979.

Pavelic J, Gall-Troselj K, Bosnar MH, Kardum MM, Pavelic K. 1996. PCR amplification of DNA from archival specimens. A methodological approach. Neoplasma 43(2):75–81.

Pearse AG. 1980. The common peptides and the cytochemistry of their cells of origin. Basic Appl Histochem 24(2):63–73.

Radonic A, Thulke S, MacKay IM. 2004. Guideline to reference gene selection in quantitative real time PCR. Biochem Biophys Res Comm 313:856–862.

Ray RA, Smith M, Sim R, Nystrom M, Pounder RE, Wakefield AJ. 1995. The intracellular PCR for small CMV genomic sequences within heavily infected cellular sections. J Pathol 177:171–180.

Ray RA, Cooper PJ, Sim R, Chadwick M, Earle P, Dhillon AP, Pounder RE, Wakefield AJ. 1996. Direct in situ RT-PCR for detection of measles virus. J Virol Methods 60:1–17.

Rubin MA. 2001. Use of laser capture microdissection, cDNA microarrays, and tissue microarrays in advancing our understanding of prostate cancer. J Pathol 195(1):80–86.
Sagawa M. 1997. Detection of K Ras point mutation by in situ PCR in cell suspensions: comparison of the indirect and direct methods. Cancer 17:181–195.
Sallstrom JF, Zehbe I, Alemi M, Wilander E. 1993. Pitfalls of in situ PCR using direct incorporation of labelled nucleotides. Anticancer Res 13:1153–1154.
Seidman R, Peress NS, Nuovo GJ. 1994. In situ detection of polymerase chain reaction-amplified HIV-1 nucleic acids in skeletal muscle in patients with myopathy. Mod Pathol 7(3):369–375.
Sepp N, Weyer K, Haun M, Zelger B, Thaler J, Faber V, Fritsch P, Grunewald K. 1996. Differentiation of primary and secondary cutaneous B-cell lymphoma by Southern blot analysis. Am J Clin Pathol 106(6):749–757.
Shibutani M, Uneyama C, Miyazaki K, Toyoda K, Hirose M. 2000. Methacarn fixation: a novel tool for analysis of gene expressions in paraffin-embedded tissue specimens. Lab Invest Feb; 80(2):199–208.
Shieh B, Lee SE, Tsai YC, Su IJ, Li C. 1999. Detection of hepatitis B virus genome in hepatocellular carcinoma tissues with PCR-in situ hybridization. J Virol Methods 80(2):157–167.
Speel EJM, Hopman AHN, Komminoth P. 1999. Amplification methods to increase the sensitivity of in situ hybridisation: play CARDs. J Histochem Cytochem 47:281–288.
Stamps AC, Terret JA, Adam PJ. 2003. Application of in situ reverse transcriptase PCR (RT-PCR) to microarrays. J Nanobiol Technol 1:3.
Stein U, Walther W, Wendt J, Schildt TA. 1997. In situ RT-PCR using fluorescence labelled primers. Biotechniques 23:194–196.
Sukpanichnant S, Vnencak-Jones CL, McCurley TL. 1993. Detection of clonal immunoglobulin heavy chain gene rearrangements by polymerase chain reaction in scrapings from archival hematoxylin and eosin-stained histologic sections: implications for molecular genetic studies of focal pathologic lesions. Diagn Mol Pathol 2(3):168–176.
Tbakhi A, Totos G, Pettay JD, Myles J, Tubbs RR. 1999. The effect of fixation on detection of B-cell clonality by polymerase chain reaction. Mod Pathol 12(3):272–278.
Teo IA, Shaunak S. 1995a. PCR in situ: aspects that reduce amplification and generate false positive results. Histochem J 27:660–669.
Teo IA, Shaunak S. 1995b. PCR in situ: an appraisal of an emerging technique. Histochem J 27:647–659.
Tokuda Y, Nakamura T, Satonaka K, Maeda S, Doi K, Baba S, Sugiyama T. 1990. Fundamental study on the mechanism of DNA degradation in tissues fixed in formaldehyde. J Clin Pathol 43(9):748–751.
Tournier I, Bernuau D, Poliard A, Schoevaert D, Feldmann G. 1987. Detection of albumin mRNAs in rat liver by in situ hybridization: usefulness of paraffin embedding and comparison of various fixation procedures. J Histochem Cytochem 35(4):453–459.
Uhlmann V. 1998. A novel and rapid in cell amplification technique for the detection of low copy mRNA transcripts. J Clin Pathol 51:160–167.
Unger ER, Hammer ML, Chenggis ML. 1991. Comparison of 35S and biotin as labels for in situ hybridization: use of an HPV model system. J Histochem Cytochem 39(1):145–150.
Urieli-Shoval S, Meek RL, Hanson RH, Ferguson M, Gordon D, Benditt EP. 1992. Preservation of RNA for in situ hybridization: Carnoy's versus formaldehyde fixation. J Histochem Cytochem Dec; 40(12):1879–1885.
vande Vijver C. 1995. In situ PCR: principle, protocols and promise. Tÿdschrift vande Belgische Vereniging ven Laboratorium technologie 22(4).
Walboomers JM, Melchers WJ, Mullink H, Meijer CJ, Struyk A, Quint WG, vander Noordaa J, ter Schegget J. 1988. Sensitivity of in situ detection with biotinylated probes of human papilloma virus type 16 DNA in frozen tissue sections of squamous cell carcinomas of the cervix. Am J Pathol 131(3):587–594.
Walker F, Bedel C, Dauge Geffory D, Lehy T, Madelenat P, Potet F. 1996. Improved detection of HPV infection in genital contra-epithelial neoplasia in HIV positive women by PCR-ISH. Diagn Mol Pathol 5:136–146.
Weirich G, Hornauer MA, Bruning T, Hofler H, Brauch H. 1997. Fixed archival tissue. Purify DNA and primers for good PCR yield. Mol Biotechnol 8(3):299–301.

Wickham CL, Boyce M, Joyner MV, Sarsfield P, Wilkins BS, Jones DB, Ellard S. 2000. Amplification of PCR products in excess of 600 base pairs using DNA extracted from decalcified, paraffin wax embedded bone marrow trephine biopsies. Mol Pathol 53(1):19–23.
Zawilinska B, Bulek K,Kopec J, Daszkiewicz E, Rojek-Zakrzewska D, Kosz-Vnenchak M. 2006. In situ RT-PCR can distinguish between productive and latent cytomegalovirus infection in the blood cells of bone marrow transplant recipients Acta Virol 50(1):67–70.
Zehbe I, Hacker GW, Rylander E, Sallstrom J, Wilander E. 1992. Detection of single HPV copies in SiHa cells by in situ polymerase chain reaction (in situ PCR) combined with immunoperoxidase and immunogold-silver staining (IGSS) techniques. Anticancer Res 12(6B):2165–2168.
Zhang S, Troyer DL, Kapil S, Zheng L, Kennedy G, Weiss M, Xue W, Wood C, Minocha HC. 1997. Detection of proviral DNA of bovine immunodeficiency virus in bovine tissues by PCR and PCR-ISH. Virol 236:249–257.

Index

I

K

L

M

N

O

P

U

V

Z

Color Plates

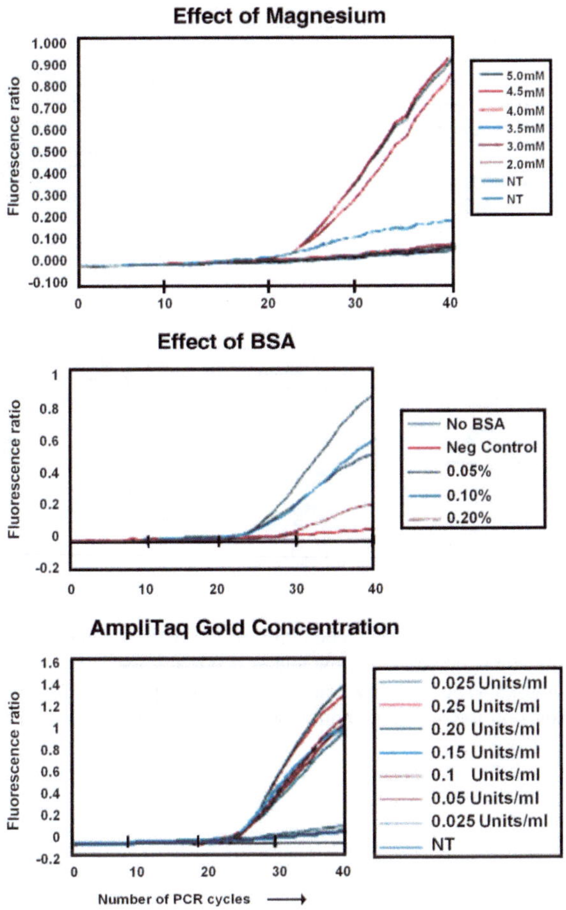

Fig. 6.4 Influence of divalent magnesium ions on real time PCR efficiency using thermostable Taq DNA polymerase. The efficiency (but not necessarily the specificity) of PCR increases with increasing Mg^{++} concentration. At approximately 10 mM Mg^{++} concentration, enzyme activity (PCR efficiency) is reduced. BSA and Taq polymerase concentration can also be two other important variables. This real time PCR test system involved the optimisation of a beta-actin specific PCR. In the figure, the X-axis indicates PCR cycle number and the Y-axis the fluorescence ratio (emission intensity of the reporter probe divided by the emission intensity of a passive reference) (Reproduced from Taylor et al., 1997. With permission from 'Oxford Journals')

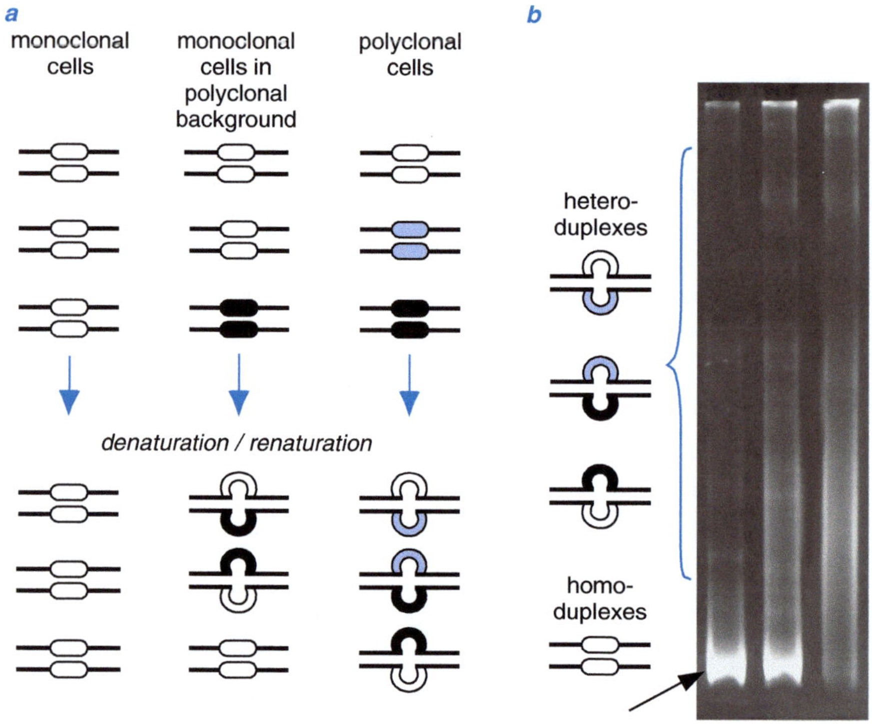

Fig. 9.6 The principle of PCR SSCP. Depending on the nature of the alleles of a given gene, various combinations of mutant and wild type alleles is possible. Such homo- and hetero-duplexes display varying electrophoretic mobility leading to banding pattern differences as displayed in panel b. The two lanes on the left show homoduplex presence (320 basepair long product highlighted by an arrow)

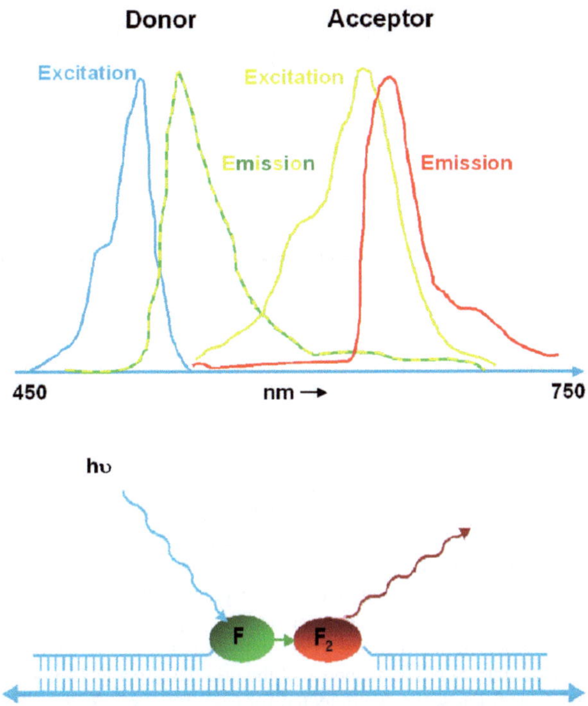

Fig. 9.16 The principle of fluorescence resonance energy transfer (FRET); Top panel: schematic representation of the overlapping spectral characteristics of quencher and donor, essential if adequate quenching of donor excitation by the acceptor is to occur. Bottom panel: Steric position of the quencher and the fluorophore in a fluoresence quenching test (Reproduced from Roche LightCycler Protocols. With permission)

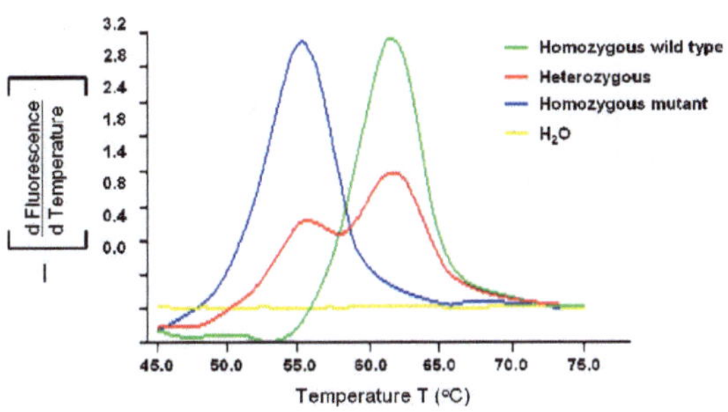

Fig. 9.18 Genomic DNA from a homozygous wild type, a homozygous mutant, and a heterozygous mutant were amplified and analysed on the LightCycler using the hybridisation probes format. The difference in melting temperatures of the samples can easily be visualised by comparing the first negative derivatives (–dF/dT) of the melting curves. The characteristic melting temperatures of the genotypes allow discrimination (Reproduced from Roche LightCycler Protocols. With permission)

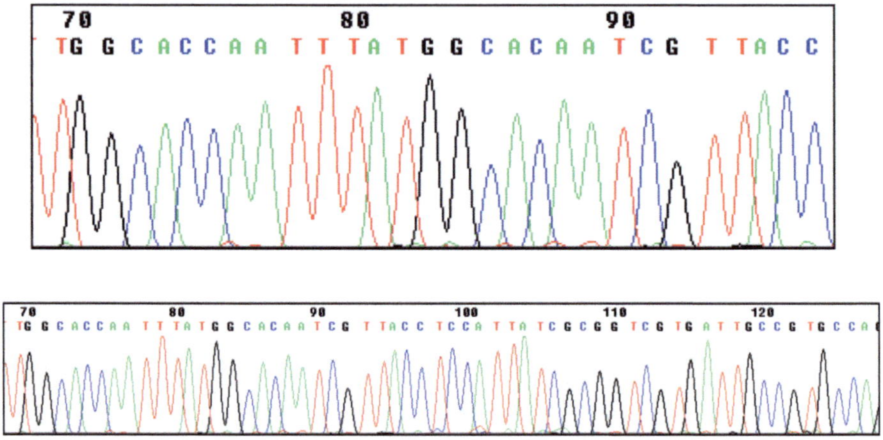

Fig. 9.19 The principle behind the DNA sequencing of method of Sanger et al. [1977]. If a dideoxynucleotide triphosphate (as opposed to the usual deoxynucleotide triphosphate) is incorporated into a growing DNA chain by a thermostable DNA polymerase, then termination of that growing DNA chain will occur at the point where the dideoxynucleotide triphosphate is incorporated. By using a mixture of four dideoxynucleotide triphosphates labelled with different fluorescent dyes in a PCR mix alongside the usual deoxynucleotide triphosphates, then a series of labelled differently sized amplimer molecules are generated which may be separated by denaturing polyacrylamide gel electrophoresis. The sequence of the original target DNA may then be determined based upon the particular fluorescent dye detected for each fragment length

a

CYCLE	AMOUNT OF DNA 100% EFFICIENCY	AMOUNT OF DNA 90% EFFICIENCY	AMOUNT OF DNA 80% EFFICIENCY	AMOUNT OF DNA 70% EFFICIENCY
0	1	1	1	1
1	2	2	2	2
2	4	4	3	3
3	8	7	6	5
4	16	13	10	8
5	32	25	19	14
6	64	47	34	24
7	128	89	61	41
8	256	170	110	70
9	512	323	198	119
10	1,024	613	357	202
11	2,048	1,165	643	343
12	4,096	2,213	1,157	583
13	8,192	4,205	2,082	990
14	16,384	7,990	3,748	1,684
15	32,768	15,181	6,747	2,862
16	65,356	28,844	12,144	4,866
17	131,072	54,804	21,359	8,272
18	262,144	104,127	39,346	14,063
19	524,288	197,342	70,324	23,907
20	1,048,576	375,900	127,482	40,642
21	2,097,152	714,209	229,468	69,092
22	4,194,304	1,356,998	413,043	117,456
23	8,388,603	2,578,296	743,477	199,676
24	16,777,216	4,898,763	1,338,259	339,449
25	33,554,432	9,307,650	2,408,866	577,063
26	67,108,864	17,684,534	4,335,959	981,007
27	134,217,728	33,600,615	7,804,726	1,667,711
28	268,435,456	63,814,168	14,048,506	2,835,109
29	536,370,912	121,293,220	25,287,311	4,819,686
30	1073,741,324	230,466,618	45,517,160	8,193,466

b

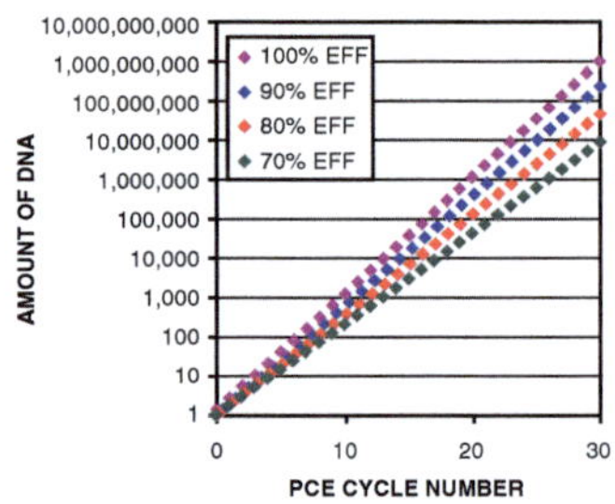

c

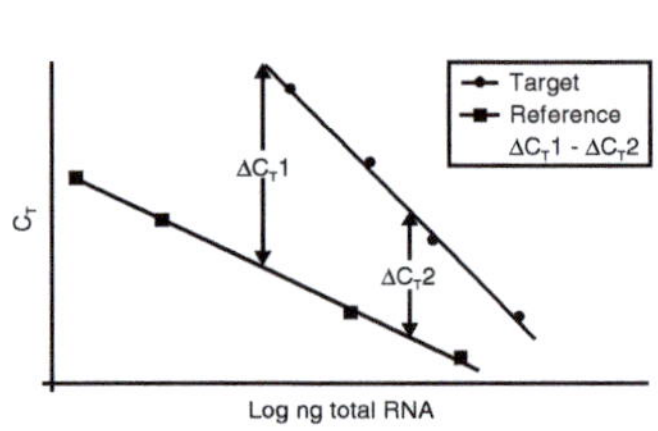

Fig. 10.6 PCR efficiency assessment (A, B). The graphs can be used to deduce the numerical replication value of each and every individual thermocycle and an average value for the entire PCR, based on a presumed replication efficiency value (from 100% to 70% from left to right in A). B shows a graphical version of the Table under A. The parallel assay in (C) allows for a comparison of the amplification efficiency of standard sample C1 with that of test sample C2 (Reproduced from Richard C. Hunt M.A. Ph.D. University of South Carolina School of Medicine. With permission)

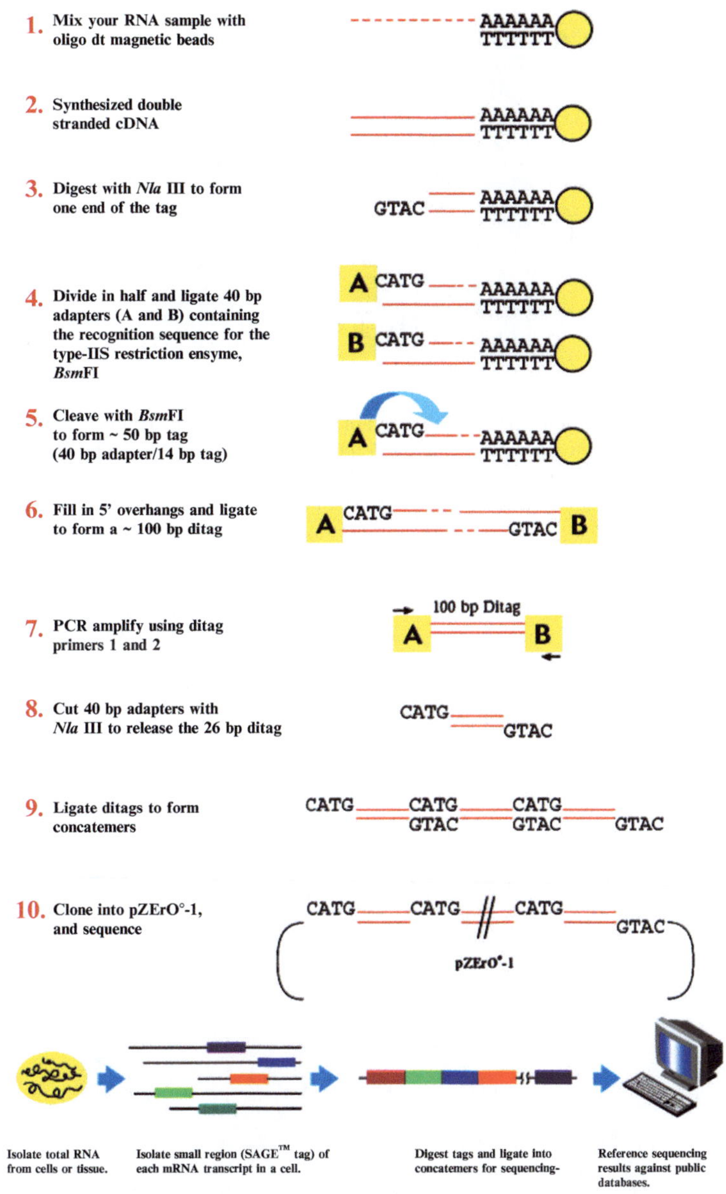

Fig. 12.17 Schematic presentation of the various steps of the Serial Analysis of Gene Expression (SAGE) protocol (see text in this chapter for more precise written explanation and www.invitrogen.com for more detail on the I-SAGE kit for Genome Wide Expression Analysis) (Copied from Invitrogen)

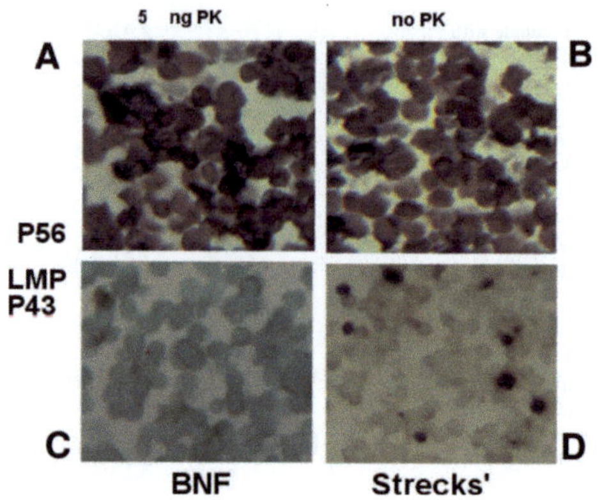

Fig. 13.6 Differences in non-specific DNA synthesis in sections of SiHa cell pellets infiltrated with paraffin at different temperatures. Panels A and B show the effect of embedding cells with a commercially available paraffin formulation of Paraplast (infiltration at 56°C), whilst Panels C and D show the effect of embedding cells in low melting point paraffin (Merck) (infiltration at 43°C). Cells were fixed in buffered neutral formalin containing 5 ng/ml phosphate buffered neutral formalin (BNF) (panels A and C) or Streck fixative (panels B and D). Non-specific DNA synthesis (after performing a standardised IS-PCR protocol without any PCR primers) is very apparent in the nuclei of cells that have been heated to 56°C (Paraplast). Much less non-specific DNA synthesis is observed when the temperature during paraffin impregnation was kept at 43°C (low melting point paraffin)

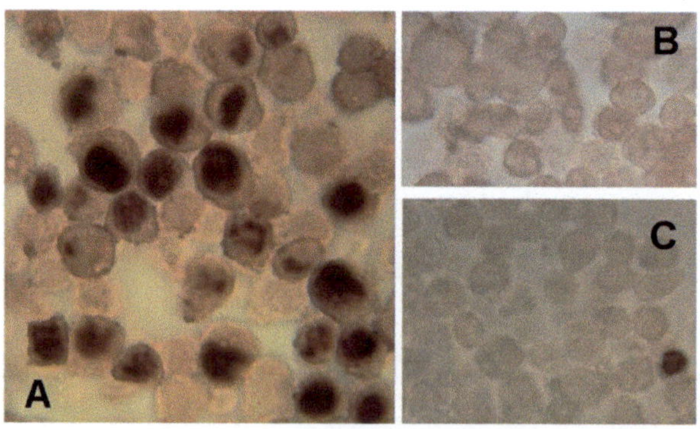

Fig. 13.18 Optimised IS-PCR using 3.5 ng of proteinase K as a pre-treatment in a wet hot- start PCR containing tRNA. The program used 17 cycles of PCR at an annealing temperature of 55°C. Panel A. IS-PCR on SiHa (HPV positive) cells using OL16 primers and showing the nuclear localisation of the amplicons (NB. Some cells do not contain the target and remain unstained). Panel B. A549 (HPV negative) control cells. Panel C. SiHa cells that have undergone IS-PCR thermocycling in the absence of PCR primers showing that sporadic cells contain nuclear artefacts of non-specifically synthesized DNA

MIX
Papier aus verantwortungsvollen Quellen
Paper from responsible sources
FSC® C105338

If you have any concerns about our products,
you can contact us on
ProductSafety@springernature.com

In case Publisher is established outside the EU,
the EU authorized representative is:
Springer Nature Customer Service Center GmbH
Europaplatz 3, 69115 Heidelberg, Germany

Printed by Libri Plureos GmbH
in Hamburg, Germany